BRILL'S STUDIES IN INTELLECTUAL HISTORY

VOLUME 66

THE CRISIS OF CAUSALITY

Voetius and Descartes on God, Nature and Change

BY

J.A. van RULER

E.J. BRILL
LEIDEN · NEW YORK · KÖLN
1995

The paper in this book meets the guidelines for permanence and durability of the Committee on Production Guidelines for Book Longevity of the Council on Library Resources.

Library of Congress Cataloging-in-Publication Data

Ruler, J.A. van.
The crisis of causality : Voetius and Descartes on God, nature, and change / J.A. van Ruler
p. cm. — (Brill's studies in intellectual history, ISSN 0920-8607 ; v. 66)
Includes bibliography references and index.
ISBN 9004103716 (alk. paper)
1. Voet, Gijsbert, 1589-1676. 2. Descartes, René, 1596-1650. 3. Causality—History—17th century. 4. God—History of doctrines—17th century. 5. Philosophy of nature—History—17th century. 6. Change—History—17th century. 7. Vitalism—History—17th century. 8. Mechanism (Philosophy)—History—17th century. I. Title. II. Series.
B4000.V64R85 1995
122'.092'2—dc20 95-17523
CIP

Die Deutsche Bibliothek - CIP-Einheitsaufnahme

Ruler, Johannes A. van:
The crisis of causality ; Voetius and Descartes on God, nature and change / by J.A. van Ruler. - Leiden ; New York ; Köln : Brill, 1995
(Brill's studies in intellectual history ; Vol. 66)
ISBN 90-04-10371-6
NE: GT

ISSN 0920-8607
ISBN 90 04 10371 6

PRINTED IN THE NETHERLANDS

To my grandmother,
Mrs. M. H. Brouwer-Van Doorn

CONTENTS

PREFACE

This study on the reception of Cartesianism is the result of a four-year fellowship as assistant-in-training at the Faculty of Philosophy of the University of Groningen. During my time in Groningen, I was much encouraged by the active support of the staff and students of that faculty, without which it would have been much harder to bring the project to fruition. It would have been quite impossible without the help of the staff of the various libraries that keep the rare books and prints referred to in this work. I should like to thank the staff of the university libraries of Groningen, Utrecht and Leiden in particular for their assistance. Further thanks are due to Peter Stork for helping me out with the Greek and to René Gude and Régine Dugardyn for sharpening my mind on some central issues of Cartesian philosophy.

I should also like to express my gratitude to the colleagues at *Filosofie Magazine* for providing a stimulating environment—nothing so comical as writing on philosophy—in which I have been able to work on other subjects than seventeenth-century philosophy alone, and to Ton & Chantal for temporarily providing a place to work in the very literal sense of the word. There have been times at which the relief and comfort of friends and relatives was of more importance than any professional support. Besides so many others, I remember my parents, who have encouraged me all along, and Jeroen & Jet, Roeland, Han & Annemarie, Karen & Peter, Régine, Pa & Ma Vlas, Klaartje & Leo, Dr. Ferguson and, of course, Jeanne for helping me through some especially hard moments. In the end, my greatest debt is to my friend and companion, Jan Vlasblom.

Three people have contributed to the appearance and readability of this book in a very direct way. The first is Arjo Vanderjagt who, besides being a great colleague, made me broaden my intellectual horizon at times when Voetius and Descartes might have seemed to be all there was. Nevertheless, one tends to stick to one's first fascination, and it is Cartesianism and the conceptual changes which—three hundred and fifty years ago—this philosophy brought about, that have occupied me so much that I wrote this book about it. It was Theo Verbeek who, during my student years in Utrecht, first aroused my interest in that Golden Age of philosophical history and who gave me the opportunity of co-operating with him in the publication of the documents relating to the *Querelle d'Utrecht*. It has been a pleasure to have worked with him since. Finally, it was a great honour to work under the supervision of John North, a man whose erudition is matched only by the extent of his respect for the work and idiosyncrasies of others—a virtue too rarely found in the academic world. It is one of his convictions that one cannot write anything of much interest without being earnestly involved in the subject. I entirely agree. Let this book be the expression of what has captivated me these last few years.

Argentière, 1 March 1995 Han van Ruler

ABBREVIATIONS

AT	[Descartes, René,] *Œuvres de Descartes*, publiées par Charles Adam & Paul Tannery, Paris (Léopold Cerf) 1897-1913 in 12 vols. / Nouvelle présentation, en co-édition avec le Centre National de la Recherche Scientifique, Paris (Librairie Philosophique J. Vrin) 1964-1971 in 12 vols.
COT	Descartes, René, *Meditations on First Philosophy*, With selections from the Objections and Replies, Translated by John Cottingham, with an introduction by Bernard Williams, Cambridge (C. U. P.) 1986.
CSM	[Descartes, René,] *The Philosophical Writings of Descartes*, translated by John Cottingham, Robert Stoothoff, Dugald Murdoch and Anthony Kenny, Cambridge (C.U.P.) 1985-1991, in 3 vols.
Dibon	Dibon, Paul, *L'Enseignement philosophique dans les universités néerlandaises à l'époque pré-cartésienne (1575-1650)*, Dissertation Leiden 1954, Amsterdam (Institut Français d'Amsterdam) 1954.
Duker	Duker, A. C., *Gisbertus Voetius*, Leiden (E. J. Brill) 1893-1915 / reprint: Leiden (Groen) 1989 in 4 vols.
Jöcher	Jöcher, Christian Gottlieb, *Allgemeines Gelehrten-Lexicon* [etc.], Leipzig (Johann Friedrich Gleditschens Buchhandlung) 1750-1751 / Reprint Hildesheim (Olms) 1960-1961.
Narratio	*Testimonium Academiæ Vltrajectinæ, et Narratio Historica quà defensæ, quà exterminatæ novæ Philosophiæ*, Rheno-Trajecti, Ex Typographiâ Wilhelmi Strickii, 1643.
Querelle	Descartes, René / Schoock, Martin, *La querelle d'Utrecht*, Textes établis, traduits et annotés par Theo Verbeek, Paris (Les impressions nouvelles) 1988.
Responsio	Regius, Henricus, *Responsio, sive Notæ in Appendicem ad Colloraria Theologica-Philosophica* [etc.], Ultraiecti, Apud Ioannem à Doorn, 1642.
Rodis-Lewis	Descartes, René, *Lettres à Regius et Remarques sur l'explication de l'esprit humain*, Texte latin, traduction, introduction et notes par Geneviève Rodis-Lewis, Paris (Vrin) 1959.
Select. Dispp.	Voetius, Gisbertus, *Selectarum Disputationum Theologicarum Partes V*, Utrecht and Amsterdam 1648-1669; I-III, Ultraiecti, Joh. à Waesberge 1648-1659; IV, Amstelodami, Jo Janss. à Waesberge 1667; V, Ultrajecti, Ant. Smijtegelt, 1669.
Thijssen-Schoute	Thijssen-Schoute, C. Louise, *Nederlands Cartesianisme*, avec sommaire et table des matières en français, Amsterdam (Noord-Hollandsche Uitgevers Mij.) 1954 / reprint, bezorgd en van aanvullende bibliografie voorzien door Th. Verbeek, Utrecht (HES) 1989.

With great effort, Christianity had learned to read the Bible through the spectacles of Aristotelian philosophy, and it had to make no less an effort to un-learn this.

R. Hooykaas

There is so much to do, however, one scarcely knows where to begin. A good deal of attention, for example, has been given to Galilei's discussion of causality and how it breaks with the past, yet to the best of my knowledge, there has been no study of the notion of causality among Aristotelian thinkers.

Charles B. Schmitt

INTRODUCTION

When in 1685 the Groningen philosopher Johannes Bertling (1626-1690) published his *Seminary of All Transnatural Wisdom* in defence of the accepted School-philosophy, it was not merely against "the fraudulent machinations of both ancients and moderns" that he fulminated. More particularly, Bertling aimed at "denouncing and overthrowing" one philosophical school in particular: Cartesianism. As Bertling tells us even on the title-page of his work,

> dangerous tenets that threaten the truth and purity of the heavenly Doctrine daily present themselves ever more often.[1]

Bertling's warnings may seem to be no more than a variation on the ever reccurring reactionary theme of accepted religious principles and ethical standards being abandoned and of things in general getting worse. Yet Bertling had special reasons in 1685 to bemoan the fate of natural theology and to reaffirm "the legitimate and true use of our serving Sophia in Sacred Theology."

This book is about what had happened some forty-five years earlier, when Bertling's teacher, the Utrecht theologian Gisbertus Voetius (1589-1676), took it upon himself to defend against Descartes' iconoclastic philosophical revolt the type of philosophy which had proved so fruitful for theology. I hope to point out why

---

[1] Johannes Bertling, *Seminarium totius transnaturalis sapientiæ duobus tomis Philosophiam Primam et integram Pneumaticam sive Theologiam Naturalem Nervosa & succincta Methodo complectens*, Groningæ, Typis Catharinæ Zandt, *Viduæ* Huysman, Provinciæ & Academiæ Typographæ Ordinariæ, Anno 1685, titlepage. Bertling is a philosopher long forgotten. "[Ein] Philosoph zu Gröningen, florirte 1661", is all that Christian Gottlieb Jöcher has to say about him in his *Allgemeines Gelehrten-Lexicon*, Leipzig (Johann Friedrich Gleditschens Buchhandlung) 1750-1751 / Reprint Hildesheim (Olms) 1960-1961, (hereafter: *Jöcher*) Bd. I, col. 1038. See however also: C. Louise Thijssen-Schoute, *Nederlands Cartesianisme*, Amsterdam (Noord-Hollandsche Uitgevers Mij.) 1954 / reprint Utrecht (HES) 1989, pp. 488-489 and p. 492, where she describes Bertling's violent anti-Cartesianism. Johannes Bertling, born in Groningen of German parents, was a pupil of both Gisbertus Voetius and Martin Schoock, two of the main characters to appear in this book. Cf. Paul Dibon, *L'Enseignement philosophique dans les universités néerlandaises à l'époque pré-cartésienne (1575-1650)*, Dissertation Leiden, 1954, pp. 183, 184 and 219; and: F. Sassen, "350 jaren wijsgerig onderwijs te Groningen", in *Groninger Universiteitsblad* 15-3 (1965), p. 72.

exactly Descartes' thought formed such a threat and why the philosophy that Voetius—and, later, Bertling—stood for, namely the late-Scholastic philosophy that had its base in the Renaissance commentaries on the works of Aristotle, was thought to be the only system appropriate to the support of Christian doctrine. It is widely affirmed that in some way, philosophy was in pre-modern times thought to be the handmaiden of theology. Yet it is not self-evident what this might mean. What, in other words, could Sophia do in order to please the masters of the faith?

An event that cannot be ignored in the intellectual history of seventeenth-century Europe, is the rise of Copernicanism. Whatever the precise offence that Galileo made against theology's rule, his claim that the Earth *realiter* revolves around the Sun, certainly explains the best-known—and most widely disputed—clash of Reason and Religion. In chapter 1, I shall deal with the question of Copernicanism and analyse its role in Voetius' denunciation of Descartes. Copernicanism, however, was not the main issue in the theologian's defence of accepted philosophical dogma. As my title indicates, I shall present the philosophical debate between Voetius and Descartes as a crisis of causality.

The concept of causality, around 1650, was undergoing rapid change. Questions such as "Why does a stone fall?" and "What makes heavenly bodies revolve in their orbits?" were of particular importance for natural philosophy. Especially from 1572 onwards—when, in answer to the appearance of the famous *stella nova*, Tycho Brahe had put forward the conjecture that the heavens were of a corruptible nature—faith in the accepted world-picture was rapidly declining. The development of a new image of the Universe and, along with it, of new ideas of causal mechanisms in Nature, was a question of a few generations only, culminating in Newton's theory of universal gravitation by the end of the century.

Theology could not but be influenced by this, especially since, in seventeenth- and eighteenth-century Europe, people still saw God not merely as the source of what exists, but also as a governer and administrator of the Universe. One spoke of God not only as *Creator*, but also as *Conservator*, and *Co-operator*.[2] Created things, whether

[2] I shall not here discuss the typically Calvinistic way of describing God not only as Creator and Director of Creation, but also as Elector of the Faithful. Even in the eighteenth century, such descriptions were common. In a Groningen dissertation *De Creato Mundo* of 1755, a student of theology defending Leibnizian optimism against deist tendencies, refers to God as "the electing, creating and governing

alive or not, were not just thought of as *creatures*, but as *instruments* of divine providence as well.

This book deals with the emergence of a philosophical system, Cartesianism, that arose in the midst of the era of "scientific revolution", and our thesis will be, that the main objection of theologians like Voetius against the so-called *New Philosophy* of Descartes and others was that the New Philosophy could not aptly describe the relation between God, the Creator, and Nature, His Creation. Along the way, various related conclusions will be drawn. Likewise, I shall argue that there was no obvious reason for Scholastic adversaries of the New Philosophy to see the philosophical innovations of the day as being revolutionary in the sense of introducing ideas which had never been known before. The sort of questions which the discipline of physics was supposed to answer simply did not leave room for such creativity. To someone like Voetius, the *novelty* of the New Philosophy lay in the unexpectedness of a re-emergence of theories contradicting those of Aristotle rather than in any "revolutionary" aspects of the theories themselves. Also, I hope to prove that, in the face of the critique of those thinkers from whom Cartesianism could expect—and in fact received—the fiercest opposition (viz. Scholastic philosophy and theology), some of the characteristics it developed, such as its *a priorism* and its "rationalism", may be explained in new ways, ways different from those that take Descartes' text at its face value.

Yet Cartesianism, even from the perspective of its Scholastic critics, will not in itself form the prime subject-matter of what follows. In fact, all aforementioned topics are made subordinate to the central idea behind this work, which is to reformulate the debate between Voetius and his pupil Martin Schoock (1614-1669) on the one hand and Descartes and his ally Henricus Regius (1598-1679) on the other, in terms of conflicting concepts of causality. This calls for caution, since there are serious philosophical problems with these terms. What,

---

God" (*besluitende, Scheppende, bestierende*). Cf. Henricus Hillers, *Bovennatuurkundig Vertoog waar in dat geen, 't welk twyffelagtig is in de vraag aangaande de geschapene Beste Werelt, opgeheldert wort*, Te Groningen, By Hajo Spandaw, Boekverkoper in de Gulden straat, in het Oude Zwarte Kruys, 1755, pp. 1-2. Like all Dutch university disputations and dissertations of the seventeenth and eighteenth centuries, this dissertation also invokes divine assistance (*Goddelyke Bystant*) on its title page, in the hope that God may allow the event of its public defence to take place as scheduled. In Latin disputations, such pious additions occur in the form of a *Deo Volente, Deo Auxiliante, Deo Assistente, Deo Concurrente*, etc.

for instance, do we mean when we use the term *cause*? The issue is, in part, a terminological one. Well-known caveats accompanying introductions to Aristotle's theory of the "four causes" for example, are a direct result of the translation of the Greek αἰτία as "cause" or *causa*. To represent Aristotle's catalogue of αἰτίαι as a series of "causes" is, however, anachronistic and does not do justice to the analytic character of Aristotle's project. For Aristotle, listing more than one type of αἰτία was a way of aknowledging that there are many answers that might be given to the question of why things are such as they are.[3] Whilst Scholastic writers apparently saw no

---

[3] This way of putting things could lead to the confusion that αἰτίαι are merely linguistic or mental items, which, for Aristotle, they are not. In order to avoid such confusion, Julius M. Moravcsik proposes to interpret Aristotle's theory of αἰτίαι to be an answer to the question "What features of parts of reality make these [parts] intelligible?" (Cf. "What Makes Reality Intelligible? Reflections on Aristotle's Theory of *Aitia*", in Lindsay Judson (ed.) *Aristotle's* Physics*: A Collection of Essays*, Oxford (Clarendon Press) 1991, p. 31.) Others, however, deny that Aristotle was looking for explanations rather than causes. Cynthia A. Freeland for instance aims to re-establish Aristotle as a "causal realist" (Cf. *idem*, pp. 49-72) on the basis of the idea that the various types of "causes" he offers are not subjective explanations depending on specific interests of those asking different "why-questions", but "causal statements [..] that obtain in the world itself." However, various issues are at stake here. On the one hand, Aristotle was indeed talking of "causal connections" that obtain in the world itself. This rules out reading Aristotle's theory of αἰτίαι in the "pragmatic" way of authors such as Bas van Fraassen whom Freeland criticises in her paper. It does not, however, imply that the modern notion of cause functions in the same way as does the Greek αἰτία. For the modern reader, it may seem paradoxical to say that Aristotle on the one hand believed in "real" causal factors obtaining in the world itself and on the other defined αἰτίαι as various ways of explaining events. For Aristotle, such a paradox did not occur. As Morasvcsik quite rightly points out, Aristotle neither knew of, nor accepted, a "Kantian [...] dichotomy between the noumenal and the phenomenal world", or, for that matter had he "Humean doubts about the reality of causal powers." Accordingly, accepting that Aristotle was concerned with the variety of "real" factors initiating change in the world itself, we may yet talk of his αἰτίαι as "four different explanatory chains", as Moravcsik does (Cf. *idem*, p. 43). For the same reason, however, I see no difficulty in describing the four αἰτίαι as different answers to questions of why things are such as they are. For Aristotle, these answers corresponded in a non-problematic way to relations obtaining in the world of things. In other words, the question of whether or not Aristotle was a "causal realist" is irrelevant precisely because it is a modern one. We may therefore, with W. Wieland, hold on to the view that we are dealing "not with four causes, but with the four senses in which we speak of causes," without danger of making Aristotle into a neo-Kantian of some sort. As Wieland states: "the formal unity of these distinct meanings [of causes] is established through a functional element, namely through the question 'Why?'" Cf. W. Wieland, "The Problem of Teleology", in

problem in transforming this into an all-embracing concept of causation for material, formal, efficient and final "causes" alike, the modern reader is more likely to restrict usage of the term "cause" to the realm of efficient, i.e. working, moving causes or processes.

This terminological problem reflects a change in the meaning of the word *cause*. Although the term links up with Scholastic and theological ideas rather than with the explanations of classical mechanics[4], the everyday use of words like "cause", "causation" and "effect" on the one hand and the acceptance of the scientific world-view on the other, has in fact led us away from an easy acceptance of the Scholastic concepts of causality. Nonetheless, I shall present the conflict between Scholastic and Cartesian physics in terms of a conflict of *concepts of causality*. For all the problems attached to it, the use of the causation-terminology will not mislead us as long as we recognize what questions philosophers and theologians intend to answer when they take recourse to the concept of *cause*. In pre-Newtonian physics, the quest for causes is—generally speaking—a search for *causal agents*. Accordingly, an essential aspect of Scholastic and Cartesian physics is that both aimed to identify the *causes* through which natural motion could be explained, rather than, say, to identify the regularities and laws governing natural motion. As a result, both are concerned with offering causes in the sense of "driving forces" behind the phenomena.

From this perspective, questions concerning falling bodies and revolving planets are merely well-known examples of a much wider range of problems relating to the origins and principles of natural

---

Jonathan Barnes, Malcolm Schofield and Richard Sorabji, *Articles on Aristotle*, Vol. 1, "Science", Duckworth, London 1975, p. 147.

[4] A very elegant attack on the use of the notion of cause may be found in the works of Bertrand Russell. The passage in which he exemplifies its ineffectiveness within the natural science deserves to be quoted in full: "All philosophers, of every school, imagine that causation is one of the fundamental axioms or postulates of science, yet, oddly enough, in advanced sciences such as gravitational astronomy, the word 'cause' never occurs. Dr James Ward, in his Naturalism and Agnosticism, makes this a ground of complaint against physics: the business of those who wish to ascertain the ultimate truth about the world, he apparently thinks, should be the discovery of causes, yet physics never seeks them. To me it seems that philosophy ought not to assume such legislative functions, and that the reason why physics has ceased to look for causes is that, in fact, there are no such things. The law of causality, I believe, like much that passes muster among philosophers, is a relic of a bygone age, surviving, like the monarchy, only because it is erroneously supposed to do no harm." Cf. Bertrand Russell, "On the Notion of Cause" (1917), in *Mysticism and Logic, and Other Essays*, London (Allen and Unwin) 1963.

change. In mythology and theology and in philosophy and science, mankind has been constantly searching for the driving forces behind natural processes. Of course, to define the whole range of myths and theories which these areas comprise as a single attempt to arrive at a "theory of causation" does not do justice to the variety of cultural and intellectual contexts in which such theories arose, nor does it, from a modern perspective, offer us any clear idea of a single notion of *causality* that more or less resembles current usage of that term. Yet by tagging a single label on to the various questions touched upon below, I do not see much danger of anachronism or of making the subject so wide as to be meaningless.

Anachronism is in fact best avoided by a refusal to confine the concept of causality to its modern bounds. Seventeenth-century philosophers neither accepted the constraints forced upon the concept by a purely mechanistic interpretation of causality nor did they accept the idea of a mathematical physics that would replace a commonsensical notion of causes altogether. Writers as diverse as Voetius, Schoock, Descartes and Regius—or Boyle and Newton, for that matter—all accepted a notion of causality according to which God could in some way be seen as the ultimate "cause" of natural motion. It will, I think, prove more interesting to see in what way they could make use of this idea, than to exclude the concept of causation beforehand as being either ambiguous or too vague.

Consequently, I shall deal with theology as much as with natural philosophy or "physics" proper. Indeed, for some of the writers we shall meet, a sharp distinction between physics and theology could hardly have made much sense. The conflict between Scholastic and Cartesian ideas of natural change was first and foremost a conflict between finalistic and non-finalistic theories of causation. It will therefore be of interest to investigate theological ideas such as the question of divine providence, which Neo-Scholastic authors had developed on the basis of a finalistic theory of natural change. After first having presented Voetius' essay "On the Natures and Substantial Forms of Things" in chapter 1, which will be the text from which our discussion begins, I offer in chapter 2 a systematic analysis of some of its key concepts. Chapter 3 concentrates on the function of physical theory within pre-modern theology generally. This is contrasted in chapters 4 and 5 with the Cartesian method that threatened to undermine the theological use of physics. Chapter 6 presents some philosophical ideas concerning substantial unity which were elaborated in Rennaissance commentaries on Aristotle's *Physics* Book

II, and which, according to authors like Voetius and Schoock, made substantial forms an indispensable part of natural explanation. Next, I shall return to Regius and Descartes and explain some aspects of Cartesian methodology in the light of these Scholastic views (chapter 7). After summarizing the Cartesian idea of the physical Universe in chapter 8, I shall, in the last chapter, offer a final account of what, according to Voetius' first formulation of 1641, was wrong with it.

CHAPTER ONE

VOETIUS' ESSAY "ON THE NATURES AND SUBSTANTIAL FORMS OF THINGS"

In December 1641, Gisbertus Voetius (1589-1676), professor of theology in Utrecht since 1636, published an essay "On the Natures and Substantial Forms of Things"[1]. It is a short work containing a very general critique of the so-called *New Philosophy* of anti-Aristotelians. Amongst these, Voetius counts both long forgotten writers such as Sebastian Basso, David Gorlæus and Nicolaus Taurellus, and celebrities as Kepler, Galileo and Descartes. The main object of Voetius' essay is to defend the notion of substantial forms and the idea that every natural object is endowed with an individual "nature".[2] Both of these notions of "form" and "nature" had suffered serious attacks from the anti-Aristotelian camp. Through the mediation of one of Voetius' Utrecht colleagues, the professor of theoretical medicine and botany Henricus Regius (1598-1679)[3], the

---

[1] This text first appeared as an *Appendix ad corollaria Theologica-Philosophica nuperæ disputationi de Iubileo Romano* [etc.], an original copy of which may be found in the Bibliothèque Nationale in Paris. The academic disputation was held on 23 and 24 December 1641. Later, the text was incorporated into the *Testimonium Academiæ Vltrajectinæ, et Narratio Historica quà defensæ, quà exterminatæ novæ Philosophiæ*, Rheno-Trajecti, Ex Typographiâ Wilhelmi Strickii, 1643, pp. 36-51. In the following, I shall refer to this work as '*Narratio*', followed by page numbers. See also: R. Descartes/M. Schoock, *La Querelle d'Utrecht*, Paris (Les impressions nouvelles) 1988, pp. 69-123. This book, to which I shall refer as '*Querelle*', contains a French translation of the *Narratio*. Thirdly, the text of the *Appendix* may be found in G. Voetius, *Selectarum Disputationum Pars Prima*, Utrecht 1648, pp. 870-881. To this edition, I shall refer as *Select. Dispp.*, followed by volume number and page numbers. In 1642, Henricus Regius replied to Voetius' *Appendix* by publishing a *Responsio, sive Notæ in Appendicem ad Corollaria Theologica-Philosophica* [etc.], Ultraiecti, Apud Ioannem à Doorn. In this little work, Regius had the complete text of the *Appendix* reprinted. I shall refer to it as *Responsio*, followed by page numbers.

[2] Cf. Theo Verbeek, "Voetius en Descartes", in J. van Oort, C. Graafland, A. de Groot, O.J. de Jong (edd.), *De onbekende Voetius: Voordrachten wetenschappelijk symposium Utrecht 3 maart 1989*, Kampen (J. H. Kok) 1989, p. 218: "Voetius' inhoudelijke bezwaren tegen de Nieuwe Filosofie gingen vrijwel alle uit van de vrees dat het verwerpen van de 'vormen' ertoe zou leiden dat de individualiteit van de afzonderlijke zijnden niet langer te begrijpen zou zijn."

New Philosophy had gained followers amongst the students of the Utrecht Academy.

It must have been the latter development which alarmed Voetius and some of the other professors most of all. As the University Senate was eager to point out to the City Corporation, the New Philosophy was in conflict with traditional teaching and aimed at overthrowing the foundations of the accepted body of learning. Furthermore, it withheld students from acquiring an adequate understanding of the technical terms occurring in the works both of famous authors and of the students' own professors, and hence from being successful in science. Finally, inexperienced youngsters might draw "false and absurd" conclusions from the teachings of the New Philosophy, which were said to be harmful to other disciplines, orthodox theology in particular.[4]

Voetius' defence of Peripatetic thought is thus, among other things, a defence of traditional relations between the various disciplines. Philosophy courses should enable students to understand and follow their professors in theology, medecine and law. Philosophy should not, in other words, go its own way. To Voetius, who besides being a leading figure in the Contra-Remonstrant movement was a respected Minister of the Church, the supposed threat of a "New Philosophy" was of course of immense importance. However, as we shall see, there are many ways in which the threat to theology might be interpreted. One of them was that the discipline of theology as Voetius saw it, was dependent on a particular body of philosophical doctrine. The introduction of a new and independent philosophy could not but damage theological authority. Accordingly, Voetius reclaims authority over those points of philosophical doctrine that violate theological boundaries. In Thesis VIII of the essay on forms, for instance, Voetius attacks the Copernican idea of an annual and diurnal rotation of the

---

[3] For Regius, see M. A. J. de Vrijer, *Henricus Regius: Een "Cartesiaansch" hoogleraar aan de Utrechtsche Hoogeschool*, 's-Gravenhage (Martinus Nijhoff) 1917. An historical account of the troubles at Utrecht University in the early 1640's may be found in Theo Verbeek's introduction to the *Querelle*, pp. 19-66 and in Theo Verbeek, *Descartes and the Dutch*, Carbondale, Ill. (Southern Illinois University Press) 1992, Introduction. See also: Thomas Arthur McGahagan, *Cartesianism in the Netherlands, 1639-1667; The New Science and the Calvinist Counter-Reformation* (Ph. D. diss. Univ. of Pennsylvania), Ann Arbor and London (University Microfilms International) 1976.

[4] Cf. *Narratio*, p. 66 / *Querelle*, p. 122.

Earth. It will prove worthwhile to take a closer look at how he deals with it.

1.1 Copernicanism

Protestant theologians were in principle at least as sensitive to the question of the Earth's rotation as were their Catholic counterparts.[5] Nevertheless, it seems that Copernicanism was not a widely discussed issue in the Netherlands prior to Voetius' criticisms of the Copernican

[5] German Protestant churchmen defended the Earth's stability independently of, and prior to, the Roman Catholic Church's denunciations of Copernicanism in 1616 and 1633. Martin Luther's condemnation of Copernicanism, even before the publication of Copernicus' ***De Revolutionibus Orbium Coelestium***, is well known. It is odd that it should in fact be Luther who, in 1539, said: "But this is how it goes these days: he who wishes to appear clever should not put up with what others esteem, but make himself something of his own." Cf. Martin Luther, *D. Martin Luthers Werke, Kritische Gesamtausgabe*; Tischreden Band IV, pp. 412-413 (= WATR 4, 4638). However, as Heinrich Bornkamm quite rightly pointed out, Luther's was only an aside, made in public and only accidentally known to us because it was written down by others. He never discussed the theory seriously: "Hätte Luther nicht so viele Eckermänner gehabt, die treulich alles aufzeichneten, was bei Tisch besprochen wurde, so wüssten wir gar nichts von dieser Gelegenheitsäußerung. Kein Wort in Schriften, Briefen oder Gutachten Luthers spricht sonst von Kopernikus oder seine Lehre, und nicht der geringste Versuch ist von ihm gemacht worden, ihre Ausbreitung in Kursachsen oder einem anderen Territorium zu verhindern. Im gegenteil, die beiden namhafstesten Kopernikusanhänger der Zeit, Rheticus und Reinhold, lehrten unangefochten in Wittenberg." Cf. "Kopernikus im Urteil der Reformatoren", *Archiv für Reformationsgeschichte* 40 (1943), p. 173. Nevertheless, as we shall see hereafter, Luther's argumentation anticipates that of Melanchthon and Voetius and is characterized by the same critical standpoint towards idle curiosity. As for Melanchthon, who rejected Copernicanism, Bornkamm stresses that physical arguments were of more importance to him than religious ones and that his position towards Copernicanism was generally mild. The first argument seems to me to be untenable for reasons to be given hereafter. (See below, notes 31, 32, 34 and 36.) As for the second argument (which also occurs in Robert S. Westman, "The Melanchthon Circle, Rheticus and the Wittenberg Interpretation of the Copernican Theory", in *Isis* 66 (1975), pp. 173-174; see below, note 36), it is certainly true that Melanchthon's reaction to Copernicanism was, at least in later years, a very moderate one. However, the fact that Lutherans proved tolerant towards the new astronomy in the sense that they did not censure the opinion or took "sonstige kirchliche Maßnahme" against it (as Bornkamm rightly points out, p. 182), does not imply that they did not have reservations about it on theological grounds. Melanchthon certainly did.

hypothesis.[6] In the essay on substantial forms, he argues that Copernicanism is in conflict with the Bible and mentions various Scriptural passages from *Joshua*, *Ecclesiastes* and the *Psalms* as a "first type of argument" against it.[7] Voetius does not stand alone, but rather delves into exegetical discussions which had for many years been going on between Copernicans and their religious adversaries in other parts of Europe—Italy in particular. The Biblical verses which he mentions all speak of a stable Earth and a moving Sun.[8] These presented the Copernicans with a problem. They somehow had to offer

---

[6] Voetius triggered off a lively discussion on the subject. Further: Rienk H. Vermij, *Secularisering en Natuurwetenschap in de zeventiende en achttiende eeuw: Bernard Nieuwentijt*, Amsterdam (Rodopi) 1991, p. 65: "Het optreden van Voetius maakte het stelsel van Copernicus in Nederland godsdienstig verdacht. [...] Voor 1650 hadden Nederlandse theologen zich zelden iets gelegen laten liggen aan het wereldsysteem." In the years 1655-1656, a "second major Cartesian crisis" was to occur, in which Copernicanism was widely discussed even though, according to Theo Verbeek, "the deeper question was scepticism". See his: "From 'Learned Ignorance' to Scepticism", in Richard H. Popkin / Arjo Vanderjagt (edd.), *Scepticism and Irreligion in the Seventeenth and Eighteenth Centuries*, *Brill's Studies in Intellectual History* 37, Leiden (E. J. Brill) 1993, pp. 36-43 esp. For a history of Copernicanism in the Netherlands, see R. Hooykaas, "The Reception of Copernicanism in England and the Netherlands", in *The Anglo-Dutch Contributiuon to the Civilization of Modern Society; An Anglo-Dutch Symposium*, London (Oxford U. P.), pp. 33-44.

[7] Voetius, *Narratio*, p. 49 / *Select. Dispp.* I, p. 880 / *Responsio*, p. 35 / *Querelle*, p. 114: "*Priorem* classem argumentorum constituemus nunc Iosus 10. v. 12. 13. 14. Ecclesiast. 1. v. 4. 5. 6. 7. Psalm. 19. v. 5. 6. 7."

[8] In the King James Version, they read as follows. *Joshua* 10:12-14: "Then spake Joshua to the Lord in the day when the Lord delivered up the Amorites before the children of Israel, and he said in the sight of Israel, Sun, stand thou still uopn Gibeon; and thou, Moon, in the valley of Ajalon. And the sun stood still, and the moon stayed, until the people had avenged themselves upon their enemies. Is not this written in the book of Jasher? So the sun stood still in the midst of heaven, and hasted not to go down about a whole day. And there was no day like that before it or after it, that the Lord hearkened unto the voice of a man: for the Lord fought for Isreal." *Ecclesiastes* 1:4-7: "One generation passeth away, and another generation cometh: but the earth abideth for ever. The sun also ariseth, and the sun goeth down, and hasteth to his place where he arose. The wind goeth towards the south and turneth about unto the north; it whirleth about continually, and the wind returneth again according to his circuits. All the rivers run into the sea; yet the sea is not full; unto the place from whence the rivers come, thither they return again." Psalms 19:5-7: "Which [i.e., the sun] is as a bridegroom comig out of his chamber, and rejoiceth as a strong man to run a race./His going forth is from the end of the heaven, and his circuit unto the ends of it: and there is nothing hid from the heat thereof. The law of the Lord is perfect, converting the soul: the testimony of the Lord is sure, making wise the simple."

a less rigid interpretation of the verses in order to make Scripture agree with their idea of a rotation of the Earth.[9] The usual way to do so was to say either that the Bible spoke metaphorically, and with the tongue of the masses, or that the Bible was not relevant for answering non-religious questions. Referring to the works of Lansberg and Kepler and to a letter by Foscarini which was published along with Galileo's *World Systems*[10], Voetius shows himself to be well aware of the possibility of avoiding a literal interpretation of the Biblical text. However, according to him,

---

[9] As Galileo put the question: "The reason produced for condemning the opinion that the earth moves and the sun stands still is that in many places in the Bible one may read that the sun moves and the Earth stands still. Since the Bible cannot err, it follows as a necessary consequence that anyone takes an erroneous and heretical position who maintains that the sun is inherently motionless and the earth movable." Cf. *Discoveries and Opinions of Galileo*, Translated with an Introduction and Notes by Stillman Drake Garden City, New York (Doubleday Anchor Books) 1957, p. 181. The book offers, amongst other things, an English translation of the *Letter to Madame Christina of Lorraine Grand Duchess of Tuscany Concerning the Use of Biblical Quotations in Matters Science*, from which the quotation is taken. In this *Letter to Christina*, Galileo tackled the Scriptural issue at length, arguing that physical conclusions should be found by making use of sense experience and necessary demonstration only. Scripture should be interpreted, says Galileo, according to physical conclusions that leave no room for doubt, but Scripture should not settle disputes of physics. The *Letter to Christina*, which was written in 1615, would only be published in Strasbourg in 1636 (Cf. *Discoveries and Opinions*, p. 171). However, according to Jerome Langford, Galileo's interference in question of Biblical interpretation heightened the tension between him and Church authorities. See Jerome J. Langford, *Galileo, Science and the Church*, Ann Arbor 1966 / Ann Arbor (The University of Michigan Press, Ann Arbor Paperbacks) 1992[4], pp. 69-78 esp., and note 18, below.

[10] Voetius, *Narratio*, p. 50 / *Select. Dispp.* I, p. 880 / *Responsio*, p. 35 / *Querelle*, p. 114. Philips Lansberg (1561-1632) was a Flemish Protestant clergyman and well-known Copernican astronomer who influenced Kepler both on astronomical and exegetical questions. See P. C. Molhuysen / P. J. Blok (edd.), *Nieuw Nederlands Biografisch Woordenboek*, Leiden (A. Sijthoff's Uitgevers-maatschappij) 1912, deel II, cols 775-782. The second edition of the Latin translation of Galileo's *Dialogi*, the *Systema Cosmicum*, Lugdunum 1641, contains both an abstract from Kepler's *Astronomia Nova* and the translation of a letter by Paolo Antonio Foscarini of 1615; both dealing with the conciliation of Copernicanism and Scriptural authority. See also: Theo Verbeek, *Querelle*, pp. 475-476, note 97.

> the exceptions to these verses [of *Joshua*, *Ecclesiastes* and *Psalms*] are weak, untheological, dangerous and [...] excessively playful and extravagant with regard to Holy Writ.[11]

Voetius warns against taking Biblical interpretation too lightly and refers to *Romans* 12:3 and *Isaiah* 66:2, to inspire intellectual modesty and consciousness of divine grace especially regarding our understanding of the works of God.[12] Voetius had already put forward this standpoint in a previous work, the *Thersites Heautoutimorumenos*.[13] There he had rejected the Copernican resources on theological grounds. Not that he would want to intrude in other disciplines. On the contrary, Voetius explains that it is the Copernicans who, "for the sake of probabilities, i.e. their probable hypotheses, violate Holy Writ."[14] As a consequence, theologians have a perfect right to interfere *in theorema de Terræ immobilitate*. In fact, they are asked even by astronomers to give a decisive answer to the question whether the Copernican world-system is to be regarded a true picture of the Universe. Again, who else but theologians would have to

---

[11] Voetius, *Narratio*, p. 50 / *Select. Dispp.* I, p. 880 / *Responsio*, pp. 35-36 / *Querelle*, p. 114. The text is more or less identical to that of Voetius' *Assertiones theologicæ de præjudiciis veræ religionis*, Ultraj. 1634, "coroll. philosophico-theol., thes. 6", quoted in A. C. Duker, *Gisbertus Voetius*, vol. II, Leiden (E. J. Brill) 1910 / reprint: Leiden (Groen) 1989 (hereafter: *Duker*), p. 32, note 1: "Coelum quiescere, terram autem motu diurno rotari, saltem ex Scriptura probari non potest; quin immo non obscure ei repugnat. Ps. 19:6, 7 et 104:5. Ecclesiast. 1:4, 5. Exceptiones, quas nonulli hic adferunt, dilutæ et absurdæ sunt."

[12] Cf. *Romans* 12 :3: "For I say, through the grace given unto me, to every man that is among you, not to think of himself more highly than he ought to think; but to think soberly, according as God hath dealt to every man the measure of faith." *Isaiah* 66:2: "For all those things hath mine hand made, and all those things have been, saith the Lord: but to this man will I look, even to him that is poor and of a contrite spirit, and trembleth at my word."

[13] The *Thersites Heautoutimorumenos Hoc est, Remonstrantium Hyperaspystes, catechesi, et Liturgiæ Germanicæ, Gallicæ, & Belgicæ Denuo insultans, Retusus* [etc.], Ultrajecti, Ex Officinâ Abrahami ab Herwiick & Hermanni Ribii, 1635, forms part of a polemic by Voetius (and Martin Schoock, 1614-1669) with a Remonstrant author—probably Jean Batelier (1593-1672)—who had criticised Voetius' rejection of Copernicanism in an academic disputation. *Vide* Theo Verbeek, *Querelle*, p. 476, note 99. Ἑαυτὸν τιμωρούμενος is the title of a play by Menander, the 'Self-tormentor'. Θερσίτης is an Homeric name, meaning 'the Audacious'. In the *Thersites*, or 'Audacious Self-tormentor', Voetius and Schoock launch a violent attack on the Remonstrant faith, in which the issue of Copernicanism occurs as only one of many points on which Voetius and the Remonstrants hold different opinions.

[14] Voetius, *Thersites*, p. 256.

decide upon the right interpretation of Scriptural texts? Who else should silence those philosophers "who proclaimed that the story of Creation abounded with marvellous absurdities"?[15] Who should see to it that students of theology learned physics? And why let astronomers deal with holy things?[16] According to Voetius, theologians had every reason to discuss "the immobility of the Earth", but he remains eager to exclude all suspicion that he would want to transgress disciplinary boundaries or outwit astronomers in astronomy:

> we only venture to defend the true interpretation of *Ps.* 19:6 and 104:5 and *Eccles.* 1:4-5, which is accepted by all Theologians and Churches. This task is entrusted to us and to every theologian alike.[17]

And, according to Voetius, the "true interpretation" is a literal one. Even if one were to accept the view that the Scriptural text is accommodated to the vulgar ear, as did Galileo[18], this would not solve

---

[15] According to Voetius, "after Simplicius, there are today many of that kind", *Thersites*, p. 257. Simplicius (VI), the Neo-Platonist commentator of Aristotle, was the classical example of an heretical critic of the Christian account of Creation and was often attacked by Protestant theologians. Voetius refers to Franciscus Junius (or François du Jon, 1545-1602), a French Calvinist who served as a chaplain in the army of William of Orange and from 1592 until his death was professor of theology in Leiden. Cf. Alfred Davain, *François du Jon (Junius), Pasteur et Professeur en Théologie 1545-1602*, Reprint Paris 1882, Genève (Slatkine) 1970. Apart from three *Prælectiones in Geneseως* and an *Analysis* of the Book of *Genesis*, Du Jon wrote a *Confutatio argumentorum* XXII *quæ olim a Simplicio, in sacram Mosis historiam de Creatione, fuerunt proposita, & hoc nostro seculo ab hominibus prophanis atheisque recocta imperitis obtrundur*. See his *Opera Theologica*, Editio postrema, Genevæ, Apud Samuelem Crispinum, 1613, in 2 vols.

[16] Voetius, *Thersites*, pp. 257-258.

[17] Voetius, *Thersites*, p. 263.

[18] Galileo compared the matter to the way in which the Bible speaks of God, picturing Him as having "feet, hands, and eyes, as well as corporeal and human affections, such as anger, repentance, and sometimes even the forgetting of things past and ignorance of those to come. These propositions uttered by the Holy Ghost were set down in that manner by the sacred scribes in order *to accommodate them to the capacities of the common people*, who are rude and unlearned." Cf. Galileo, *Letter to Christina*, in *Discoveries and Opinions*, p. 181 (translation by Stillman Drake; my italics). It has been argued that it was Galileo's interference with Biblical exegesis that led to his first conflict with the Church in 1616. Cf. Jerome Langford, *Galileo, Science and the Church*, pp. 50-78 esp. It was the publication of the *Dialogo sopra i due massimi sistemi del mondo*, that brought about Galileo's second conflict with the Roman Catholic Church. Given the vast amount of secondary literature on the backgrounds of Galileo's trial and the inescapable attempts to settle questions of liability in this examplary crisis of religious and scientific authority, it is quite impossible to give a full account of the great variety

the problem that Scripture would still offer a misrepresentation of the facts in case the Copernican theory was true.

> So in order to please the vulgar, the Holy Spirit (blasphemy begone) would lie, and change white into black, square into circular, great into small, east into west?[19]

Even though the Bible is written in terms that are intelligible to everyone, there is yet no reason for Scripture to contain falsehoods.[20] The same reply may be given to the other argument—it was also given by Galileo[21]—that the Bible does not pursue total accuracy in matters less relevant to the faith. Would it therefore, Voetius asks,

> if I understand it right, [be] erroneous, misleading, or at least troublesome?[22]

Voetius sees no reason to shed doubt on the authority of Scripture and lays emphasis on the fact that a long tradition of theologians and philosophers had rather used the Bible as a source of *natural*—as well as ethical and religious—knowledge.[23]

---

of interpretations of what went on in Rome in the year 1633. The *Dialogue* did in any case provoke the Papal countermoves. Langford presents a detailed analysis of the events and, in a postscript to the third edition, offers a critical review of the most important publications on the subject of more recent date.

19 Voetius, *Thersites*, p. 266. Note that *Ecclesiastes* 1:6 reads: "The wind goeth toward the south, and turneth about unto the north; it whirleth about continually, and the wind returneth again according to his circuits."

20 In later discussions on Copernicanism, Voetius' pupil Martin Schoock would reiterate this standpoint. See Theo Verbeek, "From 'Learned Ignorance' to Scepticism", in Richard H. Popkin / Arjo Vanderjagt (edd.), *Scepticism and Irreligion*, p. 37, where Verbeek paraphrases Schoock's view as follows: "Had the Spirit chosen to instruct the faithful "ex mente vulgi", He would have been either unable or unwilling to speak the truth, which is utterly absurd." For references to Schoock: *ibidem*, notes 21 and 29.

21 According to Galileo, conclusions of physics "*in no way concern the primary purpose of the sacred writings*, which is the service of God and the salvation of souls—matters infinitely beyond the comprehension of the common people." The quotation is from the English edition of Galileo's *Letter to Christina*, in *Discoveries and Opinions*, p. 181 (translation by Stillman Drake; my italics).

22 Voetius, *Thersites*, p. 266.

23 That this would remain Voetius' position throughout, is shown by the 1656 disputation on "Some Miscellaneous Opinions", in which the question of Copernicanism is shortly dealt with in the following way: "*Terram moveri, solem quiescere. Quæritur*, quid futurum sit, textibus scripturæ hactenus à theologis non tantum, sed & Christianis physicis atque astronomis (quorum aliquos alibi indicamus) in contrariam sententiam acceptis?" See *Select. Dispp.* IV, p. 754. In the

The words of *Joshua*, *Ecclesiastes* and *Psalms* are clear and descriptive and those of *Ecclesiastes* and *Psalms* in fact form a complete logical sequence. Besides, if verses which relate to astronomical facts were not to be taken as literally true, the same might be said for those relating to chronology. Chronological dates are often rounded off in the Bible, but, according to Voetius, this should be no reason for arguing that there is no chronology to be deduced from Scripture.[24] The same holds for astronomy. There can be no pretexts when it comes to interpreting the passages from *Joshua*, *Ecclesiastes* and *Psalms*. *Psalms* 19 is in fact a fine example of "natural theology", in which the Psalmist

> leads us from the Book of Nature and the works of God, to knowledge of our Creator and in that way inspires piety.[25]

In the whole chapter of the *Thersites* in which Voetius discusses Copernicanism, the astronomical question is presented not just as a relevant question for theology, but in fact as a theological one. For the twentieth-century reader who is not acquainted with theological interference in theoretical disciplines, but has rather been brought up with the tale of Galileo's martyrdom, it is hard to appreciate the problem which the seventeenth century had to confront as regards these questions of authority. It is clear that Voetius did not himself see his discussion of Copernicanism as an interference in astronomy. Whereas we might interpret the situation as a transgression of disciplinary boundaries by the theologian, Voetius' own view of the matter is completely the opposite. What right, Voetius asks, do

---

Thersites, Voetius mentions various Christian writers who had written in the tradition of commentaries on the book of Genesis and quotes his near contemporary Lambertus Danæus as saying that "physics is included in Holy Writ and is in some way a part of theology and subjected to it." *Thersites*, pp. 267-268. We shall return to these writers on physico-theology in Chapter 3, below.

[24] Education in chronology was one of great relevance for the student of theology, especially so since chronological matters formed an important point of incongruity between the Hebrew and Septuagint texts of the Old Testament. Another problem was that chronologies of Egyptian dynasties extended human history to times before the Flood or even before the supposed date of Creation. Cf. Anthony Grafton's excellent article on "Joseph Scaliger and Historical Chronology: The Rise and Fall of a Discipline", in *History and Theory* 14 (1975), pp. 156-185. On the implications for Protestant theology, see David S. Katz, "Vossius and the English Biblical Critics", in Richard H. Popkin / Arjo Vanderjagt (edd.), *Scepticism and Irreligion*, pp. 142-184.

[25] Voetius, *Thersites*, p. 273.

astronomers have to meddle in Scripture? Why should the opinion of nearly all theologians and philosophers, both living and dead, yield to the hypothesis of some arithmeticians and geometers, who stand alone in rejecting the universal judgement of the learned?[26] There is no indication that Voetius expects that the issue may, in the end, also be settled by other than theological considerations.

Both in the *Thersites* and in the essay on forms, Voetius gives replies to the non-Scriptural kind of arguments concerning Copernicanism. Yet his primary concern is with theology. In the *Thersites*, the non-theological part of his argumentation is no more than a short, rhetorical, reply to the arguments in favour of Copernicanism. Voetius mentions the "Copernican" argument of the unimaginable nature of a heavenly sphere revolving at the incredible speed of an estimated 2,317,829,692 German miles per hour. He retorts that the estimated speed of the Earth (225 Gmph for the daily and 740 Gmph for the annual rotation) is just as incredible and that it is contrary to fact that an arrow would be carried away in an eastward direction at the speed of nearly 4 miles a minute.[27] In the essay on forms, Voetius' position is again the standard Scholastic one, in which "counterexamples" to the idea of a moving Earth are given on the basis of ballistic and gravitational considerations such as the weight falling back perpendicularly on the Earth's surface, the cannon ball shot from different directions and the stability of buildings, trees, the waters of the seas, and other things which, according to anti-Copernicans, would be lost in space if the Earth were to move.[28]

In the essay on forms, the "natural reasonings" take up the larger part of the argument. This is probably due to the fact that Voetius had

---

[26] Voetius, *Thersites*, p. 259. Voetius argues (quite rightly) that nearly all philosophers and theologians reject the Copernican viewpoint; *Thersites*, p. 256, p. 259 and p. 267. An exception is made for the 'thersites heautoutimeros' or 'audacious self-tormentor' whom Voetius is addressing throughout the work; see above, note 13.

[27] Voetius, *Thersites*, pp. 260-261.

[28] Voetius, *Narratio*, p. 50 / *Select. Dispp.* I, p. 881 / *Responsio*, p. 36 / *Querelle*, p. 114. See also the *Tersites*, pp. 260 ff. A fine analysis of the various mechanical problems involved, of the way in which adversaries of the Copernican theory posed them and of their eventual solution by Galileo, Kepler and Newton may be found in I. Bernard Cohen, *The Birth of a New Physics*, London (Heinemann) 1960 / London (W. W. Norton & Co.) 1985 / London (Penguin) 1992. Another very lucid source is Jacques Gapaillard, *Et pourtant elle tourne, Le mouvement de la Terre*, Paris (Éditions du Seuil) 1993, which describes the history of the idea of the Earth's movement from ancient times up to the present day.

been provoked to answer to the arguments of his colleague Regius, who had presided over a disputation in which the Copernican viewpoint was defended.[29] But in any case, Voetius' writings on Copernicanism do not convey a great interest in physics or astronomy. Neither, then, would it be fruitful to dwell on his restatement of stock-arguments from pre-Galilean physics. What makes his discussion of Copernicanism interesting is that it shows us where Voetius' priorities lie. His reaction to Copernicanism is very well summarised in the long quotation from Melanchthon which he takes up in the *Thersites*.

Melanchthon criticizes the vanity of writers who haughtily, or "for the love of novelty", invent new ideas such as that concerning the stability of the Sun and the "eighth sphere".[30] He goes on to say that the right thing to do is "to respectfully embrace what God has revealed to be true." Although some may laugh at the divine testimonies with an appeal to physics,

> we nevertheless think it honorable to devote philosophy to the heavenly sayings and, given the obscurity of the human mind, to consult divine authority whenever we can.[31]

Next, Melanchthon goes over the passages from *Psalms*, *Ecclesiastes* and *Joshua*, concluding that these convey the truth so clearly, that there is no reason to be "distracted by the tricks of those who think it is a virtue to throw the sciences into disorder."[32] With Scripture in hand, nothing more could be desired, even if one were to abstain from all physical and mathematical reasonings.[33]

As far as appreciation of mathematical studies is concerned, Melanchthon and Voetius are at opposite extremes.[34] Yet Voetius

---

[29] Cf. *Duker* II, p. 149, app. XLIX.

[30] As Emil Wohlwill pointed out in 1904, Melanchthon scrapped this particular phrase in the second edition of the *Initia doctrinæ physicæ*. See Robert Westman, "The Wittenberg Interpretation", p. 173 and *ibidem*, note 31.

[31] *Thersites*, p. 269 and Philip Melanchthon, *Initia Doctrinæ Physicæ*, Carolus Gottlieb Bretschneider (ed.), *Coprus reformatorum*, Vol. XIII, *Scripta Phil. Melanthonis ad Historiam Profanam et Philosophiam spectantia*, Halle (C. A. Schwetschke und Sohn) 1846, col. 216.

[32] *ibidem*, cols 216 and 217.

[33] Cf. Voetius, *Thersites*, p. 269: "Etenim si probæ examinemus sacrum codicem, haud difficile est ex eo omnium fundamentorum Astronomiæ [...] veritatem evincere & obtinere, adeo ut si cunctis Physicorum & Mathematicorum rationibus abstinendum foret, in his solis nihil ultra desiderari possit."

[34] Cf. Robert S. Westman, "The Wittenberg Interpretation". Westman shows that Melanchthon not only became more moderate in later years (see above, note 5),

finds in Melanchthon the same rejection of idle curiosity, the same incentive to piety and the same standpoint as regards the self-sufficiency of Biblical interpretation. As was said, in the essay on forms, Voetius warns against taking Biblical interpretation too lightly and refers to *Romans* 12:3 and *Isaiah* 66:2 to inspire intellectual modesty.[35] He might just as well have left it at that. No physical argument was in any case to be expected that might prove the theologian wrong and the Copernican opinion to be more than just a mathematical hypothesis. The issue of Copernicanism thereby illustrates the self-confident position of the Neo-Scholastic theologian whom Voetius here represents: a position which was marked by an intention to save the unity of truth and which was upheld by a system of education in which theology, as a "higher" discipline, held an unquestioned right to supervise philosophical theory. It is to these topics that I shall now turn.

1.2 Physics, Theology and the Unity of Truth

The question of Copernicanism does not seem to fit well into the general line of argument of Voetius' essay on forms. Neither the concept of "nature" nor that of "substantial form" occurs in the *Thesis* which deals with Copernicanism. Yet Voetius' rejection of the

---

but that he was in fact very interested in the Copernican theory, and at points accepted certain of its aspects: "Thus he praises Copernicus' lunar theory because it is 'so beautifully put together' [...] In several places he uses Copernican data for the solar apogee and for the apogees of the superior planets. And in the second edition of his *Initia doctrinæ physicæ* [...] he tones down the negative allusions to Copernicus" (*idem*, p. 173). Finally, supporting the mathematical studies of his pupils, Melanchthon, the *Præceptor Germaniæ*, also stood at the basis of the spread of Copernican ideas throughout Germany. In these respects, Melanchthon's attitude is, therefore, very different from that of Voetius, who, in the *Thersites*, makes an effort to humiliate mathematicians—saying that the entire community of theologians and philosophers are "no less rational beings than arithmeticians and geometers" (*ibidem*, p. 259)—and, in the Cartesian debates, warns against the magical conclusion that numbers could have physical efficacy. (We shall see what he means by this in Chapter 6, below.) As for Melanchthon, Robert Westman, in "The Wittenberg Interpretation", p. 173, justifyably says that: "[the] customary dichotomy of "pro" and "anti" Copernican [...] becomes less than adequate as a description of Melanchthon's views and those of his disciples". Nevertheless, this is only so with respect to the question of mathematical vs. physical interpretations of Copernicanism. As far as the physical interpretation is concerned, Melanchthon has the same difficulties as Voetius on theological grounds.

[35] See above, note 12.

Copernican hypothesis is typical both for the main line of argument in the essay and for his standpoint as regards physics in general.

As we saw, in attacking Copernicanism, Voetius' principal concern is with saving the authority of Scripture. This, as he puts it in the *Thersites*, is his proper task (*sparta*) as a theologian. From that point of view, a questioning of the accepted geocentric world-order could only be interpreted as idle curiosity. This is also how Melanchthon saw it: Why regard the hypothesis of the Earth's rotation as an actual fact? Like Melanchthon, Voetius might well have accepted the possibility that the Copernican system might have advantages as a mathematical hypothesis[36], but since the Bible in fact teaches us the stability of the Earth, the physical question is settled.

Thus, the "natural reasonings" against Copernicanism only have the task either of supplying additional or circumstantial evidence for what is already known by the authority of Holy Writ—as in the essay on forms—, or—as in the *Thersites*—of giving counter-arguments to the idle hypotheses of philosophers who speak with double tongue, that is, against the Christian truth. Apart from that, the "first type of argument", i.e. the Scriptural evidence, settled the question anyway and in fact leaves no room for further discussion.

---

[36] Cf. Robert Westman, "The Wittenberg Interpretation", p. 169: "Whilst the historical origins of this dissociated methodological viewpoint may be traced back to an earlier split between natural philosophy and mathematical astronomy in the Middle Ages and to Osiander's unique role in affixing his anonymous letter to *De revolutionibus*, the origins of its institutional entrenchment and promulgation must be sought in Melanchthon's Wittenberg circle." And, *idem*, pp. 173-174: "[Melanchthon] recognizes that the Earth's motion *could* be interpreted as a real one, and he explicitly rejects this possibility." This was also what Voetius' disciple Martin Schoock did. He too argues that, as long as one does not conclude that there are real eccentric spheres or epicycles in the heavens, one may introduce all sorts of eccentric orbits and epicycles in order to save the celestial phenomena. Cf. Martinus Schoockius, *Admiranda Methodus Philosophiæ Renati Des Cartes*, Vltraiecti, Ex Officina Joannis van VVaesberge, 1643, p. 238 / *Querelle*, p. 301. Schoock's point is to argue that a mathematical model never has the force of a physical theory concerning the true nature of things. Theo Verbeek interprets Voetius' tolerance of Jacob Ravensberg's (1615-1650) disputation on Copernicanism as proof that Voetius accepted the defence of the Copernican view as long as it was thought to be a mathematical hypothesis only. Cf. Verbeek, "Voetius en Descartes", in J. van Oort *et al.*, *De onbekende Voetius*, pp. 212-213. It should however be said that Ravensberger inclined to the Tychonian system, which was consistent with Biblical views and to which Voetius was generally well-disposed.

This way of reasoning calls into question the status of the "natural reasonings" in Voetius' writings. There is of course a long and complicated history of viewpoints on the use of natural knowledge within theology and of the relevance of philosophy for the faith, especially in the context of the so-called problem of double truth.[37] At the end of the sixteenth century, sceptical and otherwise anti-Scholastic ideas were widespread and were often accepted in both Protestant and Catholic circles. Voetius not only aimed to call a halt to these anti-rationalist schools, but was, on the other side, also supposed to defend the faith against rationalist tendencies, such as those of the Socinianist and Remonstrant thinkers.

It is therefore no surprise that we find the first three of Voetius' collected disputations devoted to the difficult issue of searching the right equilibrium between reason and faith.[38] In the first disputation, "Of Human Reason in Matters of Faith", Voetius states that there are two main controversies: one with the Socinians and the other with some modern Catholics. The Socinians are said to be akin to the sceptics and the Pyrrhonists.[39] They hold reason to be the highest religion, saying that one can decide on the possibility or impossibility of articles of faith only by reason, and that nothing should be believed that seems contrary to it.[40] Voetius analyses the matter by introducing various distinctions concerning the notions of reason and faith. I shall not discuss these. What matters is that he claims that all truth, all articles of faith and all its conclusions are, in the last instance,

[37] A complete exposition of the development of the problem of double truth leading from the fourteenth-century revival of Thomism and the fifteenth and sixteenth-century discussions between the various branches of Aristotelianism in Italy to Spanish and German Neo-Scholasticism of around 1600, may be found in Charles H. Lohr's excellent chapter on "Metaphysics" in *The Cambridge History of Renaissance Philosophy*, Cambridge (C. U. P.) 1988 / 19902, pp. 598-638 esp.

[38] Voetius, *Select. Dispp.* I. The first of these is the disputation "De Ratione Humana in Rebus Fidei", which was held in February 1636, in the first days of Utrecht University. The second is the "De Theologia Scholastica" of February 1640 and the third is a disputation of July 1636 on the question "Quousque Se Extendat Autoritas Scriptvræ", to which I shall return later.

[39] The fact that Voetius mentions scepticism and Pyrrhonism in relation to Socinianism, seems to offer an argument in favour of Richard Popkin's project of including dogmatic writers such as Isaac la Peyrère and Spinoza into the sceptical tradition on the basis of their criticisms of Biblical truth. Cf. Richard H. Popkin, *The History of Scepticism from Erasmus to Spinoza*, Berkeley (University of California Press) 1979. See also: Richard H. Popkin / Arjo Vanderjagt (edd.), *Scepticism and Irreligion*, pp. 3-4 esp.

[40] Voetius, "De Ratione Humana in Rebus Fidei", *Select. Dispp.* I, p. 1.

dependent on the "external principle of faith", which is God's Word. Human reason in its present state of Fall, is, on the other hand, poor and blind and hence incapable of scrutinizing divine mysteries.

It would be wrong, however, to say that Scripture is the only principle of faith. Some French Catholics unjustly attribute this position to the Protestants.[41] Against them, Voetius argues that Reformists *do* accept the natural light as well as Biblical authority. The issue indicates the vulnerability of a standpoint which tries to keep a balance between the two extremes of critical reason and blind faith. However, as is often the case in theological disputes, the polemic leaves plenty of room for a variety of intermediate standpoints, especially since Voetius' formulation of the Catholic reproach makes refutation easy. Voetius lists a whole series of arguments in favour of the use of reason in theology and faith. The interesting point is that questions of "natural theology" are immediately left out. For with regard to these, the question does not arise. Viewpoints that pertain to natural theology are by definition to be examined and proved in both ways: "primarily from Scripture, secondarily by the natural light."[42]

In seventeenth-century Protestant thought, "natural theology" may refer to that part of metaphysics which was called *metaphysica specialis*, or *pneumatology*. The discipline deals with "the nature, properties and activities of spiritual being",[43] that is to say, with God, angels and human souls. The occurrence of these subjects within the discipline of metaphysics is significant in that it shows that the Christian metaphysician believes certain doctrines of revelation, such as the immortality of the soul, to be knowable also through natural means, viz. through the study of metaphysics.[44] Therefore, such

---

[41] Voetius' adversaries here are the so-called "Verronians". Franciscus Veron (d. 1649) was a Jesuit priest who committed himself to the task of converting his Huguenot compatriots. He is reported to have written a *Veronian Method for Convincing Heretics*. Cf. *Jöcher* IV, cols 1543-1544.

[42] Voetius, *Select. Dispp.* I, p. 7.

[43] Charles H. Lohr, *The Cambridge History of Renaissance Philosophy*, p. 629. See also Brian P. Copenhaver and Charles B. Schmitt, where they discuss the *Natural Theology or Book of Creatures* of Raimond Sebond: "*Natural* (as distinct from *revealed*) theology infers God's existence and attributes from creatures by analogical and other rational arguments", in *Renaissance Philosophy* (*A History of Western Philosophy* 3), O.U.P., Oxford 1992, p. 252.

[44] On natural theology in our period, see also: John E. Platt, *Reformed Thought and Scholasticism: the Arguments for the Existence of God in Dutch Theology 1575-1650*, Leiden (E. J. Brill) 1982.

doctrines may be known through both ways, or, as Voetius says, "primarily from Scripture, secondarily by the natural light."

In a larger sense, "natural theology" might refer to any way of reasoning by which natural knowledge and revealed religion concur.[45] It is in this way that Voetius uses the concept in the context of Copernicanism. As we saw above, Voetius claims that *Psalms* 19 offers a fine example of "natural theology", in which the psalmist

> leads us from the Book of Nature and the works of God, to knowledge of our Creator and in that way inspires piety.[46]

This means that truths of physics may also count as examples of truths which may be proved both ways: either by revelation or through the "natural light".[47] And again we may point to Melanchthon, who illustrated the issue of the double way to truth with an example belonging not to pneumatology, but to physics proper. As Charles Lohr writes, giving an example of truths accessible to reason and confirmed by revelation, Melanchthon "appealed to the fact that the universal experience of men that the Earth is stationary, while the Sun moves, is confirmed by scripture."[48] Here the Copernican issue is even taken as exemplary for those questions which can be known both

---

[45] Apart from divine attributes, God's *works* might also be the point of focus in the *theologia naturalis*, as it had already been in the *Natural Theology* of Raimond Sebond (see above, note 43). It seems that under the influence of the Newtonian physico-theology, the concept of "natural theology" was later to be used in physical rather than metaphysical contexts. See J. W. Buisman, *Tussen Vroomheid en Verlichting*, Zwolle (Waanders) 1992, p. 28: "De achttiende-eeuwse geschriften spreken vaak over rede en openbaring als twee zelfstandige en met elkaar overeenstemmende kenbronnen, waarbij de eerstgenoemde ook competent werd geacht op het terrein van de zogenaamde "natuurlijke theologie", i.e. de leer aangaande de eigenschappen en werken Gods [...]. Uiteindelijk evolueerde dit optimisme bij enkelen zelfs tot de overtuiging dat de rede de openbaring geheel kon missen."

[46] Voetius, *Thersites*, p. 273.

[47] See also note 23, above, where I referred to Voetius' quotation of Lambertus Danæus, who said that "physics is included in Holy Writ and is in some way a part of theology and subjected to it." Voetius, *Thersites*, pp. 267-268.

[48] Charles H. Lohr, *The Cambridge History of Renaissance Philosophy*, p. 623. See also Melanchthon: "Grata est autem bonæ menti confirmatio veritatis, ubi intelligit ad naturalem noticiam accessisse divinam vocem. Ratio deprehendit terram stare immotam, et solam moveri. Sed cum idem divinitus traditum audimus, firmius adsentimur." Melanchthon, Philippe, *Liber de anima*; *Scripta Phil. Melanthonis ad Historiam Profanam et Philosophiam spectantia*, in Carolus Gottlieb Bretschneider (ed.), *Coprus reformatorum*, Vol. XIII, Halle (C. A. Schwetschke und Sohn) 1846, col. 152.

through revelation and natural reasoning. Voetius' standpoint is the same as Melanchthon's in this matter and his reaction to Copernicanism is exemplary in the same sense as it is for the German theologian. Their principal argument is that of the unity of truth.

Voetius' way of dealing with the question of Copernicanism is important in that is shows us how to read his 1641 essay against the New Philosophy where it discusses other topics. Saving the unity of truth is in fact the *fil rouge* of the essay "On the Natures and Substantial Forms of Things". Right at the start of *Thesis* II of the essay, where Voetius presents his first "preliminary note or doubt" to the followers of the New Philosophy, he asks them to consider

> whether they can give a satisfactory account of the conciliation of [their] opinion with Holy Writ. For truth agrees with truth and Christian philosophers will rather a thousand times profess their learned ignorance, than inflict even the tiniest sort of prejudice upon divine truth.[49]

The point at issue is not Copernicanism, but the rejection of substantial forms—a question to which we shall return in the next chapter. What matters here, is that, like questions of metaphysics, questions of physics may also have a double frame of truth, both Scriptural and natural. According to Voetius, this is indeed the case. In the third of his *Selected Disputations*, Voetius puts forward the question of the authority of Scripture and its reach.[50] There are, according to Voetius, two ways of going astray: one *in excessu*, by ascribing too much to divine inspiration, the other *in defectu*, by denying important Biblical information the status of being divinely inspired. The first category is easily dealt with: we should not overestimate the value of the marginal notes which scribes have added in the margins of the Old Testament.[51] As to the other category, Voetius lists various examples, such as that of the Anabaptists, who only accept the New Testament, and that of authors who attribute

---

[49] Voetius, *Narratio*, p. 39 / *Select. Dispp.* I, p. 872 / *Responsio*, p. 11 / *Querelle*, p. 105. On the function of "learned ignorance" in Voetius' writings, see Theo Verbeek, "From 'Learned Ignorance' to Scepticism; Descartes and Calvinist Orthodoxy", in Richard H. Popkin / Arjo Vanderjagt (edd.), *Scepticism and Irreligion*, pp. 31-45.

[50] "Quousque Se Extendat Autoritas Scriptvræ", *Select. Dispp.* I, pp. 29-48.

[51] Voetius, *Select. Dispp.* I, p. 36: "In excessu peccant & nimis [Scripturam] extendunt, qui τὸ *keri* seu lectiones marginales V.T. faciunt autheticas, vel solas excluso & extruso τὸ *ketib* [...], vel una cum *ketib*."

Biblical passages to fallible human effort instead of to the infallible and immediate revelation of the Holy Spirit.[52] Just as often, Voetius takes the opportunity to refute the Socinians and their allies (i.e. those of the Remonstrant creed) who are typical for doubting the authority of Scripture on the basis of human reasonings. Most significantly, however, he refers to his Remonstrant adversary of the *Thersites*,

> who denies that the *Scriptores sacros* [...] are authentic in those matters which are not absolutely necessary for salvation.[53]

Though not naming Copernicanism, Voetius here repeats his earlier argumentation against the followers of Copernicus and Galileo. His point is, that taking Scripture literally is as much an axiom of natural philosophy as of natural theology. Scripture is not merely concerned with metaphysical questions such as the immortality of the soul, but with physical Nature too. For this reason, adherents of the New Philosophy should first consider whether their theories are consistent with Scripture.

This emphasis on the unity of truth, even outside the context of the *metaphysica specialis*, leaves one wondering whether it is in fact the *object* of natural philosophy to vindicate Scripture, in which case the main task of physics would be exegetical. Extreme though this position may seem, it is exactly what Voetius holds to be the case. As he had said in his *Sermon on the Usefulness of Academies and Schools*, a sermon given on the occasion of the opening of Utrecht University:

> If these things [i.e., the secrets of Nature] are taught in the academies and schools, what else does one do but explain Scripture and take from it the doctrines of things natural? Moreover, if a student of physics examines many things in such a science and aims to become experienced in it, what else is he doing but admiring and considering the works of almighty God, and at the same time lending a helping hand to the theologian and indeed to all readers and lovers of Holy

---

[52] Voetius mentions Castellio who did not take seriously what St. Paul wrote on "election" in *Romans* 9. Voetius, *Select. Dispp.* I, pp. 32 and 40. Castellio, or Castalio is the Latin name for Sebastien Châtillon (or Chasteillon), was a Basel professor of Greek and the author of a book *On Predestination and Justification in Chapter 9 of the Letter to the Romans*. Chatillon, a fierce critic of Calvin and Beza, was later to be regarded a precursor of Arminianism. Significant in this respect is the fact that his theological works appeared in a Dutch translation in Haarlem in 1613. Cf. *Jöcher* I, cols 1737-1738.

[53] Voetius, *Select. Dispp.* I, p. 40.

> Writ (who, night and day, find their meditation and joy therein) in order to understand even better so many of its chapters and proverbs.[54]

Scripture thereby acquires a central place even outside the boundaries of theology. According to Voetius, any intellectual undertaking is necessarily connected with an attempt to deepen our knowledge of Creation and hence of Holy Writ.[55]

It is for his emphasis on the authority of Scripture in particular, that Voetius is ranked amongst the key figures of the *Nadere Reformatie*, or "Further Reformation", a pietist movement within the Dutch Reformed Churches for which the primacy of theology was a central doctrine.[56] But whereas the doctrinal aspect of Voetius' emphasis on Scripture and his place in the movement of the *Nadere Reformatie* are well documented, it is much less well appreciated how

---

[54] Voetius, *Sermoen van de Nuttigheydt der Academien ende Scholen mitsgaders der Wetenschappen ende Consten die in de selve gheleert werden* [etc.], Tot Utrecht, Bij Ægidius ende Petrus Roman, Druckers van de Universiteyt, Anno 1636, p. 28.

[55] See also S. van der Linde, "Gisbertus Voetius in zijn conflict met Jean de Labadie", in *Opgang en Voortgang der Reformatie: Een keuze uit lezingen en artikelen van Prof. Dr. S. van der Linde*, Amsterdam (Ton Bolland) 1976, p. 264: "In dezelfde lijn ligt wat Voetius in deze oratie [sc. of March 1636] uiteenzet aangaande de *bijbel*. Die is hem het Woord van een heil, dat alles te boven gaat. Maar hij weet dat Woord evengoed aan te prijzen als fundamentboek voor alle wetenschappen, die daar oriëntatie en inspiratie hebben te vinden." According to Voetius, this applies to astronomy in particular: "Soo dat niet sonder reden eenighe treffelijcke Meesters in dese conste geerne bekent hebben datmen uyt de Schriftuere de waerheyt van alle Astronomische fundamenten kan ende behoort te bewijsen / ende dat de Schriftuere de veylichste haven is om daer uyt inde Zee van de Astronomische speculatien henen te vaeren." Voetius, *Sermoen van de nuttigheydt der Academien ende Scholen*, p. 32. See also De Vrijer, *Henricus Regius*, p. 206.

[56] See Rienk H. Vermij, *Secularisering en Natuurwetenschap*, p. 60 and the further literature mentioned, p. 60, note 10. In *Opgang en Voortgang der Reformatie: Een keuze uit lezingen en artikelen van Prof. Dr. S. van der Linde*, students of Professor Van der Linde emphasize Van der Linde's view of the *Nadere Reformatie* as differing in important aspects from English Puritanism and German Pietism, *op. cit.*, p. 7. It was with regard to the fact that revealed religion seems to leave no room for natural theology otherwise than in trying to convince the unbeliever, that in our days the Calvinist theologian Professor A. A. van Ruler has dealt with "the other side of natural theology", i.e., with the meaning which "general (i.e. natural) revelation" might have—not for the heathen, but for the Christian himself. Van Ruler's answer—that natural theology inspires gratitude for being created *tout court*—seems to me however, to be too existentialist an interpretation to be of use for understanding Voetius. See A. A. van Ruler, *Theologisch Werk* Deel VI, Nijkerk (G. F. Callenbach) 1973, pp. 41-48.

Voetius' standpoint with regard to physics and the unity of truth was intertwined with contemporary views on university schooling. In the next section, I shall present Voetius' position in relation to pedagogy.

1.3 Didactical matters

As stated in section 1.1, the Senate of Utrecht University protested against the introduction of novel theories, particularly on the grounds that alternative theories obstructed succesful education. The proper function of philosophical theory was to proffer the basic apparatus for the higher faculties. If students were to neglect the traditional subject-matter of their propedeutic years, further educational development would be hampered. Given the fact that the higher faculties (especially theology and medicine) based their teachings to a considerable degree on Aristotelian terminology, the substitution by novel theories of the "received" philosophy would imply more than just a change of courses. The reaction of the Utrecht Senate thus gives us insight into their view of a unified body of science.

The revival of scepticism at the end of the sixteenth century may have brought about a sense of relativism as regards the schools of philosophical thought,[57] but such an outlook was certainly not shared by university professionals like Voetius. Far from leaning towards a relativist standpoint, they saw the *philosophia recepta* rather as a fixed body of learning and a necessary tool for doing "science" in the higher faculties. Philosophical theory was supposed to have a certain intrinsic objectivity.

This is mirrored in the way in which philosophy was taught. Professors based themselves on commentaries and compendia of logic, physics and metaphysics.[58] These, it should be noted, were not held to comprise just "Aristotle's philosophy" as against the philosophies of Plato or Epicurus. Contrary to our way of seeing things, in seventeenth-century universities philosophy was not taught relative to

---

[57] For the revival of scepticism in the sixteenth century, see Richard H. Popkin, *The History of Scepticism from Erasmus to Spinoza*, Berkeley (University of California Press) 1979.

[58] An aspect sharply criticised by Descartes in the fourth section of his 1643 *Letter to Voetius*, "On Voetius' knowledge and his use of books", AT VIII-II, pp. 40-42 / *Querelle*, pp. 350-351. On Descartes' criticism of Voetius' dialectics, see: K. van Berkel, "Descartes in debat met Voetius. De mislukte introductie van het Cartesianisme aan de Utrechtse Universiteit (1639-1645)", in *Tijdschrift voor de Geschiedenis der Geneeskunde, Natuurwetenschappen, Wiskunde en Techniek* 7-I (1984), pp. 4-18.

philosophers. What was taught was the *subject* of philosophy. Although it was generally accepted that this philosophy derived mainly from Aristotle, the content of the subject was not thought to depend on its accidental source.[59]

Since philosophy was a preliminary to higher education, the introduction of alternative philosophical or physical theories was judged to be quite impracticable. This is part of what is intended in the verdict against the New Philosophy of Utrecht University's Senate. In the case of theology, which in its Neo-Scholastic form was highly dependent on Aristotelian theory, the need for a thorough education in traditional terms was of the utmost importance. To Voetius and other Neo-Scholastics—were they Catholic or Protestant—the success of teaching theology thus depended on the success of teaching philosophy. Supplying a certain technical jargon was one way in which philosophy acted as the "maidservant" of theology. But there were other ways in which theology was most vulnerable to philosophical or scientific change.

At the end of the previous section, I claimed that Voetius' standpoint with regard to physics and the unity of truth was intertwined with contemporary views on university schooling. We may now see why. Since, according to Voetius, the Bible was supposed to embody a great number of "natural", "physical", or "philosophical" truths as well as purely religious ones, philosophical systems were judged according to their success in securing the unity of truth.[60] This would mean that that philosophical theory was to be accepted which in its "secular" way matched a Biblical or religious viewpoint. Accordingly, Voetius' Essay "On the Natures and Substantial Forms of Things" is full of arguments connecting theological and Biblical

---

[59] Likewise, a Neo-Scholastic philosopher as the Leiden professor Burgersdijk would not teach from Aristotle, Ramus, or Bacon. What he did was to teach a subject—e.g., metaphysics—using mainly Spanish contemporary sources. See my "Franco Petri Burgersdijk and the Case of Calvinism within the Neo-Scholastic Tradition", in E. P. Bos and H. A. Krop (edd.), *Franco Burgersdijk (1590-1635): Neo-Aristotelianism in Leiden*, Amsterdam / Atlanta GA (Rodopi) 1993, p. 54 esp.

[60] It has been argued that disciplinary fragmentation and loss of theological commitment are important criteria for distinguishing between pre-modern and modern philosophy. See Beverley Southgate, "'Torn between Two Obligations': The Compromise of Thomas White", in Tom Sorell (ed.), *The Rise of Modern Philosophy: The Tension between the New and traditional Philosophies from Machiavelli to Leibniz*, Oxford (Clarendon Press) 1993, pp. 107-127, and my forthcoming review of this book in the *Archiv für Geschichte der Philosophie*.

views with the notions of the accepted, and mostly Aristotelian, body of learning.

We have already seen what this led to in the case of the discussion over the Copernican hypothesis. Since Scripture settled truth, the unity of truth could only be saved by parallel reasoning in philosophy. This is another way in which philosophy was to be the "maidservant" of the higher discipline of theology: apart from offering technical vocabulary, it was also supposed to demonstrate and deduce those "natural truths" which were thought to be in accordance with Scripture and theological dogma.[61] And this servitude was expressed in the educational hierarchy as well: theology could prescribe what philosophical doctrines were to be taught.

The University Senate spoke of the undermining character of novel philosophies only in general terms. The theologians themselves were more specific. Besides the fact that the rise of the New Philosophy led to quarrels incompatible with the study and practical training of their students and that the young Utrecht academy would be suspected of "paradoxical modernisms", the Faculty of Theology pointed out specific philosophical doctrines incompatible with Christianity. They said, first, that the denial or negligence of certain ideas of metaphysics and logic could hardly be avoided once new ideas were made acceptable and that scepticism would be the result. Moreover, the new ideas of physics which were put forward by Henricus Regius, the Professor of Medecine, were of great consequence for questions of substance and accident, substantial and accidental union, the unity of body and soul, the efficacy of quantity etc. According to the theologians, these questions were of more concern to theology than to medecine. Further, it would be intolerable to let students of theology accept ideas which were incompatible with the *physica sacra* of

---

[61] The "maidservant"-metaphor was brought forward by later Voetians especially against Cartesianism and Copernicanism, as the Utrecht minister Jac. du Bois wrote, philosophy, being only the maidservant of theology, should be silent when the mistress speaks: "De Philosophie is maar een dienaresse van de Theologie / de dienaresse nu moet swijghen daer de vrouwe spreeckt." J. du Bois, *Naecktheyt van de Cartesiaensche Philosophie, Ontbloot in een Antwoort op een Cartesiaensch Libel, Genaemt Bewys* [etc.], T'Utrecht bij Johannes van Waeberge, 1655, p. 10, cited also in De Vrijer, *Henricus Regius*, p. 209 and Theo Verbeek, "From 'Learned Ignorance' to Scepticism", in Richard H. Popkin / Arjo Vanderjagt (edd.), *Scepticism and Irreligion*, p. 41.

Moses, as interpreted by Church Fathers, Scholastics and recent Christian authors.[62]

Copernicanism was a theory which according to many theologians was not in line with Scripture, but the *physica sacra*, or Mosaic physics dealt with different questions entirely. Not only was the Christian tradition of studying Nature in relation to Scripture founded long before Copernicus: it also took its origin primarily in the Book of *Genesis*, in which no reference to the relative positions of the Earth and Sun occurs.[63] One of my main objects in the rest of this book will be to present other topics that form part of a Christian physics. For the moment it should be enough to have pointed to the fact that for the zealot of the Further Reformation, a certain educational hierarchy was as self-evident as was a reluctance to let philosophy develop on its own. The intertwined character of the questions of double truth, of disciplinary "authority", and of university organisation, was not a matter of coincidence.[64] Though appreciated differently, this intertwinedness was equally acknowledged and discussed by authors as diverse as Voetius and Galileo.[65]

---

[62] *Narratio*, p. 26 / *Querelle*, pp. 96-97. See also: Theo Verbeek, "Voetius en Descartes", in *De onbekende Voetius*, pp. 203 ff.

[63] Apart from being often interpreted—especially in relation to Galileo's trial—as the most important point of conflict between theology and physics, Copernicanism has also been the central focus of some studies of the conflict between Voetius and Descartes. So for instance by Vermij, who discusses Copernicanism, whilst skipping over all other points of conflict for the reason that they are "mainly esoteric questions of metaphysics". Rienk H. Vermij, *Secularisering en Natuurwetenschap*, p. 61. Even Theo Verbeek, who has studied the *Querelle d'Utrecht* in a great variety of aspects, discusses only Copernicanism where he comments on the propositions of the *physica sacra*. In the passage referred to however, Voetius and the other members of the Faculty of Theology in fact discussed Copernicanism apart from the "sacred physics of Moses". Cf. Theo Verbeek, "Voetius en Descartes", in *De onbekende Voetius*, pp. 210 ff.

[64] Therefore, the Senate's verdict against the New Philosophy is not as "superficial" as it was thought to be by Regius' biographer De Vrijer. See *Henricus Regius*, p. 38. I agree with De Vrijer that matters of character were of less importance than "the background of differences of principle" (*idem*, p. 33). To my mind however, the Senate's verdict in fact points to differences of principle and should therefore not be taken as a superficial judgement of the conflict between Voetius and Regius.

[65] In the *Letter to Christina*, Galileo discusses the view that makes theology "the queen of all the sciences". According to Galileo, some theologians claim that "[theology] need not bend in any way to accommodate herself to the teachings of less worthy sciences which are subordinate to her". They arrogantly send off scientists to re-do their work when it does not match the "absolute authority" of

All of this may give us a clearer view of the general idea behind Voetius' essay "On the Natures and Substantial Forms of Things". It is directed against both anti-Aristotelian novelties and non-Aristotelian alternatives in philosophy which Renaissance learning had helped to rescue from oblivion. According to Voetius, such alternatives were not to be welcomed. And Copernicanism was only one of many problematic viewpoints. A host of other ideas and new ways of "invention" were criticised by Voetius in the various *Theses* of the short work. Most of the ideas that Voetius rejects are related to corpuscular philosophies. Voetius criticizes them whilst vindicating the Scholastic conception of the individuality and of the individual efficacy of natural substances. His standpoint, in short, is a restatement of Scholastic physics, which it will be my task to present in the following chapters as a philosophy that was thought to safeguard causality in accordance with the theology of a God-Creator governing the world. I shall first, however, set the stage by adding a historical note on events in Utrecht.

1.4 The Utrecht Crisis and the Essay on Forms

The writing of the essay on forms was motivated—or rather provoked—by the attitude of Voetius' colleague Henricus Regius. Being a correspondent of Descartes', and someone generally well-disposed to the emergence of new, mechanical theories in medicine and biology, Regius, at the time a junior professor of medicine and botany, had arranged to give a weekly course on some specific problems of physics.[66] Though soon aware of the unorthodox character of his

---

theological opinion. Contrary to what Galileo says, someone like Voetius would in fact welcome scientists to bother him with questions of physics. Nevertheless, Galileo's analysis offers an elegant and clear description of the position of someone like Voetius in the eyes of an adversary. What divides Voetius and Galileo is that Galileo accepted the idea that the new scientific investigations forced scholars to redefine the various forms of intellectual and religious authority, whilst Voetius—and many other divines—tried to hold on to accepted views of Biblical and disciplinary authority, which had no bearing on any notion of scientific progress. However, confident of the fact that he would be able to demonstrate the Earth's movement, Galileo himself gives Scripture absolute precedence over those physical theories which are only probable, leaving it to "wise theologians" to secure the unity of truth by showing that *demonstrated* physical truths do not contradict the Scriptures. Cf. Jerome Langford, *Galileo, Science and the Church*, pp. 72 ff.

[66] *Narratio*, p. 12 / *Querelle*, p. 86 and p. 463, note 19. See also: De Vrijer, *Henricus Regius*, p. 18 ff. In his *Letter to Father Dinet*, which was published together with the second, Elzevier, edition of the *Meditations* in 1642, Descartes

teachings, the other professors are reported not to have taken the courses of their colleague very seriously at first.[67] It was only when he took the side of one of his students at a public disputation and thereby behaved "in a rather inconvenient manner" towards the *præses*, professor Senguerdius, that a conflict began to fester.[68] The spread of new ideas within the Utrecht academy, however, is said to have started especially from the day on which Regius introduced the Cartesian theory of the circulation of the blood, which he did in a public disputation of 10 June 1640.[69] A host of other unorthodox theses were to trouble the more reluctant professors in the coming year. The final outburst however, came on 8 December 1641, when Regius had put forward for discussion the thesis that "man is an accidental being", an *ens per accidens*, which is to say that man is not a single unity of mind and body.[70] The public disputation broke out into chaos[71] and reactions were soon to follow. It was the senior professor of medicine, Guilielmus Stratenus (or Willem van Straten, 1593-1681), who was first to attack his colleague in a rather *ad hominem* way.[72] It was to be

---

says that, having read Descartes' *Dioptrics* and *Meteors*, Regius had written a complete course on Physiology on his own. This was the draft of what was later to become Regius' chief work on physics, the *Fundamenta Physices* of 1646. Cf. AT VII, p. 582-583 / *Querelle*, p. 141 and p. 484, note 49. Further: Theo Verbeek, "Regius's *Fundamenta Physices*", in *Journal of the History of Ideas* 55 (1994), pp. 533-551.

[67] *Narratio*, p. 13 / *Querelle*, p. 86.

[68] Arnold Senguerd, or Senguerdius (1610-1668) was appointed professor of philosophy on 11 June 1638, toghether with Regius, Martin Schoock (1614-1669) and Arnold van Goor. See *Querelle*, p. 463, note 16. The subject of the disputation was the magnet, of which Regius denied that it had an occult, attractive, quality. See *Narratio*, p. 14 / *Querelle*, p. 86.

[69] *Narratio*, p. 14 / *Querelle*, p. 87. The idea of the circulation of the blood was first defended by William Harvey in 1628. It was (wrongly) reinterpreted by Descartes in Part V of the *Discourse on Method* of 1637, AT VI, pp. 46-55. I shall discuss these and other texts involving the circulation of the blood and the movement of the heart in section 5.3.2, below.

[70] *Narratio*, p. 22 / *Querelle*, p. 93.

[71] Students of medicine made it impossible for a student of theology to defend the old philosophy. The tumult did not even stop when Regius himself, who, as *præses*, had the task to keep order, tried to calm his followers amongst the student population. *Narratio*, pp. 22-23 / *Querelle*, pp. 93-94.

[72] The corollaries to his disputation of 22 December 1641 may be read as a very personal comment on Regius' ideas on medicine and on his capacity as a botanist. See *Narratio*, pp. 24-25 / *Querelle*, p. 95. Stratenus had been nominated as Utrecht's first professor of medicine on 3 March 1636, right at the opening of the university. See *Querelle*, p. 462, note 8.

Voetius who, "at the students' demand and at the suggestion of certain colleagues"[73], was to launch a general attack on the New Philosophy by writing his essay on substantial forms.

The story of what has become known as the "Utrecht Crisis", has in recent years developed into a very well documented history. In various studies, Theo Verbeek of Utrecht University has reconstructed the plot of what went on in those troublesome early years of the academy in which, as Verbeek himself has pointed out, "Cartesian ideas were first taught".[74] In what follows, I shall not go into many historical details of Voetius' clashes with Descartes and Regius, for which I depend almost wholly on published materials by Verbeek and others. In accordance with what I have said in the Introduction, I shall take it to be my object here to analyse the conflict with regard to the philosophical argumentation, and in particular the idea of causality.

1.5 Conclusion

Let us therefore return to the subject-matter of Voetius' essay. As was said, Voetius holds that the Aristotelian philosophy is more in accordance with Holy Writ than are the philosophies of those who criticize Aristotle. Accordingly, he not only argues against the impiety of inventing theories which are not in line with Christian thought, but also defends some basic notions of Aristotelian philosophy on purely theological grounds.[75] Likewise, he starts off by offering Scriptural justification for the existence of substantial forms by arguing that "permanent natures, faculties and distinct species of things" are mentioned in various verses of *Genesis* and *Proverbs* and warning the reader that

> Christian philosophers will rather a thousand times profess their learned ignorance, than that they inflict even the tiniest sort of prejudice upon divine truth.[76]

---

[73] *Narratio*, p. 35 / *Querelle*, p. 103.

[74] See Verbeek's dedication of his *Descartes and the Dutch*: "To Utrecht University, the first university in the world where Cartesian ideas were taught."

[75] Theo Verbeek captures this standpoint well, when, commenting on Voetius' Aristotelianism, he writes: "Voetius is not surprised that everything theology needs happens to be found in Aristotle." TheoVerbeek, *Descartes and the Dutch*, p. 7.

[76] *Narratio*, p. 39 / *Select. Dispp.* I, p. 872 / *Responsio*, p. 11 / *Querelle*, p. 105: "Vide Gen. 1.11.21.22.24.25. Proverb. 30.24.25 26. 26 25 [= Gen. 1:11, 1:21-22 and 1:24-25 and Prov. 30:24-28]. Ubi permanentes naturas, facultates, & species rerum distinctas innui putamus". On the Voetian idea of "learned ignorance", see

The drift of the argument is that one should take care not to contradict the Scriptures and that if the Bible tells us that the Natural world is ordered according to distinctions between sorts, then our philosophical theory should have to make room for such distinctions. According to Voetius, the Aristotelian concept of substantial form does justice to the Biblical distinction of species and permanent natures and faculties of things, whilst the Cartesian and other new philosophies fail to make room for such distinctions.

This raises a number of questions. In what way, for instance, does the concept of *species* occur in the Bible? What are its relations to the notion of substantial form and why was the latter of such importance to Voetius? To answer these questions, we shall have to take a closer look at the idea of a *Physica Reformata* as conceived by Voetius and its relations to Theology. We shall start, however, by taking a closer look at the key concepts of "nature" and "substantial form" which go to make up the contents of Voetius' essay and lie at the heart of his ideas on biological species and specific propagation.

Theo Verbeek, "From 'Learned Ignorance' to Scepticism; Descartes and Calvinist Orthodoxy", in Richard H. Popkin / Arjo Vanderjagt (edd.), *Scepticism and Irreligion*, pp. 31-45.

CHAPTER TWO

SPECIES AND FORMS

According to Gisbertus Voetius, the doctrine of substantial forms is one of the philosophical doctrines which are in keeping with Holy Writ. As the title of his 1641 essay indicates, it is the concepts of "nature" and "substantial form" in particular that Voetius aims to vindicate against the modernists of his day. But what exactly are "substantial forms" according to Voetius? Voetius refers to the Bible, mentioning the 30th chapter of *Proverbs*, amongst others, as a passage in which "permanent natures" occur. Presumably, the Biblical reference to natures, faculties and distinct species of things is meant to prove the correctness of the philosophical notion of forms. The verses of *Proverbs* 30 however, in which ants, conies, locusts and spiders are alluded to, are rather poor in that respect. Their major significance is a moral one[1] and does not seem to lie in a representation of the animal world as a realm regulated by biological diversity, let alone to indicate that the variety of species is founded on a metaphysical concept of class distinctions. Accordingly, these texts do not help us much to appreciate Voetius' notion of substantial forms. Ants differ from spiders and these again from human beings, but, the moral issue apart, *Proverbs* offers no conclusion as to the question in what way they differ.

It was Descartes who, in his *Letter to Voetius* of 1643, pointed this out to Voetius, arguing that

> in the whole body of Scripture there is no verse that you might not just as well have cited, since in every one of them there occurs some corporeal object to which you ascribe a substantial form. However, these prove as little on your behalf as do those in which snow is mentioned on behalf of people who say that snow is black.[2]

---

[1] The four types of animals are given as examples of "little things" that, nevertheless, are "exceedingly wise" in comparison to the human race, which, by contrast, is governed by vanity, adultery and a general disrespect for hierarchical relations.

[2] *Epistola ad Celeberrimum Virum D. Gisbertum Voetium*, 1643 / AT VIII-II, p. 62 / *Querelle*, p. 362.

In other words, neither a reference to different objects, nor the occurrence of class-names in the Scriptures may, according to Descartes, count as proof for the proposition that the differences between species and kinds depend on the existence of substantial forms. The quotation from *Proverbs* was, however, the easier one to solve. The verses of *Genesis*, to which Voetius also refers, are of rather more relevance to the question of biological species.

2.1 Mules and Monsters; the Question of Specific Propagation

Genesis 1:11, 1:21-22 and 1:24-25 all tell of God's creating vegetation and animal-life on Earth. Again, various biological species are named. This time however, it is explicitly stated that God created them (and that they multiply) *in species suas*, or each *secundum genus suum.*[3] Moreover, He created them *in genus suas*, that is to say, each "after its kind". Obviously, a natural biological order is hinted at, in so far as each type of plant and animal is said to belong to one of the unique species created in the beginning by God.

In his *Letter to Voetius*, Descartes does not go into the *Genesis* passage, but contents himself with referring to what he calls the "ingenious" answer of Henricus Regius to these matters. Early in 1642, Regius had entered into an involved correspondence with Descartes about how to reply to Voetius' essay on substantial forms. From Endegeest, the French philosopher sent his Utrecht friend an outline for a possible answer[4] in which he proposed to cite the text of the verses mentioned in full.[5] Regius did so, but, apparently independently of Descartes, added that he agrees with Voetius that "permanent natures, faculties and distinct species of things" are referred to in the Biblical passages. Regius however, like Descartes, denies that they prove anything in favour of substantial forms:

---

[3] In the Vulgate text of *Genesis* 1:11, the herb is said to yield seed and the fruit tree to produce fruit *iuxta genus suum.* According to Genesis 1:21, 24 and 25, God created the animals of the sea *in species suas*, every winged fowl *secundum genus suum*, the living creatures of the Earth *iuxta species suas*, and cattle, creeping thing and beast each *in genere suo.*

[4] Descartes' correspondence with Regius was published by Geneviève Rodis-Lewis as Descartes, *Lettres à Regius et Remarques sur l'explication de l'esprit humain*, Texte latin, traduction, introduction et notes par Geneviève Rodis-Lewis, Paris (Vrin) 1959, hereafter to be referred to as '*Rodis-Lewis*', followed by page numbers. See also: Th. Verbeek, *Descartes and the Dutch*, pp 18-19.

[5] Descartes to Regius, January 1642, AT III, p. 502 / *Rodis-Lewis*, p. 86.

> these [natures, faculties and species] in fact originate from the mere disposition of corporeal matter according to motion, rest, position, shape and size of the parts, which alone were imparted to created things by God. Nor can even the slightest proof of another source be given. As a consequence, no substantial form is to be attributed to them.[6]

Regius' reply is identical to Descartes' in the sense that he holds a Biblical reference to various species not to be a reference to the philosophical notion of substantial forms. The verses mentioned do not prove the existence of such forms. Moreover, Regius adds that the kinds and species mentioned in the first chapter of *Genesis* may be defined in terms of "mechanical principles" such as the size, movements, shape and position of their parts.[7] He does not, however, go into any detail, nor does he explain what is meant by the Biblical statement that all species were created and propagate each "after its kind". In other words, although animal diversity may be explained in mechanical terms, Regius does not make clear how the concepts of *genus* and *species* are to be accounted for.[8]

If we turn to Voetius however, to see what he has to say regarding the concepts of *species* and *genus*, the result is equally disappointing. As far as I know, Voetius himself does not explicitly deal with the meaning of these terms and their use in *Genesis* 1 either, although he commented on the book of *Genesis* in a series of ten academic disputations *De Creatione*.[9] These are all very elaborate pieces of

---

6 Regius, *Responsio*, pp. 14-15.

7 Although Descartes does not go into the question in his *Letter to Voetius*, he did write to Regius that nobody doubts that the prophets and apostles were ignorant of the notion of substantial forms and that the *species* and *genera* of *Genesis* might refer to accidental or modal differences only. Descartes to Regius, January 1642, AT III, pp. 501-502 / *Rodis-Lewis*, pp. 85-86 / CSM III, p. 207.

8 Regius generally showed a tendency to keep to Scholastic terminology, even when he had no use of it. See hereafter, section 7.1.

9 They were held in the autumn of 1638, Voetius' student Lubbert Spruit performing the task of *respondens* in each of the ten disputations. In university disputations *sub præside*, the student acting as *respondens* could exercise his rhetorical abilities and thus prepare himself for a possible disputation *pro gradu*, by which he would graduate. The student was responsible for the text of his graduation dissertation, whereas the *præses* (mostly a professor, but sometimes a graduate or senior student as well) was probably held responsible for the *disputatio sub præside*. The authorship of the latter however, remains a source of controversy. In her *Collegia en Colleges. Juridisch onderwijs aan de Leidse universiteit 1575-1630 in het bijzonder het disputeren*, Proefschrift Leiden 1990, Groningen

writing[10], in which the story of *Genesis* is told following the order of events of the first Six Days.[11] The question of the distinctness of species is nowhere brought up as a separate problem of discussion. Yet in order to see why Voetius interpreted the Scriptural passages as proof for the existence of substantial forms, it will prove worthwhile to take a closer look at another question concerning species which Voetius addresses in the *De Creatione* disputations.

2.1.1 Voetius in Genesim

In the eighth part of *De Creatione* the question is put "whether God then [i.e. on the Sixth Day] produced all species." Voetius answers in the affirmative[12], referring to *Genesis* 2:1-2:

> Thus the heavens and the Earth were finished, and all the host of them. And on the Seventh Day God ended his work which he had made; and he rested on the seventh day from all his work which he had made.

The Scriptural conclusion seems clear enough, but from the early Christian fathers onwards, commentators on the book of *Genesis* had been troubled by some very specific questions with relation to biological species. One of these was the problem of how to account for—as Voetius puts it—"the adulterous species" of mules, leopards, lynxes, *tityri* and the like. These animals are—or were thought to be—cross-breeds: the mule of horse and ass, the leopard of lion and panther, the lynx of wolf and deer and the *tityrus* of sheep and he-

---

(Wolters-Noordhoff / Egbert Forsten) 1990, Margreet Ahsmann offers a very detailed study of juridical education at Leiden. She argues that the Leiden situation differs from that in Germany, where the professor was wholly responsible for the text. See Margreet Ahsmann, *Collegia en Colleges*, p. 311-323. The student in the present disputations, Lubbertus Spruit, was to be Voetius colleague as a minister of the church in Utrecht from 1644 to 1651, before which he served in Blauwkapel. See on him: Duker, A. C., *Gisbertus Voetius*, Leiden (E. J. Brill) 1893-1915 / reprint Leiden (Groen) 1989, III (1914), pp. 39 and 114 esp.

[10] Leaving out the four appendices, the ten disputations take up no less than 250 pages of the first volume of Voetius' collected disputations. See *Select. Dispp.* I, pp. 552-802.

[11] Voetius thereby consciously places himself in the tradition of the *physica hexæmera*, or "physics of the Six Days", i.e. the tradition of theological commentaries on the book of Genesis, *Select. Dispp.* I, p. 726. See also below, section 3.1.1.1.

[12] Voetius, *Select. Dispp.* I, p. 726.

goat.[13] Various interrelated questions arose: were such creatures begotten *præter naturam*, i.e. contrary to nature? Was it lawful for man to breed them?[14] And, in relation to the present problem: were these species also created on the Sixth Day?

---

[13] *Select. Dispp.* I, p. 726. Authoritative "proof" for these biological views could be found in innumerable commentaries on *Genesis* and handbooks of natural history. Voetius quotes a long passage from the *Historia sacra animalium* of Wolfgang Frantze, or Franzius (1564-1628), in which conditions for the breeding of hybrids are summed up. First, Frantze explains, one can only interbreed species that do not differ much in size and have more or less equal periods of gestation. Moreover, only the most bawdy types (*salacissima*) such as dogs, cows, he-goats, pigs and asses lend themselves for the job. Finally, the animals must be mature and the time must be right, preferably in spring. But cross-breeding is in any case an artificial thing, mostly brought about by owners who stimulate the copulation of animals of different species because they expect to get stronger animals from it. Franzius, *op. cit.*, VVittenbergæ, Sumtibus Zachariæ Schureri & Iohannis Gormanni, 1616^2, p. 320 and Voetius, *Select. Dispp.* I, p. 727. Frantze was a Lutheran divine and *Professor Historiarum* in Wittemberg. Cf. *Jöcher* II, col. 719.

[14] These questions were associated with the texts of *Genesis* 36:24 and *Leviticus* 19:19. In *Genesis* 36:24, it was, according to some translations, told of Anah that he "invented mules", instead of "finding waters in the desert" as other versions have it. *Leviticus* 19:19, despite the obviously literal meaning of the prohibition it contains—"Thou shalt not let thy cattle gender with a diverse kind"—was mostly interpreted in a figurative manner. Pererius, a Spanish neo-Scholastic theologian whom Voetius refers to, quotes both Aquinas and his commentator Cajetan, or Thomas de Vio (1469-1534). Aquinas and Cajetan defended a figurative interpretation of *Leviticus* 19:19 on the basis of I *Corinthe* 9:9-10, which was a usual remedy against a literal interpretation of commands of the Old Testament that were hard to follow. In I *Corinthe* 9, the Apostle says that "it is written in the law of Moses, Thou shalt not muzzle the mouth of the ox that treadeth out the corn. Doth God take care of oxen? Or saith He it altogether for our sakes? For our sakes no doubt, this is written: that he that ploweth should plow in hope; and that he that thresheth in hope should be partaker of his hope." According to Aquinas, *Leviticus* 19:19 is equally addressed to us. Notwithstanding its Mosaic origin, Aquinas holds the text, amongst other things, to be a warning against letting vulgar people mingle with heathens and Jews. Even the literal interpretations Aquinas offers do not relate to a prohibition of cross-breeds. According to St. Thomas, the literal meaning could be either a denunciation of Egyptian divinations, a prohibition against unnatural sexual intercourse, or a general adminishion to avoid whatever occasions passionate lust. Aquinas, *Summa Theologica* I-II, qu. 102, art. 6 *ad octavum*. Cajetan claims that *Leviticus* 19:19 *cannot* relate to cross-breeds since these were common in ancient Isreal. He accordingly offers a "parabolic" interpretation of the verse, arguing that it contains a caution against accepting any novelty that goes against reason. Cf. Benedictus Pererius, *Commentarii et disputationes in Genesin*, Tomus IV, Colonia Agrippina, Ex officina Antonij Hierati, 1606, p. 47. The only writer who seems to have interpreted *Leviticus* 19:19 in a literal way is Abbot Rupert of Deutz, or Rupertus Tuitiensis (c. 1090-1135), who holds the act of

This last question had been adressed by Aquinas in the first book of the *Summa Theologica*, to which Voetius refers.[15] Aquinas defends the text of *Genesis* 2:2, saying that

> nothing totally new has afterwards been made by God, which was not in some way present in the works of the Six Days.[16]

Of course, other creatures only appeared afterwards, but, according to Aquinas, all were there on the Seventh Day in some sense or other. Woman, for example, was "materially" present in Adam's rib, whilst all the individual beings which are generated every day were *causaliter* present in the first individuals of their respective species. Finally, there are complete species which only later come into being, such as some of the small animals which spontaneously arise out of rotting-processes[17] and cross-breeds. These are said also to have pre-existed *causaliter*, since they "pre-existed in some active virtues" of things or species which went before. In this way, the species "mule" pre-existed *causaliter* in the species of horse and ass, and was hence—albeit indirectly—included in the Creation of the Six Days.

Voetius himself mentions various standpoints[18] and concludes that on the Sixth Day "all perfect species" were produced. These, however, do not include cross-breeds, which do not form a perfect species, "if indeed we do not want to call them monsters."[19] According to Voetius,

---

reproduction between animals of different kinds to be *contra naturam*. See Pererius, *In Genesin*, p. 48, and Rupertus Abbas Tuitiensis (c. 1090-1135), whose *Libri XLII de operibus sanctæ Trinitatis*, was printed in Cologne in 1528. See *idem*, Pars I, chapter 57, no paging. Pererius denounces the view. See *In Genesin*, p. 48: "At enimverò mea longè diversa est sententia, senseo equidem istiusmodi generationem animalium naturalem esse."

[15] *Select. Dispp.* I, p. 726. The reference to Aquinas however, reading "*Thomas* I. *qu*. 37. *art*. I.", should read I, qu. 73, art. 1.

[16] *Summa Theologica* I, qu. 73, art. 1 ad 3: "Ad tertium dicendum quod nihil postmodum a Deo factum est totaliter novum, quin aliqualiter in operibus sex dierum præcesserit."

[17] According to Aquinas, the animals which arise spontaneously from the rotting of inanimate materials or plants were indeed created within the Six Days. Bugs that arise from the rotting of higher species however, were only created *potentialiter*. The idea (which is St. Augustine's) is that it is in conflict with the "first arrangement of things", that something less noble would arise from a more noble substance. Aquinas, *Summa Theologica* I, qu. 72, art. 1, ad 5. Augustinus, *De Genesi ad litteram* III, C. 14.

[18] Voetius confronts Aquinas' opinion with that of Pererius, who in fact follows Aquinas on nearly every point. See e.g. note 14, above.

[19] Voetius, *Select. Dispp.* I, p. 727.

the fact that the offspring of cross-breeds are either sterile, like mules, or, if not sterile, then attracted to either one of the species from which they were propagated, proves that they form no "third nature" which is "specifically distinct" from those of their parents. They may look like a mixture of both, but, according to Voetius, the mixture does not constitute a distinct species:

> Just as those dogs which we call "leopards" or "lions" etc. are in fact dogs, [although] their external shape and perhaps other things may recall a certain similarity with a leopard or a lion.[20]

Voetius seems to argue that a leopard-like dog is a dog in the shape of a leopard, but still a dog. Now the comparison between leopard-like dogs and cross-breeds is rather troublesome in that the dogs are in fact said to be of a genuine species, whilst cross-breeds are not. Voetius' point is obviously to argue that external and other similarities are no proof in favour of the existence of distinct mixture-species. Thus, it is implicitly argued that accidental forms do not always point out substantial forms in any obvious way. If we see that a certain animal has the accidental shape of a mule, that in itself does not guarantee that we should ascribe a "new" substantial form to them which is totally distinct from the existing class of substantial forms. We should not, in any case, accept the mule as a genuine biological species. It is "just" a mixture of horse and ass.

Many years later, in the autumn of 1660 and the spring of 1661, Voetius held a second series of academic disputations *De Creatione*, this time in the form of lists of specific problems and direct answers. On 5 December 1660, number 87 was discussed: "whether all brutes were created within the Six Days, even poisonous, harmful and infected ones?"[21] The answer is affirmative, but the question led Voetius to say a few things on related topics, such as that, as Voetius puts it, "concerning mules". He quotes the protestant theologian Lambertus Danæus—or Lambert Daneau (1530-1595)[22]—, who wrote that such cross-breeds as mules

---

[20] Voetius, *ibidem*. I have no idea what sort of dogs Voetius is here referring to.

[21] Voetius, *Select. Dispp.* V, Ultrajecti, Ex Officinâ Antonii Smytegelt, Bibliopol., 1669, p. 191.

[22] On Daneau see Paul de Félice, *Lambert Daneau (de Baugency-sur-Loire) Pasteur et professeur en théologie 1530-1595*, Paris (Librairie G. Fischbacher) 1882 and Olivier Fatio, *Méthode et théologie: Lambert Daneau et les débuts de la scolastique réformée*, Thèse Genève 1974 / Genève (Éditions Droz) 1976.

> are only a sort of mixture and temperament of two kinds [*genera*] of animals, not, however, some sort [*species*] of animal in the strict sense which is entirely different from the whole class [*genus*] of breeds.[23]

Voetius agrees to this and says that it is in line with the first verses of Genesis 2, where "all species of things" are alluded to, "without exception".[24] At the same time, he quotes—with approval—the *Commentaria et Disputationes in Genesim* of the Spanish Jesuit theologian Benedictus Pererius—or Benito Pereyra (1535-1610). Pererius had recapitulated the Thomistic standpoint, offering even more distinctions among ways of being created: *materialiter, causaliter, virtualiter, potentialiter* and *secundum similitudinem speciei.*[25]

Voetius refers to both Pererius and Danæus, but these writers in fact represent two different positions as to the question whether mules were also created on the Sixth Day together with the rest of the animal world. First, there is the Thomistic position, which makes the creation of cross-breeds in some way or other a part of God's initial Creation. Secondly, there is the position of Danæus, who held that cross-breeds form no distinct species of their own and were hence *not* included in the Creation of the Six Days. The question might be raised, which of these two seemingly inconsistent positions Voetius ultimately took. The answer is: both. Although he offers the various arguments without further comment, Voetius' position may be resumed as follows. When God had completed His works on the Seventh Day, there was no mule on the face of the Earth, even though all species were already there. As Pererius argues, the mule was, so to say, con-created on the Sixth Day, but it does not, as Danæus says, form a species of its own. This option "saves" the Scriptural text, which is Voetius' primary intention.

The question of mules also tells us a bit more about the concept of *species*. Neither Aquinas nor Pererius had seen any harm in using the term *species* when writing on cross-breeds. Aquinas speaks of *species*

---

[23] Lambertus Danæus, *Physica Christiana Pars Altera, sive de rervm creaturarvm natvra. Qvæ in sex tractatvs pro diervm, quibvs Deus ipse operatus est, numero diuiditur, in Opvscvla Omnia Theologica*, Genevæ, apvd Evsthativm Vignon, 1583, p. 359.

[24] Voetius, *Select. Dispp.* V, p. 191.

[25] Pererius, *In Genesin* I-IV, pp. 53-54.

novæ with regard to organisms that have only later come into being[26], whilst Pererius says that "many species" arise even today. However, both may here use the term in a rather non-technical sense. As we have seen, Voetius himself uses the phrase "adulterous species" to denote cross-breeds, even though he in fact denies them the status of a "perfect species". The difference lies in the use of the adjective. Pererius too, when commenting on a quotation from the *Hexæmeron* of Basil of Cæsarea[27], uses the phrase "the more perfect animal species" to refer to those types of animals which God created on the Sixth Day, arguing that the "less perfect" were produced later.[28] Accordingly, we might say that mules were not "actually" produced, but only *causaliter* (or *potentialiter*) in the *more perfect species* of horse and ass.

The idea of distinguishing genuine species from imperfect derives in part from Aristotle. I shall present some of the relevant themes of Aristotelian natural teleology in the section 2.2. First, however, let us take another look at the question of the mules, especially in comparison with monsters.

2.1.2 The Non-Species of Mules and Monsters

Voetius is quite clear about the fact that cross-breeds are not perfect species. As we have seen above, in the first series of disputations *De Creatione* of 1638, he even gives a clue as to what cross-breeds positively are. According to Voetius, they are rather like monsters. Monsters are, like hybrid animal "species", also excluded from God's initial act of Creation of the Six Days.[29] The occurrence of such

---

26 Aquinas, *Summa Theologica* I, qu. 73, art. 1 ad 3.

27 The nine *Homilies on the Hexæmeron* of St. Basil of Cæsarea (330-379) were a prime source for the hexæmeric tradition. A nice edition of the work is: Basile de Césarée, *Homélies sur l'Hexaéméron*, texte grec, introduction et traduction de Stanislas Giet (*Sources Chrétiennes* N° 26 *bis*), Paris (C. E. R. F.) 1968.

28 Pererius, *In Genesim*, p. 47: "[...] one can also say that at the time those animal species were not produced 'actually', but only *causaliter* and *potentialiter*, since God made the more perfect animal species [...] from which [...] the less perfect are afterwards produced through the force of natural causes." The passage in fact deals with the animal species which were thought to spontaneously arise from organic material in rotting processes, but the argument holds also for cross-breeds, which were usually mentioned alongside these as examples of animals created causaliter within the works of the Six Days. The term *potentialiter* is even introduced by Pererius especially with regard to cross-breeds.

29 In the 1660 disputations, Voetius quotes the protestant theologian Hieronymus Zanchius (1516-1590). Arguing that animal *genera* are non-eternal, since they were

deformities in the course of Nature led Voetius to answer various theological questions with respect to man-born monsters in the 1638 disputations. It was asked whether monsters are human and whether, if they live, they should be baptized, or rather be killed.[30] As for monsters that are not born of woman, Voetius lists and criticizes many accounts of mythic and exotic animals.[31] Only in the 1660 disputations does he return to the question of whether monsters were created in the beginning. His answer is, of course, negative, but his argument is interesting enough to quote in full:

> I retort that monsters have no causes *per se* or proper [and] definite [causes] (for causes *per accidens* are no causes); for [they are] failures[32] and deviations[33] of nature.[34]

In other words, monsters are caused "accidentally". They have no "proper" and "definite" causes of their own. Presumably, the fact that monsters do not belong to a definite species of their own and the fact that they have no definite causes, amount to the same thing. Here, the example of specific propagation serves as a metaphor for causality in general. Genuine causality is species-bound.

Monsters then, are "accidents" in the causal chain of specific propagation. But what about mules? If the same holds for cross-breeds, then the fact that cross-breeds are sterile is indicative of their

created in time, viz. on the Fifth and Sixth Day, Zanchius adds: "Genera dico animalium: quoniam monstra novum genus non constituunt: sed in unoquoque genere peccata sunt naturæ." Hieronymus Zanchius, *Opera Theologica*, Tomus III: *De Operibus Dei intra Spacium Sex Dierum Creatis*, Exudebat Stephanos Gamonetus, 1605, I, lib. 7, cap. 2, th. 1, p. 475 and Voetius, *Select. Dispp.* V, p. 191.

30 *Select. Dispp.* I, pp. 750-751. Voetius' answer to the first question was affirmative, provided that there was reason to believe that the monster was supplied with a human soul. As for the second problem, it might in some cases be worth waiting until the monster had grown up. It should in any case not be killed, "where there is even the least possible doubt". Presumably, Voetius means to say that man-begotten monsters should not be killed when there is even the slightest possibility that the monster is in possession of a human soul.

31 Voetius, *Select. Dispp.* I, p. 727.

32 Ἁμαρτήματα, a term literally taken from Aristotle, *Physics* II 8, 199^{b}4, and occurring in the same sense for instance in Plato, *Gorgias*, 479 a 8.

33 Παρέκβασις, "digression", here in the sense of "aberration"; a concept Aristotle uses mainly in his ethical and political works. See Bonitz, *Index Aristotelicus*, Berlin 1870 / reprint Graz (Academische Drücke und Verlagsanstalt) 1955, p. 568.

34 Voetius, *Select. Dispp.* V, p. 194.

non-specific character—which seems to be the most important reason for Voetius' refusal to accord to them the status of a distinct species.[35] Cross-breeds are exceptions, or even deviations from the general course of Nature. That is why they are said to be "rather like monsters".

Voetius defends the concept of substantial form, arguing that the Bible refers to "permanent natures, faculties and distinct species of things". We have seen that in the case of the animals which were created on the Sixth Day of Creation, he draws a sharp dividing-line between *species*—which were then created and which propagate "each after their kind"—and those animals which are non-specific "failures" of Nature, and which are said not to have any proper and definite causes and to have come into being only "accidentally".

This way of linking the concepts of biological species, specific natures and accidental deviances of Nature to a general theory of causation, ultimately goes back to Aristotle—Book II of the *Physics* in particular. It is there that we find Aristotle developing a theory of causation on the basis of an idea of "natures" resembling biological kinds. It is there also, that Aristotle struggles with the concept of "accidental causation", linking this to the example of monstrous births. Despite many similarities, however, the Scholastic idea of species differs from its Aristotelian forerunner. It will not be my main object here to present the Aristotelian line of argument in full, but rather to take out those ideas which may throw light upon, or which are worth contrasting with, Scholastic doctrine.[36] Let us therefore turn to Aristotle.

2.2 Aristotelian Teleology

Voetius denies mules the status of a "perfect" species. The Greek term for which *perfectus* (perfect, complete) is the traditional translation is

---

[35] Voetius, *Select. Dispp.* I, p. 727: "Aut enim sunt plane steriles, ut muli; aut si generant, pulli eorum in alterutram speciem inclinant; nec tertiam quondam naturam ab utrâque specificè distinctam constituunt."

[36] In recent years, students of the Aristotelian corpus have offered many arguments for not reading Aristotle's philosophical works too easily as an abstract of his biological practice. It is a result of the works of David Balme in particular that the Aristotle of the biological works is nowadays presented as rather different from the essentialist philosopher he has so often been taken to be. In the following, I shall go into various discussions relating to Aristotle's position, but since my aim is to draw attention to Scholastic additions to Aristotelian theory, Aristotle's own position will tend to appear in a somewhat negative way.

τέλειος, a term with evident affinity to τέλος[37]. Τέλος is, as a technical term of Aristotelian philosophy, usually translated as "end", or "fulfilment"[38]. Being easily associated with τέλος, the term τέλειος thereby also acquires a specifically "finalistic" connotation. Nevertheless, in many passages, the term may well be translated as, for instance, "complete" instead of "perfect" and may justifiably be said to comprise a non-finalistic sense of perfection. Terms like "finalistic" and "non-finalistic", however, are themselves ambiguous. In chapter 3, we shall come accross various senses in which they might be used, according to the various kinds of "finalism" in Scholastic explanations of natural change. But first I shall start with Aristotle, and give a short survey of his account of *teleological* explanation.

2.2.1 Perfection and Natural Teleology in Aristotle

Aristotle uses the adjective τέλειος ("perfect", "complete") in several different ways.[39] In an ethical context, it appears for instance where he

---

[37] See P. J. Du Plessis, *Teleios: The Idea of Perfection in the New Testament*, Dissertation Kampen (NL) 1959, Kampen (Kok) 1959, p. 73: "The outline so far has confirmed our impression that basically, the adjective is to be reviewed from its nominal root. The distinct qualities of *telos* are therefore to be kept in mind when considering the circle of meanings characteristic of *teleios*."

[38] The use of the latter term is defended by Christopher Kirwan in his edition of the *Metaphysics*, *Clarendon Aristotle Series*, Oxford (Clarendon) 1971, p. 167: "The word for 'fulfilment' is '*telos*', from which '*teleios*' derives. The traditional translation 'end' surpresses this connection, and obscures the point that life's end, or death, teleutê, is a telos in a subsidiary sense only [a point Aristotle makes in Met. 1021^{b}29]." Note that Kirwan also prefers "complete" to "perfect" for teleios in general, although he typifies the three distinctive meanings which Aristotle gives of *teleios* as 'entire', 'perfect', and 'complete' respectively, see hereafter, note 39.

[39] The most elaborate description of the term's meaning occurs in *Metaphysica* IV 16, 1021^{b}12-1022^{a}3. It is disputed amongst scholars whether in this passage Aristotle offers three or, rather, four types of perfection. H. Tredennick (in his Loeb edition of the *Metaphysics*, Cambridge Mass. and London (Harvard / Heinemann) 1980, p. 267) identfies four kinds of perfection. I am inclined to accept the view of W. D. Ross (*Aristotle's Metaphysics* (ed. W. D. Ross), Volume I, Oxford (Clarendon Press) 1924 (=1966^4), pp. 331-332) and Christopher Kirwan (*Aristotle's Metaphysics*, p. 167) who suppose Aristotle to hints at three types of perfection only. According to Aristotle, to say that something is perfect can imply, firstly, that no portion is to be found outside of it. See instance *Metaphysica* IV 16, 1021^{b}13-14: "the complete time of each thing is that outside which there is no time to be found which is part of that time." (I here follow Kirwan's translation.) Secondly, something can be called τέλειος when it cannot be surpassed in excellence "relative to its genus." This for instance occurs when we call a doctor a

investigates the question of what "perfect happiness" (τελεία εὐδαιμονία) is.[40] It is concluded that, for mankind, perfect happiness lies in the activity of contemplation. This happiness is supposedly "perfect" because it is a happiness resulting not just from activating any one of our virtues, but from activating our highest virtue, which is the virtue that is most typical for us, viz. contemplation. In this sense, contemplation is the ultimate "end" or "aim" (τέλος) of human activity.

Aristotelian philosophy is full of such "ends". Another passage of the *Nicomachean Ethics*, where Aristotle deals with the nature of pleasure (ἡδονή),[41] may give us an idea of the role that the concept of "end" or "perfection" plays in other contexts. Aristotle puts forward the view that pleasure is not a quality in progress. In other words: one either experiences pleasure, or one does not. At a later moment, the pleasure will either still be there or not, but there is no process of pleasure "building up". Another way of saying this, is that at all moments, pleasure is "perfect" or "complete". Aristotle's point is, in itself, obvious enough. Hardly anyone would, when asked whether he or she was experiencing pleasure, answer that his or her pleasure was, at that moment, only coming-to-be. Not being in a process of coming-to-be, pleasure is always "perfect", and from this, Aristotle draws the conclusion that "[pleasure] is not a form of motion (κίνησις)."[42]

"perfect" doctor. It even applies in the case of bad things, since we may, for instance, call somebody a "perfect" thief. Thirdly, something may be said to be τέλειος when it has reached its τέλος, which is to say, its fulfilment, or end. According to Kirwan, these three senses of τέλειος "correspond pretty exactly to the English 'entire', 'perfect', and 'complete'", cf. Kirwan, *ibidem*.

[40] Aristotle, *Ethica Nicomacheia* X 7. The phrase τελεία εὐδαιμονία occurs at 1177^a17.

[41] Aristotle, *Ethica Nicomacheia* X 3, 1174^a13 ff.

[42] Aristotle, *idem*, 1174^a18. I here follow H. Rackham's translation of the Loeb-edition (Cf. Aristotle, *The Nicomachean Ethics*, with an English translation by H. Rackham, Cambridge, Mass. / London (Harvard University Press / William Heineman) 1975^9, p. 519) in giving "form of motion" for κίνησις. W. D. Ross for instance uses "movement" instead, which is taken over in the Revised Oxford Translation (Cf. *The Complete Works of Aristotle: The Revised Oxford Translation*, Vol. 2, Princeton (University Press) 1914, p. 1856). There is no other reason, than that I should want to make clear that Aristotle is here talking in very general terms. As to τέλειος, the Revised Oxford Translation consistently—and justifiably—translates it as "complete" instead of "perfect". However, since we are here in the process of getting to grips with the technical meaning of τέλειος, I shall, for the moment, stick to "perfect", which has the advantage of being more closely connected with the Scholastic concept of *perfectio*. As it will turn out, however, I

By drawing this conclusion, Aristotle hints at a very general aspect of his theory of natural change and its relation to the concept of "perfection". For why is pleasure not a form of motion? It is because, according to Aristotle, motion always involves a process of change, culminating in a certain perfection. In pleasure, perfection of pleasure is *eo ipso* attained. Therefore, "[pleasure] is not a form of motion."

In all cases in which perfection is *not* instantaneous however, there *is* motion.[43] Whereas pleasure is perfect at any given moment, the process of building a house is—like any other motion—only perfected when its end has been reached. However, Aristotle not only refers to the final product. One may also say that the motion is perfect over "the whole" of its duration.[44] We could paraphrase this by saying that since in the entire process of building a temple none of the (in themselves imperfect) parts of the process is lacking, the whole process is perfect for building a temple. In other words: a complete process is a perfect process, just as well as the end—or aim—of the process of building a temple, is a complete temple. We should be aware, however, that in Aristotelian terms such statements are in fact tautological. The perfection of a process simply lies in the fact that the process *is* complete. The end (τέλος) of the process and of the temple is their completion or perfection.

In Aristotle's physics as well as in his ethics, the concept of τέλος thus functions as the "end" to which processes are directed. Natural processes are perfected only when the process is complete or when the τέλος is reached. Because of this, Aristotle's account of motion and of processes of natural change is often labelled *teleological*. Aristotle is easily misread on this point. We must keep in mind that for him, the teleological account of natural processes was a question of empirical fact and not a theoretical axiom or an *a priori* conviction. Experience

---

shall in fact make use of the terms "complete", "completeness" and "completion" also in order to explain Aristotle's use of τέλειος.

[43] Aristotle immediately goes on to contrast the perfection of pleasure with the imperfection of processes of change. His concept of κίνησις, motion, is so universally applicable that in doing so, he allows himself to step over from the ethical to the physical domain. Likewise, pleasure is contrasted with the building of a house. Cf. *idem*, 1174^{a}19-29.

[44] Aristotle illustrates this by elaborating upon the example of building. Fitting together the stones for a temple or laying its foundations are only part of the whole process of building a temple. In respect of the building of the temple, they are therefore "imperfect" processes only. However, building the temple in its entirety is in itself a perfect process, since, as Aristotle says, "nothing is lacking for what is proposed", *Ethica Nicomacheia* X 3, 1174^{a}25-26.

tells us for instance that plants and animals grow in order to reach their full growth. When this process is completed, they reach their end, or state of completion, i.e., perfection.[45]

The biological paradigm may be taken to be exemplary for all natural processes[46]: there is motion or change whenever a natural object is in the process of reaching a state of perfection or completion. In the *Physics*, Aristotle therefore counts the "end" (τέλος) among the factors by which natural processes of change may be explained. The concept of τέλος as it occurs in *Physics* II may be brought in connection with Aristotle's account of τέλειος in *Nichomachean Ethics* X. Just as building is "a form of motion" which is perfect, or "complete" when the building is there, or when the "whole" process of building is completed, so it is with natural processes in general.

With this in mind, we may return to the example of monsters. They form an exception within teleological processes. Monsters do not seem to develop to their proper "end". In fact, ordinary language intreprets "monstrous" as the complete opposite of "perfect". According to Voetius, hybrids are also rather like monsters. And Voetius refers to Zanchius, who held that

---

[45] In his excellent article on "The Problem of Teleology", W. Wieland gives a clear overview of the misunderstandings surrounding Aristotle's concept of teleology. One of his points is to argue that, in zoology, the teleological view is not so much an obstruction to scientific investigation as rather "a *guideline* for the exploration of the particular", which in fact involves observation rather than theory. W. Wieland, "The Problem of Teleology", in Jonathan Barnes, Malcolm Schofield and Richard Sorabji, *Articles on Aristotle*, Vol. 1, "Science", London (Duckworth) 1975, p. 152.

[46] A note is due here on the works of David Balme, whose studies of Aristotle's biological works have resulted in a reassessment of the relation between these and the philosopher's metaphysical, physical and logical treatises. See for instance D. M. Balme, "The Place of Biology in Aristotle's Philosophy", in Allan Gotthelf & James. G. Lennox (edd.), *Philosophical Issues in Aristotle's Biology*, Cambridge (C. U. P.) 1987, pp. 9-20. Yet what is actually at stake here is the question which part of Aristotle's work was of relevance to, or influenced which other part. There is no doubt but that, as Professor Balme has shown, Aristotle's biological works in many ways offer a refinement of his more general and earlier works. This however, does not prohibit the verdict that the biological paradigm—albeit a rudimentary one in comparison with Aristotle's later biological works—was at the heart of his earlier theory of causation.

monsters do not constitute a new kind, but are failures in respect of all individual kinds of nature.[47]

For the Christian reader, the use of the verb *peccare* (to fail, to go wrong) can hardly have failed to reveal the particular moral connotation which the question of monsters had.[48] Nevertheless, the argument is founded on purely philosophic grounds. Aristotle himself had dealt with monsters in *Physics* II.

2.2.2 *Chance and Spontaneity*

In *Physics* II 8, Aristotle puts forward the claim that monsters are "failures" (ἁμαρτήματα) of purpose in Nature.[49] The quotation is taken from the passage in which Aristotle develops the comparison of art and Nature. Rejecting explanations that make use of the concepts of "luck" and "chance", he compares failures in Nature with failures in art. The materialist might argue that there are no "ends" in Nature at all, but that chance brings about creatures of various possible formations and that Nature selects those which are most successful in their organic functioning. This seemingly "evolutionist" argument is discussed by Aristotle from the point of view of Empedocles, who held that in the period when "Love" was gaining on "Strife", a variety of limbs and organs were formed and randomly connected.[50] According to Aristotle however, teleology may just as well account for deviances occurring in Nature. He compares Nature with art, in which there are mistakes and failures too, as when writers make faults and doctors give the wrong prescriptions. These failures are accountable to intervening factors influencing the process which thereby misses its

---

[47] Voetius, *Select. Dispp.* V, pp. 191-192 / Hieronymus Zanchius, *Opera Theologica*, Tomus III, col. 475. In the context in which the passage occurs, genera are said to be non-eternal, since all were created at a certain period in time, i.e. on the Sixth Day of Creation. It is there that Zanchius adds: "Genera dico animalium: quoniam monstra novum genus non constituunt: sed in uno quoque genere peccata sunt naturæ."

[48] See e.g. Aquinas' and Cajetan's metaphorical interpretations of Biblical verses in which cross-breeds occur, as referred to in note 14, above.

[49] *Physics* II 8, 199^{b}4.

[50] *Empedocles: The Extant Fragments*, edited, with an introduction, commentary and concordance, by M. R. Wright, New Haven and London (Yale University Press) 1981, pp. 49-56 esp. For a detailed account of Empedocles' theory and Aristotle's interpretation of it, see Richard Sorabji, *Necessity, Cause, and Blame: Perspectives on Aristotle's Theory*, Ithaca (Cornell University Press) 1980, pp. 176-177.

aim. The same may occur in the case of monsters, where defects in the sperm drive the teleological process away from its intended result.

As against Empedocles, Aristotle says that it was the seed which was formed first and from which species are derived. In present-day terms, we could reconstruct this argument as saying that natural selection only occurs at the stage of animals already formed. Materialists like Empedocles may well argue that there once were monstrous births—some of which proved fit for survival. But these monsters must also have come about from seeds and not spontaneously.[51] They too, pass through a teleological sequel of productive stages. It is, however, deficient in respect of the τέλος.

The occurrence of monsters might in the first instance seem to point to an efficaciousness of chance or spontaneity in Nature. Aristotle however, argues in exactly the opposite way: since monsters are by definition exceptional occurrences within a regular pattern, the *chance*-occurrence of monsters points out that the *regular* pattern of Nature is in fact a teleological one. This reflects the general line of *Physics* II 8, in which Aristotle opposes teleology to chance and spontaneity only to conclude that the concepts of chance and spontaneity are in fact dependent on the teleological outlook.[52] The

---

[51] The adherent of "natural selection" might retort that this does not make a difference; seed may be formed spontaneously as well and in a later stage of development be tested for survival. It may be for this reason that Aristotle's argument that seed comes first, has been regarded as an "incidental criticism of Empedocles' theory, not relevant to Aristotle's main argument." W. D. Ross (ed.), *Aristotle's Physics*, Oxford (Clarendon) 1966[4] (= 1936), p. 530. Indeed, when interpreted in terms of natural selection, Aristotle's second argument is quite irrelevant. It is irrelevant for the question whether animal species are formed by chance. Seed may be formed by chance as well. At the same time, the argument is not at all irrelevant for the question of chance versus teleology on a phenomenological level. Aristotle tries to save teleology by pointing to the fact that, in the formation of animals, seed comes first. If seed comes first, the Empedoclean hypothesis of chance mutations must be rejected on account of its being fabulous. For why should one accept that there once were non-teleological chance mutations? We do not come across such mutations in Nature, but rather see specific kinds growing from specific seeds. If the primal combinations were also formed from seed, the teleological picture accounts for them as well. On the differences between Empedocles' theory and modern evolutionist accounts, see Richard Sorabji's chapter on "Ancient and Modern Theories of Natural Selection: their relation to purpose", in *Necessity, Cause, and Blame: Perspectives on Aristotle's Theory*, Ithaca (Cornell University Press) 1980, pp. 175-181.

[52] Aristotle argues that the concepts of chance and contingency are used in teleological contexts only. See W. Wieland, "The problem of Teleology", p. 144: "something is fortuitous only if it *could* come about also for the sake of an end

question of monsters is in fact a good argument for teleology: since monsters are exceptions, they may be interpreted as deviants from the teleological norm. In such a way, chance occurrences may generally be explained as exceptions to the normal course of events. Instead of being used in scientific theory as an independent factor of change, chance and spontaneity thereby become part of the teleological framework. Monsters in fact offer only the more reason for taking goal-directedness to be a necessary part of our accounts of natural processes.[53] They are deviants from the teleological path, which, in itself, is fixed.

Another way of putting this, is to say that Nature has a fixed, determinate end (τέλος) for the process of growth which is missed in case something goes wrong. We then speak of a chance occurrence, or, better, of spontaneous deviations.

2.3 The Metaphysics of Species

If, as Aristotle says, monsters are chance-occurrences within a teleological framework, there seems every reason to argue, with Voetius' contemporary Zanchius, that

(197^{a}35)." Wieland offers a perspicuous account of the interrelatedness of chance and teleology in *Physics* II 8. See *ibidem*, pp. 143-145.

[53] Of course, Darwinists might argue that species are selected on account of their being able to survive, whilst their formation is randomly caused by efficient factors of a biochemical process. It would not be an easy task to reconstruct the answer Aristotle might have given to such an option. Aristotle takes it upon himself to criticize the possibility of spontaneous organisation, which he offers as an exemplary theory of non-teleological necessity (*Physics* II 8, 198^{b}29-34). According to such a view, all natural forms would be formed by coincidence, some being able to survive, others perishing for being ill-formed. Empedocles' theory of monstrosities that were formed in an earlier stage of the universal cycle is here given as an example. Next, Aristotle restates the teleological view, claiming that monsters are exceptions analoguous to chance occurrences *within* a teleological framework (*idem*, 199^{a}33-199^{b}7). Finally, arguing that seed precedes the formation of the animal, he refers to Empedocles again, saying that what came first were seeds, not animals (*idem*, 199^{b}8-9). Thus, there is no discussion of the possibility of animal-formation occurring in random varieties. But then, there does not seem to be any reason for such a theory of natural selection for Aristotle. What is implicit in his argument is the fact that there is no empirical ground for one-time varieties. The only varieties that do occur, are monsters. These, however, are, as we have said, exceptional by definition.

> monsters do not constitute a new kind, but are failures in respect of all individual kinds of nature.[54]

This, however, leads to new questions. How is it that Nature offers a restricted number of types only? In what way are species fixed? It is on these points that Aristotle and the Scholastics go their separate ways.

2.3.1 Mules and Monsters: Exceptions to Natural Sequences

In case the end of a process of animal development is not achieved, its monstrous outcome must be brought about by intervening factors. The chance occurrences of monsters do not, in other words, fit the causal path of standard teleological processes. They are brought about by accidental factors intervening and occasioning a "failure" in the natural way of things.

An important aspect of this account is that if monsters are exceptional, the normal path must be confined to the species developing in the usual way. In other words, there is a collection of fixed teleological paths of development in Nature, each path corresponding to a certain species. Monsters follow none of these natural paths in the usual sense, but, because of external influences, follow one of them in an exeptional, deficient way. If one takes each natural path to correspond exactly to the development of a healthy individual of a certain species—in other words to one type of animal—monsters may be said not to form a different species of their own. It is for this reason that Zanchius says that

> monsters do not constitute a new type (*genus*), but are failures in respect of all individual types of nature.

Arguing that "mules are rather like monsters", Voetius in turn tries to exclude mules from the normal trajectories of teleological development in a similar way. Since the teleological paths have a fixed end for every species, exceptional occurences within one path results in a defective examplar of a species, that is, a monster. Mules are exceptions too: not because they are faulty results of a fixed sequence of developmental stages, but because they do not belong in any of the teleological paths that Nature has to offer. They do grow into something which is the *terminus* of the growing-process, just as

---

[54] Voetius, *Select. Dispp.* V, pp. 191-192 / Hieronymus Zanchius, *Opera Theologica*, Tomus III, col. 475.

monsters do. But since mules develop outside of fixed natural sequences, their growth is exceptional and, according to Voetius, equally "accidental" as is the growth of monsters.[55]

Voetius' theory of monsters is linked to a genuinely Aristotelian idea of teleological development. But what about mules? Are they the result of "external" or "internal" factors of change? In order to answer this, let us return to Aristotle once more.

2.3.2 Fixing the Number of Species

According to Aristotle, monsters are the result of defects of the seed, which, in the light of normal development, is a "spontaneous", external factor influencing the internal course of things. The teleological path is nonetheless intact: since there is no random development of things natural, but rather a fairly constant development from seed to the natural—and "best"—end, monsters are only exceptions to the teleological rule and no proof for independent chance-factors in Nature.[56]

Since Aristotle prefers "form" rather than "matter" as that to which the term "nature" (φύσις) is best applied and identifies φύσις with the

[55] As Voetius sees it, monsters have only accidental causes. See above, note 34.

[56] At *Physics* II 6, $197^{b}32$-37, Aristotle discusses the concepts of spontaneity and chance and explains that things that come to be contrary to nature (παρὰ φύσιν) are better described as occurring spontaneously (ἀπὸ ταὐτομάτου) than by chance (ἀπὸ τύχης). As such, this is a matter of linguistics only. Aristotle is investigating the use of ἀπὸ ταὐτομάτου and ἀπὸ τύχης, which differed in that the latter phrase was used in context of benefit and profitability, whilst the former refers to accidentality without any connotation of luck or fortune. However, even as a "spontaneous effect", the case of things coming to be contrary to nature is, according to Aristotle, somewhat ill-described. For what is said to arise "spontaneously" is usually brought about by external factors, whereas things produced by nature, but "contrary to it", are what they are because of factors that are "internal". The question is, what Aristotle has in mind here. It has been argued by Ross in his commentary on *Aristotle's Physics* (p. 542) that the reference is to spontaneous generation, which occurs in a natural way, but which is at the same time "contrary to nature" in the sense that, in spontaneous generation, the teleological trajectory from seed to full-grown animal is absent. This interpretation is in any case preferable to the obvious one that Aristotle is here thinking of monsters. In the latter case, the distinction between external and internal factors would in fact be faulty. As we have seen, according to Aristotle, monsters are the result of defects of the seed, which, in the light of normal development, is a "spontaneous", external factor influencing the internal course of things. Hence monsters would be the exact opposite of a thing produced by internal factors.

"substance" (οὐσία)[57] or with the form of a thing as it is given in its definition (λόγος)[58], the concept of nature seems to be given a classificatory function also. This, in other words, would mean that natures stand for species and that natural teleological development is restricted to types, species, or "natures", each developing in a definitive way. Monsters are exceptions in that they miss the natural end of one of such teleological paths. The question is: in what way is an internal process of change fixed to a nature? Is there a finite set of natures to which a finite set of teleological paths applies, or is it possible to find different sorts of development, to every one of which there corresponds a certain nature, thus adding nature to nature?

In the first alternative, the set of "natures" is fixed. I shall call this the "essentialist" view: Nature as a whole has only a limited number of "natures" or species of things to offer. In the second alternative, the concept of "nature" is wholly dependent upon the variety of natural processes we come accross in an empirical investigation of Nature. Nature as a whole is then thought to bring about whatever teleological trajectories of organism-development we observe. Each of these we may, by definition, regard as a species. This we might call the "nominalistic" or "empirical" concept of nature.

The fact that Aristotle does not regard monsters as a distinct species does not put him on the "essentialist" side. For monsters are defined relative to a certain teleological trajectory which is regularly observed, whatever the status of this trajectory—metaphysically speaking. Moreover, despite centuries of "realist" interpretations, Aristotle's biological works clearly show that he was disinclined to make use of the concept of "species" for explaining individual forms at all.[59]

---

[57] Aristotle, *Physics* II 1, $192^{b}33$.

[58] Aristotle, *Physics* II 1, $193^{a}30$-31.

[59] As Pierre Pellegrin concludes in "Aristotle: A Zoology without Species": "[...] Aristotle does not think of generation as the transmission of *specific* characteristics, to which are added individual variations; for him, the species level has no priviliges and thus is not consistent. The individual may have attributes of larger and larger groups: Socrates-man-animal. In *GA* iv $3.767^{b}32$, [...] the term *genos* has in fact a *logical* function and could designate any degree of generality at all beyond the individual. Only the context shows us that it here designates what we would call a species. Thus there is not in this text, and not in similar texts, an affirmation that what we call a species has any priviliged role in the hereditary transmission of characteristics." Alan Gotthelf (ed.), *Aristotle on Nature and Living Things: Philosophical and Historical Studies*, Pittsburgh, Penn. (Mathesis Pulications) / Bristol (Bristol Classical Press) 1985, p. 111. See also David M. Balme,

Voetius' comparison of mules with monsters, on the other hand, indicates to what extent the Scholastic theologian tied the empirical fact of animal diversity to a metaphysics of class distinctions. Where for Aristotle the theory of development to specific ends was based upon the observable regularity of biological patterns, the Scholastic theologian saw the teleological trajectories in such a way as to exclude certain animals from the fixed classes of species.

Moreover, Voetius identifies the Biblical use of *genus* and *species* as corresponding to the philosophical concepts. The separate question of whether animal diversity may or may not function as an argument in favour of a metaphysical realism of kinds, is presented the other way around: a metaphysical realism of kinds functions as an argument to exclude mules from the class of earthly creatures. When, on the Sixth Day of Creation, God created all species of living things, no mules were to be found on the face of the Earth. Interpreting the Aristotelian notion of φύσις and species (οὐσία)[60] in a metaphysical sense, the words of Scripture could be backed with philosophical theory. God created "permanent natures, faculties and distinct species of things"[61]. To match the Scriptural evidence, we have an essentialist notion of "nature" that explains such permanent species as natural kinds.[62]

"Aristotle's Biology was not Essentialist", in *Archiv für Geschichte der Philosophie* 62 (1980), p. 1: "Here I argue that in the *G. A.* Aristotle holds that the animal develops primarily toward parental likeness, including even non-essential details, while the common form of the species is only a generality which 'accompanies' this likeness. In the *P. A.* he argues for teleology with the question "What benefits an animal of this kind?", not with the question "What benefits *all* animals of this kind?" He treats species as merely universals obtained by generalisation. While it is true that species-membership may help to explain the features of individuals, this is not because a species is an efficient cause of individual formation, but because individuals in like circumstances are advantaged by like features." See also *idem*, p. 7: "It is often remarked that [Aristotle] has only the one word *eidos* for both form and species. But he does not need a technical word for species, since he does not hypostasise it into an entity or absolute; its status is merely that of a universal. True, his attitude too it is not nominalistic: the lowest universal especially, the infima species, has an objective validity [...] But from the fact that it is objectively valid, it does not follow that it need also be formally precise or even unchanging."

[60] On Aristotle's use of οὐσία for "species" and "form" alike, see above, note 59.

[61] See above, chapter 1, note 76.

[62] According to Aristotle, the species is not necessarily eternal, although there may be groups of organisms in which there is an eternal succession of individuals

2.3.3 Substantial Forms

We may now assess what Voetius means by the term "substantial form". What matters most to Voetius when referring to *Genesis* 1, is the fact that the philosophy of "natures" matches the Biblical text. This is not to be seen as a coincidence. On the contrary, for Voetius, it was the unity of truth which was at stake. As Voetius himself says in connection with this in a passage which we have already once quoted:

> truth agrees with truth and Christian philosophers will rather a thousand times profess their learned ignorance, than that they inflict even the tiniest sort of prejudice upon divine truth.[63]

As is the case in the question of the mule, divine truth might leave some of the details unanswered, yet the central point is clear: God in the beginning created all creatures after their kinds and had finished His works within the Six Days. So that this picture could be offered philosophical support, the Scriptural concept of species was reinforced by a metaphysical concept of "natures". Close to this notion stands the other—genuinely Scholastic—notion that Voetius mentions in the title of his essay: the notion of substantial form.

If, as the Bible says, all beings are created according to class distinctions, every natural species must be endowed with an "essence" that makes it belong to a particular class. This is what the notion of substantial form stands for: an individual essence. Scholastic authors accordingly divided the whole natural order of things into classes of different species, each individual being defined by a specific—although for its metaphysical character somewhat elusive—"substantial form". Indeed in Voetius' time, these classificatory and definitional functions of the substantial form were assumed to be necessary in order to describe the natural world. Renaissance commentators on Aristotle quoted the Philosopher by defining the substantial form as the "what it is". The famous Jesuit philosophers of Coimbra, or *Conimbricenses*,

---

procreating themselves. (See also below, note 88.) As James Lennox argues, it would seem plausible that Aristotle did hold the view that *forms* are eternal, since Aristotle argued "that the form characteristic of a species does not come to be or pass away." "Nonetheless", Lennox continues, "[Aristotle] did not claim this, and for good reason. The form is not the sort of thing that can be eternal, even though it is the basis for predicating eternality of a natural kind and of its members." James G. Lennox, "Are Aristotelian Species Eternal?", in Alan Gotthelf (ed.) *Aristotle on Nature and Living Things*, p. 89. As we have seen, for Voetius on the other hand, natures and species of things are "permanent".

[63] See above, chapter 1, note 76.

for instance, whose works were widely read and recommended by Voetius himself[64], said that the substantial form was the *ratio quidditatis*, in other words, the "ground for the what-ness"[65] of a thing. They further explained this as

> that in which the essence of every natural compound is particularly contained, or what completes (*absolvit*) the essence of a thing and its definition, and distinguishes it from other things.[66]

Accordingly, in his defence of the concepts of nature and form, Voetius asks how the adherents of the New Philosophy, which does away with forms, can explain and defend

> the distinction between the entity of substance and accident. For in their theory, there cannot be any substantial difference between a wolf, a sheep, a whale, an elephant, a snake a stone, a tree, a turnip, an aconite, wheat, the Sun, the Moon [and] the Earth.[67]

Of course, the New Philosophers would reply, as did Regius,[68] that these substances differ essentially on account of the motion, shape, position and size of their parts and that no further metaphysical distinction of species is needed. According to Voetius, that in itself does not take away the difficulty that substances would in that case not be distinguished from other substances, any more than substances are from accidents, or accidents amongst themselves. All differences would only be accidental.[69] This is clear enough, but it is no more than

---

[64] In his outstanding study on Renaissance Aristotelianism, Charles B. Schmitt characterizes the Coimbra commentaries together with those of Julius Pacius as "perhaps the most important of the bilingual, annotated editions [of Aristotle's texts] of convenient format". Charles B. Schmitt, *Aristotle and the Renaissance*, Cambridge Mass. and London (Harvard U. P.) 1983, p. 44. Voetius might well have thought the same. See below, chapter 6, note 46 and chapter 9, note 80.

[65] *Quiditas* (or *quidditas*) was a scholastic term designating the essence of a separate entity. As the example proves, *quidditas* remained acceptable despite humanistic criticism of such non-classical terminology by Renaissance authors as Lorenzo Valla. Cf. N. Kretzmann, A. Kenny and J. Pinborg (edd.), *The Cambridge History of Later Medieval Philosophy*, Cambridge (C. U. P.) 1982/1990[5], p. 816.

[66] *Commentari*[*um*] *Collegii Conimbricensis* (...) [*i*]*n octo Libros Physicorum Aristotelis Stagiritæ* I, Lvgdvni, Svmptibvs Ioannis Baptistæ Bvysson, 1594 / reprint Hildesheim (Georg Olms Verlag) 1984, p. 180.

[67] Voetius, *Narratio*, p. 39-40 / *Select. Dispp.* I, p. 872 / *Responsio*, p. 11 / *Querelle*, p. 106.

[68] See above, note 6.

[69] Voetius, *Narratio*, p. 39-40 / *Select. Dispp.* I, p. 872 / *Responsio*, p. 11 / *Querelle*, p. 106.

a restatement of the case. The question is: what is so problematic with the acceptance that substances differ only accidentally from each other?

We may find an answer in the works of the philosophers of Coïmbra, who immediately added to their description of the "form" a "clearer" (*apertius*) definition. According to the *Conimbricenses* the form was best defined as

> a simple, substantial act, constituting, with matter, a single thing in itself.[70]

It is explained that "substantial" is added in order to exclude forms which are merely accidental, and that the substantial form is said to constitute, with matter, a single thing in itself, in order to exclude *assisting* forms such as the intelligences by which the heavens are moved. I shall postpone discussion of these notions to chapter 6, below. For now, let it be noted that the fact that only those forms are meant which constitute single unities with matter, should not be taken to mean that forms could not exist separately from matter. In that case, the Conimbricenses add, the rational soul would not be a form when it dwells outside of matter, while, on the contrary, the soul is "by its essence, which cannot be taken from it, the act and form of the body."[71]

Both the example of the soul and the fact that the substantial form is presented as an *actus* are indicative of the way in which Scholastic doctrine emphasized not so much the classificatory function of the form as much as its causal function. In the chapters to follow, I hope to point out that the concept of substantial form, as something which, together with matter, constitutes a "single thing in itself", was a feature of Aristotelian philosophy that was of the utmost importance to Voetius—far more important than, for instance, the Aristotelico-Ptolemaic order of the Universe. For the Neo-Scholastic theologian, Aristotelian philosophy offered a third way out between materialist ideas on the one hand and vitalist conceptions introducing "assisting" forms on the other. In this respect, the concept of substantial form was crucial. Let us therefore take a closer look at it and examine its function in the Scholastic theories of causation.

---

70 Voetius, *ibidem*.

71 Conimbricenses, *In Physicam*, p. 181.

2.4 *The Internal Motor of Natural Objects*

In the example of monsters, Voetius and Zanchius attributed the accidental occurrence of disfigured organisms to accidental "causes".[72] The desirable end being missed, they would say that external, accidental influences "caused" the monster. This way of putting things comes close to Aristotle's, but again, Scholastic and Aristotelian views are not wholly identical.

2.4.1 *Four Types of* Αἰτίαι

In *Physics* II, the end, τέλος, or completion, aquires the status of a causal factor in physical processes. The reason is that in *Physics* II, Aristotle is investigating the ways in which we give answers to questions in which we ask for the "why" of things. Here and elsewhere, giving answers to such questions is considered to be the proper object of science.[73] If we consider why, for instance, someone walks, we may answer that he does so for his health. The τέλος (in this case, health) explains to what end or for what reason something is done. According to Aristotle, such ends may be found in Nature also. If one were to ask, for instance, why such and such motions accompany the formation of an embryo, the answer might be given that this is in order for it to grow. Growth is thus an end, a τέλος for this motion, but it is only an intermediate end. The foetus will develop further and further until it reaches its ultimate end, or completion: a full-grown organism of a certain type. This is the τέλος that ultimately explains the preceding processes of change and motion.

In the *Physics*, Aristotle not only identified the end (τέλος) as what may be sought in a why-question. As is only too well-known, he gave four αἰτίαι, or "causal determinants"[74]. They have become known as the material, the formal, the efficient and the final αἰτίαι (A1, A2, A3 and A4). According to Aristotle, all of these may be used in order to explain natural processes. However, in some cases, some of these factors may coincide.[75] For, as Aristotle has it, the "what" (A2) and

---

[72] See above, note 34.

[73] Aristotle, *Physics* II 3, 194^b17-20; *Posterior Analytics* I 14, 79^a23-24.

[74] I take the translation of αἰτίαι as "causal determinants" from Wicksteed and Cornford in the Loeb-edition of the *Physics*. See Aristotle, *The Physics*, ed. P. H. Wicksteed and F. M. Cornford, Vol. I, Cambridge, Mass. / London (Harvard University Press / Heinemann) 1980^4 (= 1929), p. 131 (195^a3 ff.). See also my Introduction, note 3, above.

[75] *Physics* II 7, 198^a24-27.

the "that for the sake of which" (A4) may be the same. And that is exactly what happens in growth. The growth of the child is explained by the end or completion at which it aims, but the completion itself is nothing but the full-grown individual. In other words, the "end" (A4), that is, being full-grown, is identical to the "what" (A2) of having acquired all specifying characteristics of a human being.

In *Physics* II 7, the search for the right number of causal determinants (αἰτίαι) is brought to an end. Aristotle concludes that there are four. In particular cases they may coincide, but the student of Nature should take account of all four of them if he is to explain why from these, something will necessarily result. In 199ª30-32 on the other hand, Aristotle literally says that the formal αἰτία is identical to the final determinant, i.e. the end or the "that for the sake of which". He thereby makes the same point as when he says that the final (A4) and formal (A2) determinants of natural processes may, at times, coincide. Moreover, as is said in 198ª24-28, the efficient factor (A3) may also coincide with the τέλος (A4) and the "what" (A2). This may be explained in the following way. Since man is born of man,[76] man initiates the growth of other human beings, his descendants.[77] Thus growth, for instance, is not only to be explained by its end—a mature human being coinciding with the specific "what", namely man—but also with the initiating factor in conception: man. Hence, in growth, all factors apart from the material factor (A1) coincide.

2.4.2 *The Concept of "Nature" in* Physics *II*

When explaining the end, or completion, as a causal determinant, Aristotle says that the τέλος is the best (τὸ βέλτιστον)[78] for a thing. In terms of Aristotle's project of finding causal determinants, this would mean that to say that all natural processes may be explained by

---

[76] *Physics* II 7, 198ª26-27.

[77] According to Aristotle, formal features are wholly dependent on the male parent, the female providing only the embryonic material, or, as David Balme puts it: "Aristotle argues that the male contribution to the foetus is form and movement only, nothing somatic." David M. Balme, "The Place of Biology in Aristotle's Philosophy", in Allan Gotthelf / James G. Lennox (edd.), *Philosophical Issues in Aristotle's Biology*, Cambridge (C. U. P.) 1987, p. 18. Note however that, apart from the teleological description, Aristotle accepts a corresponding efficient explanation of the movements of the semen also. See Balme, *ibidem* and Richard Sorabji, *Necessity, Cause, and Blame: Perspectives on Aristotle's Theory*, Ithaca (Cornell University Press) 1980, p. 160.

[78] *Physics* II 3, 195ª24.

pointing out the end or completion to which they are directed, is the same as to explain them as a striving "for the best". In other words, not every last stage of development is a teleological end, but only that which is good for something.[79]

The examples are, again, of a biological type. Leaves are for instance "good" because they protect the fruit.[80] And so generally,

> [given] the necessary limitations of heat and environment, each animal form is the best possible: that is, the form which brings it the most functional advantage, what Aristotle calls "the useful".[81]

From Aristotle's examples and his use of the biological paradigm in general, we may conclude that the predicates of ἀγαθόν, βέλτιον and βέλτιστον refer to the relative successful *functioning* of things which are said to be "good", "better" or even "best". This would still mean that every τέλος is "best", but also that we should reserve the term τέλος for well-developed and properly working things only. Aristotelian teleology would hence interpret processes not only in terms of a completion, but also in terms of the functionally successful product of the operation.[82] Things are, in other words, formed *for the*

---

[79] Aristotle points to the example of death: he does not agree to the poetical expression that calls death the end for the sake of which one is born. In that case, teleological explanation is used in an inadmissible way. Aristotle, *Physics* II 2, 194ª30-33. August Meineke mentions the anonymous fragment at *Fragmenta Comicorum Græcorum*, Berlin (Georg Reimer) 1857 / reprint Berlin (Walter de Gruyter & Co.) 1970, volume V-I, p. 123. Aristotle not only identifies the final determinant—the "that for the sake of which"—as the "best", but gives this as a reason for the fact that all other (material, formal and efficient) determinants are drawn to the end as to their "good".

[80] Susan Sauvé Meyer, "Aristotle, Teleology, and Reduction", in *The Philosophical Review* 101 (1992), p. 791: "[According to Aristotle,] parts of natural organisms develop because of the good ends they serve [...] For example, an animal's teeth develop flat in the back of the mouth and sharp in front because such teeth are suitable for biting and chewing food [...]; plants grow leaves because they are good for covering fruit, and roots grow down rather than up, because they are for nourishment, which is below." For a survey of the various types of purposive explanation in Aristotle, see also: Richard Sorabji, *Necessity, Cause, and Blame: Perspectives on Aristotle's Theory*, Ithaca (Cornell University Press) 1980, pp. 155 ff. Apart from parts of living organisms which are thought essential, Sorabji also distinguishes *luxury* organs, which are "partly or wholly for the sake of the *good*" and formative processes as candidates for teleological description.

[81] David Balme, "Aristotle's Biology was not Essentialist" in *Archiv für Geschichte der Philosophie* 62 (1980), p. 11.

[82] See also Charles H. Kahn, "The Prime Mover and Teleology", in Allan Gotthelf (ed.), *Aristotle on Nature and Living Things*, p. 197: "It will be convenient

sake of working in a fine way instead of being formed *for the sake of* completion only.[83]

Perfection then, is nothing but the full development of form in a way which is "good" or "best" for a thing. This is their end or τέλος. At the same time, *nature* (φύσις) is identified as the end (τέλος) and the "that for the sake of which".[84] This is nothing else but a restatement of the idea that final and formal factors coincide. If a process reaches its end, it acquires a state of perfection. *Physics* II starts off with Aristotle's well-known division of things that exist. There are things that are by nature (φύσει) and things that are otherwise. By nature are animals and their parts, plants and the elements of earth, water, air and fire. These are distinguished from things that are not by nature because of their having an inner principle of change, or as Aristotle puts it, they have

> in themselves a principle (ἀρχή) of motion and rest.[85]

As he explains further on,[86] this means that nature may in one sense be regarded as matter that has in itself the principle for change, or—and better—as shape or form, which is the actualisation of matter.[87] Whichever of the two—matter or form—one takes to be the "nature" of the thing, the point of *Physics* II 1 is, that in things that are "by nature", matter is actively orientated toward form, without—as in art—there being an agent that applies form to matter. In other words,

---

to sketch Aristotle's general pattern of teleological explanation to see how it applies to the cases we have illustrated. The following schema is borrowed from Andrew Woodfield [...] *A is (or occurs) for the sake of B* may be analysed as: (i) B is good (for the relevant subject); (ii) A contributes to, or is necessary for, B; (iii) Therefore, A occurs (the relevant subject has or does A)." The original may be found in A. Woodfield, *Teleology*, Cambridge (C. U. P.) 1976, p. 206.

[83] Wieland does not distinguish between these two interpretations of teleology, seemingly since both are equally inconsistent with the universal cosmic goal which Wieland rightly rules out as being part of Aristotle's natural teleology. The idea that leaves are formed in order to give shelter to the fruit *etc.*, does however add another layer of final explanation to the teleology of processes leading to an end. The latter type of teleology might explain the growing of leaves by pointing at the definition of the end-result (a tree), without making the usefulness of fruit-shelters part of a final explanation. Nevertheless, Weiland is right in concluding that "*agathon* ('good') or *beltiston* ('best') is just a predicate which necessarily belongs to every *telos* as such." W. Wieland, "The Problem of Teleology", p. 149.

[84] Aristotle, *Physics* II 8, 199^a30-32.

[85] Aristotle, *Physics* II 1, 192^b14.

[86] Aristotle, *Physics* II 1, 193^a28-31.

[87] Aristotle, *Physics* II 1, 193^b6-8.

Nature works of itself and natural processes are directed towards ends that are achieved because natural things have internal principles of change.

2.4.3 *The Problematic Nature of Mules*

With this in mind, let us for a moment return to the question of mules as it was discussed by Voetius. From the fact that a similar coincidence of causal factors was accepted by Scholastic authors commentating upon Aristotle's text, we may understand how Voetius came to exclude mules from the category of genuine species and why he connected this with the fact that mules could not propagate their kind. Voetius' standpoint on the question of mules may then be reconstructed as an example of Aristotelian teleology in the following manner. Processes "aim" at perfection, but perfection only occurs in cases in which there is a formal "what" (A2) to be arrived at. There seems to be nothing wrong with functionally well-developed mules. What is curious about them is that mules do not form a lasting species, let alone an "everlasting" one.[88] In the "essentialist" interpretation of the concept of species however, this means that there is no formal "what" for the mules. Hence there is neither a definitive end (A4) to the process, nor a genuine propagatory "cause", since in all propagation these factors coincide. The "cause" must therefore be "accidental".

The "end" of a mule being of an exceptional character, so is their "what", and *vice versa*. This is a pseudo-Aristotelian way of expressing what Voetius and Zanchius mean when they claim that mules and other hybrids do not form a species of their own and are accidentally caused.[89] With Aristotle's idea of the coincidence of causal determinants in mind, let us take a closer look at the example of specific propagation and its relation to the *forma substantialis*.

---

[88] "Many species discussed in detail in Aristotle's biology lack the appropriate sort of causation to be eternal; moreover, Aristotle conceived of hybrids produced by individuals similar but not of the same kind." James G. Lennox, "Are Aristotelian Species Eternal?", in Alan Gotthelf (ed.), *Aristotle on Nature and Living Things*, p. 89. See also above, note 62.

[89] As Voetius puts it in the Neo-Scholastic terminology of his day: "monsters do not have causes that are *per se*, proper [to themselves] and definite (for causes *per accidens* are no causes)." Voetius, *Select. Dispp.* V, p 194. See also above, note 34.

2.4.4 *Causes for* Αἰτίαι

As the case of mules and monsters has shown, the division of Nature into natural kinds and species equals a division of Nature into sharply distinguished teleological trajectories. Each species follows a specific developmental path. Commenting upon the passage of *Physics* II 7, Aquinas for instance introduces the concept of a "generating form" acting as the *causa movens* in the process of generation. This form is of the same species ("man") as the "generated form", which is the individual brought about by generation and which, at the same time, functions as the end, or *causa finalis* at—and by—which the process is directed.[90] We might call the generating form the C3, the generated form the C2 and the "final cause" the C4 of the process.

In various publications, Fr. James A. Weisheipl has pointed out that Aquinas follows Aristotle closely by distinguishing between the generative cause, or αἰτία, and the principle, or ἀρχή of natural objects. In generation, it is a *cause* by which something is brought about, just as in animals that can move themselves, their own "nature" (φύσις) is the *cause* of deliberate motion. All other natural objects however have been endowed with an intrinsic nature which is not a cause, but a *principle* of spontanuous action and which, once the natural object is brought about, will always act without any further cause as long as there are no impediments.[91] Aquinas argues that the *causa efficiens* is a "perfecting" cause (for instance) when it

---

[90] *Physics* II 7, 198^{a}25-27; *In Physicorum* II, lectio 11, Maggiòlo, p. 117.

[91] See for instance Weisheipl's collected series of articles in James A. Weisheipl, O. P., *Nature and Motion in the Middle Ages, Studies in Philosophy and the History of Philosophy* Volume 11, Washington D. C. (The Catholic University of America Press) 1985, p. 65: "When Aristotle defines "nature as "the principle and cause of motion and rest, etc." [...], he clearly means that every "nature" is at least a *principle* (*archê*), but it can also be a *cause* (*aitia*), as in animals that move themselves by parts, as in running, flying, swimming, and the like. In all other cases, even in animals that move themselves, all other motions come from "nature" as from a *principle*, a source, of motion, as falling down, floating, growing up, being alive with blood circulating, and possessing specific traits." Weisheipl argues this case especially against historians of science from William Whewell down to Ernst Mach, Pierre Duhem, Alexandre Koyré, Anneliese Maier, Stillman Drake and E. J. Dijksterhuis, all of whom interpreted Aristotle's and Aquinas' as maintaining that bodies need an internal cause in order to act in the way they do. The discussion centres around the concept of the free fall of heavy bodies, the medieval *impetus*-theory and the concept of inertia. Weisheipl points out that Aristotle and Aquinas have been unjustly represented as having obstructed the modern concept of inertia by their allegiance to the maxim that everything that is moved is moved by another. In fact, the maxim *Omne quod movetur ab alio*

induces a substantial form in generation.[92]

Such an induction of form is to be regarded as an induction of the nature, or formal factor of natural processes. For instance, the parent is the efficient cause of the nature of the child. It is this nature, or substantial form, by which the child can act as an efficient cause itself. It is this nature also by which the child may grow. However, in growth, no further αἰτία is needed, since the child will simply grow for the fact that it is a human being. Likewise, according to Aristotle, in all non-animate objects, no "cause" is needed in order to explain their motion. They simply behave on account of an internal "principle", which is to say, on account of their having a certain nature. The cause of such natural action is identified as that by which the object was brought about. The object does not itself however, need a further cause to explain its action.

Aquinas followed Aristotle on this point.[93] Yet, being attributed the title of "cause" in the particular cases of generation and deliberate action, the formal factor, substantial form, or nature, was mostly interpreted as a cause in all cases of natural motion alike. In other words, the *nature*, or *substantial form* was, in Scholastic philosophy, mostly seen as the "cause" of motion and change in non-animate objects also. What started out in Aristotle as one of the factors needed for explaining natural processes, was reified into a "substantial form" which, being induced into a natural object, could be held responsible for all its actions.[94] The subtle differences between Aristotelian and

---

movetur, does not bear on, for instance, the free fall of bodies at all. Heavy bodies, according to Aristotle and Aquinas, will, as long as there are no impediments, always fall downward without there being any further *cause* for this motion. Cf. James Weisheipl, "Galileo and the Principle of Inertia", in *idem*, pp. 49-73. See also: "The Principle *Omne quod movetur ab alio movetur* in Medieval Physics, in *idem*, pp. 75-97 (= *Isis* 56 (1965), pp. 26-45).

[92] Aquinas, *In Physicorum* II, lectio V, Marietti , p. 92.

[93] Aquinas, *In Physicorum* II, lectio I, n. 4: "In corporibus vero gravibus et levibus est principium formale sui motus, quia sicut alia accidentia consequuntur formam substantialem, ita et locus, et per consequens moveri ad locum; non tamen ita quod forma naturalis sit motor, sed motor est generans, quod dat talem formam, ad quam talis motus consequitur." Quoted also in Weisheipl, *Nature and Motion in the Middle Ages*, p. 65, note 1. For Aristotle's account of the movement of objects to their natural places, see Keimpe Algra, *Concepts of Space in Greek Thought*, Leiden (E. J. Brill) 1995, esp. pp. 195 ff.

[94] This interpretation corresponds to a shift that scholars have detected in Aristotelian terminology, the history of which has been traced back to John Philoponus (V-VI), or John the Grammarian. It involves a reinterpretation of other

Scholastic conceptions of formal causality are important and have too often been ignored. Aristotle explained movement teleologically by drawing attention to the end or completion of a process, which is to be identified with the achievement of form. It is based on an analysis of the—empirically observable—factors which have to be taken into account in attempting to describe natural processes. We may describe the resulting picture of natural motion which Aristotle gives in terms of a perfection. It is a perfection *to* the form, explaining natural motion in terms of the actualisation of something potential. This picture however, differs radically from the Scholastic idea, which ascribes to the (substantial) form a causal efficacy by which it generates the process itself as an internal motor of action. Natural processes are thereby aimed at a perfection not *to* the form, but *by* the form—the form being the initiator of motion within natural objects.[95] Natural change is, accordingly, explained by referring to the substantial form as the internal cause, or internal motor of the process.

Accordingly, apart from being the individual instance of a specific nature, the substantial form was also attributed a "causal force", explaining natural teleological action.[96] Moreover, in Medieval and

---

aspects of Aristotelian physics, such as the concepts of nature (φύσις) and the relation of matter and form. See Helen S. Lang, *Aristotle's Physics and Its Medieval Varieties*, New York (State University of New York Press) 1992, p. 123: "For these commentators, as for Philoponus, too, the role of form as actuality, and the subordination of matter as actively oriented toward form, has been replaced by a notion of an intrinsic mover." Lang analyses *Physics* II, VII and VIII and some Medieval commentaries on these texts and convincingly points out incongruities between the Aristotelian point of view and those of his commentators which have hitherto mostly gone unnoticed. The explicit theory that in natural movement, the form is the mover or *motor coniunctus* that accompanies the body which it informs as an efficient cause, has been traced back to Avicenna. Cf. James A. Weisheipl, O. P., "The Specter of the Motor Coniunctus in Medieval Physics", in *Nature and Motion in the Middle Ages*, pp. 99- 120 (= A. Maierù and A. P. Bagliani (edd.), *Studi sul XIV secolo in memoria di Anneliese Maier*, Roma (Edizioni di Storia e Letteratura) 1981, pp. 81-104).

[95] My description of the difference between Aristotelian and Scholastic accounts as a difference between perfection *to* and perfection *by* the form is a variation on Helen S. Lang's "[For Aristotle] form is the form *of* a thing, rather than form *in* a thing". Helen S. Lang, *Aristotle's* Physics *and Its Medieval Varieties*, p. 137.

[96] According to Helen S. Lang, Aristotle had no need of such an internal motor since he made no sharp division between passive and active functions of matter and form respectively: "Here we arrive at a key issue: for Aristotle, in natural things to be moved does not imply a passive principle. Matter (or potential), which is moved by form (or actuality), is moved precisely because it is never neutral to its mover: matter is aimed at—it runs after—form. Because of the active orientation of the

Renaissance writings, the internal-motor picture was regarded as genuinely Aristotelian. Voetius in any case reproaches the champions of the New Philosophy that in their opinion there is no

> intrinsic motor in created substances, or substantial principle of motion which is internal and proper [to the thing in question].[97]

I shall come back to this passage in chapter 6, below. Presently, let us resume what we have gathered so far.

2.5 Conclusions

By vindicating the concepts of *nature* and *substantial form* against the New Philosophies of Descartes and other anti-Aristotelians, Voetius defends the concept of form both as *species* and as *intrinsic motor*. The first covers an essentialist interpretation of observable animal diversity. The second is linked to a pre-scientific notion of causality, attributing to natural objects a power of action "from within". Voetius claims, moreover, that the concept of substantial form is consistent with Biblical ideas. To quote Theo Verbeek:

> Voetius is not surprised that everything theology needs happens to be found in Aristotle's work. He and most of his contemporaries interpreted it as no more than the articulation of common sense.[98]

The essentialist notion of species and the idea of natural substances being causally responsible for their action, are two ideas expressing common sense-interpretations of Nature. They are not Aristotle's, but they were at least supposedly derived from his works. The point, however, was not that Aristotle was thought to come closer to common sense than other philosophers did. It was rather, according to Voetius, that *theology* had need for Aristotle. In the next chapter we shall see why.

---

moved towards its mover, no third cause is required to combine matter and form. They go together naturally: form constitutes a thing as natural, and matter is aimed at form." Helen S. Lang, *Aristotle's* Physics *and Its Medieval Varieties*, p. 26.

97 *Narratio*, p. 40 / *Select. Dispp.* I, p. 873 / *Responsio*, p. 12 / *Querelle*, p. 106.

98 Theo Verbeek, *Descartes and the Dutch*, Prologue, p. 7.

CHAPTER THREE

FINALIST PHYSICS

Lambertus Danæus and Hieronymus Zanchius, both near contemporaries of Voetius,[1] are classified in Voetius' reading-guide for students of 1644 as "systematic writers on Physico-Theology".[2] The reference is a positive one: Voetius claims that they are in fact the most comprehensive and professional writers in the field.[3] Their importance to Voetius is probably due to the fact that Danæus and Zanchius were among the first Protestant scholars to define the Reformers' standpoint as regards natural philosophy. In the first section of this chapter, I shall discuss their idea of a "Reformed Christian Physics", which in fact was based on older writings on *physico-theology*, such as those of the *hexæmeric* tradition of Church Fathers and Scholastics.

Protestant writings were not the only rock on which Voetius was to build his defence of the old body of physics against Cartesianism. Not only would Church Fathers and Medieval Scholastics remain an important source; in the essay on substantial forms, Voetius also refers to various "modern" Catholic commentators on Aristotle. These *recentiores* of the Neo-Scholastic tradition in fact served the Utrecht theologian as welcome authorities on nearly every page of his voluminous *Selected Disputations*. I shall accordingly examine some of the physical and metaphysical views of this "commentary tradition" in relation to Aristotle's *Physics*.

---

[1] On Danæus (1530-1595), see above, chapter 2, note 22. Hieronymus Zanchius (or Girolamo Zanchi, 1516-1590) studied philosophy and theology at Padua, converted to Protestantism and later taught philosophy and theology in Heidelberg. He seems to have taken a more or less conciliatory position in the theological strife of his times. Cf. *Jöcher* IV, cols. 2147-2148.

[2] Voetius, *Excercitia et Bibliotheca, Studiosi Theologiæ*, Rheno-Trajecti, Apud Wilhelm Strick, 1644, pp. 383-384 / Ultrajecti, Ex officina Johannis à Waesberge, 1651, p. 370: "*Physico-Theologi*: qui sunt 1. *systematici*, qui vel peculiari syntagmate, ut *Danæus*; vel in cursu theologico physicam scripturariam explicant, ut *Zanchius*."

[3] That is to say, together with Amandus Polanus, who is also mentioned. Polanus (1561-1610) was a German protestant theologian and writer of, amongst other things, a *Summa Doctrinæ Christianæ* and a book against the notorious Contra-Reformist authority Cardinal Bellarmine.

The resulting picture is that of a physics in which the idea of *final causation* occupies a central place in a variety of ways. As an illustration of how the concept of final causation was put to use, I shall offer some examples of physical explanations as they were taught to Dutch universtity students in Voetius' day.

3.1 The Physica Sacra

Some long quotations from the Church Fathers precede the proper text of Lambertus Danæus' *Physice Christiana.*[4] St. Basil, St. Ambrose and St. Cyril[5] are all quoted as defending a Christian view of nature which is said to outweigh and transcend the knowledge of the Greeks.

3.1.1 The Early Traditions

Right at the start, St. Basil is quoted as questioning the knowledge of the ancients. "The wise men of the Greeks", he says,

> argued variously concerning the nature of things and—the first [explanation] always being ousted by the next—no theory (*ratio*), no fixed, stable and unshaken opinion wholly remained among them. We may therefore without any difficulty refute their points of view, for because of their mutual disagreement they themselves bring about their own downfall.[6]

The fact that the contradictory reasonings of the Greeks never led to a single viewpoint as regards the nature of things is linked to the fact that the Greeks lacked knowledge of Nature's ultimate "cause", God:

---

[4] Lambertus Danæus, *Physica Christiana, sive, Christiana de rervm creatarvm origine et vsv disputatio*, Genevæ, Apud Evstathivs Vignon, 1588[3].

[5] Danæus quotes from the second chapter of the first of preachings on Creation by St. Basil of Cæsarea, or Basil the Great (c. 329-379), the text of which may be found in Basile de Césarée, *Homélies sur l'hexaéméron*, Texte Grec, Introduction et Traduction de Stanislas Giet, Paris (Éditions du C. E. R. F.) 1968, pp. 92-97. Danæus also offers quotations taken from the first three chapters of the first day of the *Exameron* of St. Ambrose of Milan (c. 334-397) (*Exameron*, in *Sancti Ambrosii Opera* I (*Corpus Scriptorum Ecclesiasticorum Latinorum* Vol. XXXII-I), ed. Carolus Schenkl, Prague / Leipzig / Vienna (F. Tempsky / G. Freytag) 1897, I 1,1-I 3, 11, pp. 3-9), and from St. Cyril of Alexandria's (c. 376-444) work against the Roman Emporer Julian. Cf. Cyrille d'Alexandrie, *Contre Julien* I, Livres I et II. Introduction, Texte Critique, Traduction et Notes par Paul Burgière et Pierre Évieux, Paris (Les Éditions du CERF) 1985, Lib. II, 569 C 15-576 A 11, pp. 232-243.

[6] Danæus, *Physica Christiana*, p. 6 / Basile de Césarée, *Homélies sur l'hexaéméron*, pp. 92-93.

> For those who were ignorant of God did not grant that there is a conscious cause presiding over the origin of things: they brought forward and accepted things that seemed succesively to come up and suit their first ignorance.[7]

Two points attract our attention immediately. First, St. Basil argues that, not knowing the divine source of being, the physical explanations of the gentiles were of necessity *ad hoc* accounts of natural change. Second, there is a critique of what we might call the "materialist interpretation" in these opening lines. Not only is it so that—not knowing God—the pre-Christian philosophers necessarily came up with unwarranted and incomplete principles of things; they also lacked an accurate idea of the *type* of cause that may be brought in as principle. The Christian knows it is a *causa mentis compotis*, that is to say, a *conscious* cause. Scripture in other words offers us both a way out of the philosophers' quarrels and, by offering an intelligent source of Nature, excludes the possibility that randomness or chance might account for natural processes.

The type of "scepticism" as regards natural knowledge which is involved in this denunciation of the ancients, reflects the mistrust with which writers such as St. Basil faced theories that were not backed by a supernatural revelation of truth. How much happier is the Christian who does not have to depend on the contingency of doubtful theorizing! Having Holy Writ at one's disposal, one is not prone to be left in spiritual confusion: the acceptance of "one way, one truth"[8] puts an end to the abhorrent situation of a variety of incompatible doctrines.

3.1.1.1 The Physica Hexæmera

Scripture, the first chapter of *Genesis* in particular, was thus taken to be the proper starting-point for explaining Nature. In chapter 2, we have seen how Voetius took up the problem of biological species in the context of a commentary on *Genesis* 1. Voetius said that there were many ways in which to study creatures according to their various species, but that he preferred the way "common to all theologians and

---

[7] *ibidem.*

[8] Cf. *Physica Christiana*, *Præfatio*, p. 20, where Danæus argues that, with Christians, one cannot with equal calm excuse discord and dissension. For there is only *one* "way of truth", to be found in God's Word: "In Christianis autem non eadem æquitate potest discordia dissensioque excusari, quia veritatis una est via, cuius in verbo Dei certissima, si modo ad illud attenderent, videbant fundamenta."

chronographers", which is to divide the works of Creation into six days.[9] Voetius thus puts himself in the tradition of, as he himself calls them, *operis hexæmeri scriptores* or *scriptores hexæmeri*. The term *hexæmeron* refers to the "[work] of the six days" (ἡ ἑξαήμερος), i.e. to God's work on the first six days when He created the world. It may have been introduced by Philo of Alexandria (25/20 BC - c. 50 AD).[10] It was under the influence of Philo in any case that Basil wrote his famous nine *Homilies on the Work of the Six Days*.[11] Ambrose also wrote a Latin *Hexaëmeron* which, together with St. Augustine's *De Genesi ad litteram* and *De Genesi contra Manicheos*, formed the most authoritative sources within the tradition.

It was this tradition which Voetius and the members of his faculty must have had in mind when they warned against the acceptance of ideas incompatible with the *physica sacra* of Moses, as interpreted by Church Fathers, Scholastics and recent Christian authors.[12] In fact, the term *physica sacra* or *physica mosaica*, rather than referring in a general way to a body of physical ideas taken from Scripture (such as, for example, the stability of the Earth), refers more specifically to the Mosaic writings of the Pentateuch, *Genesis* in particular. The conclusions of natural philosophy which were taken from this part of the Bible did not at all include the stability of the Earth, which question only implicity occurs in the book of *Joshua*. What Moses taught, apart from moral law, is comprised in the common-sensical and essentialist doctrines which we discussed in the former chapter, and certain notions of creation, conservation and divine providence.

3.1.1.2 Finalistic Priorities

As already explained, Danæus opens his *Physice Christiana* with passages from Basil's *First Homily on the Hexameron*, Ambrose's *Hexameron* and Cyril's book *Against Julian*. St. Basil is quoted as saying that the Greeks lacked knowledge of Nature's ultimate "cause", God. On the authority of Scripture, the Christian knows this to be a conscious cause instead of a blind, "materialist" one. Rejecting

---

[9] Voetius, *Select. Dispp.*, I, p. 597. See also above, chapter 2, note 11.

[10] Cf. *Lexicon für Theologie und Kirche* vol. V, Freiburg (Herder Verlag) 1960, p. 315.

[11] Basil was also influenced by Theophilus' *Apology* and by Origins homilies on the book *Genesis*. Cf. Stanislas Giet's introduction to: Basil de Césarée, *Homélies sur l'hexaéméron*, Paris (Éditions du CERF) 1968, pp. 49-56.

[12] See above, chapter1, note 62.

various materialist and atomist views,[13] Basil concludes that those who held them

> surely spun a spider web, [accepting] such very slight causes and meagre principles of heaven and Earth.[14]

According to Basil, the reason for their ignorance in these matters is that the gentile philosophers never heard the opening line of *Genesis* 1, which comprises all that is needed to disabuse them of their misconceptions: *In principio fecit Deus coelum & terram* (as it reads in Danæus):

> In the beginning, God created the heaven and the Earth.

For the Christian, this is enough to refute the materialist idea that the Universe is

> without any government and administration and brought about by chance or fate.[15]

Accepting providential government, on the other hand, the Christian is safeguarded from delusion in matters physical. This is also what is hinted at in the second quotation taken up by Danæus, which is from Ambrose. Equally emphasising the workings of God in Nature, Ambrose points to the fact that, in the New Testament, God promised "to administer the world up to its consummation."[16]

The hexaemeric tradition thus demanded that the work of Creation, the *opificium mundi*, be described and analysed in accordance with the Word of providence. Making this statement into an enlarged motto of his *Physice Christiana*, Lambertus Danæus shows the same devotion.

---

[13] "[Quidam] ad materiales suppositiones confugerunt, vniuersíque causam ad elementa mundi redegerunt atque accomodarunt. Quidam individua & insecabilia corpora & moles, atque meatus omnium quæ conspiciuntur, continere naturam asseruerunt: & ortus occasúsque rerum concursionibus nunc ac coagmentationibus, nunc disceptationibus corpusculorum illorum insecabilium fieri, validiorémque connexionem eorundem diuturniorum corporum durationis caussam esse." Danæus, *Physica Christiana* , p. 6 / Basile de Césarée, *Homélies sur l'hexaéméron*, Texte Grec, Introduction et Traduction de Stanislas Giet, Paris (Éditions du CERF) 1968, pp. 92-93.

[14] *Physica Christiana*, pp. 6-7/ Basile de Césarée, *idem*, pp. 94-95.

[15] *ibidem*. My translation is taken from Danæus. Note however, that Basil uses only τύχη for chance (*casus*) and fate (*fortuna*). Cf. Basile de Césarée, *Homélies sur l'hexaéméron*, p. 94.

[16] Danæus, *idem*, p. 12 / St. Ambrose, *Exameron*, *Sancti Ambrosii Opera* I (*Corpus Scriptorum Ecclesiasticorum Latinorum* Vol. XXXII-I), p. 9.

3.1.2 The Protestant Tradition

Accordingly, Danæus too, in his preface to the *Physice Christiana*, criticises the multitude of philosophical opinions.[17] The conclusion is that one cannot rely upon the study of Nature without knowing Scripture. This leaves Danæus with a problem, for in that case, what would be the point of treating questions of physics at all? The dilemma is a particularly pressing one for someone who, like Danæus, takes it upon himself to write a *Christian Physics*. Accordingly, chapter 2 of the *Physice Christiana* deals with the question whether physics is a worthy subject for the Christian to deal with.[18]

The answer is, of course, affirmative. It is in fact already implicit in chapter 1, where Danæus divided the discipline in two distinct parts. First, he says, there is "universal" physics, which treats of the heavens, the elements and their function as principles of what follows from them and of

> the common and general way and scheme (*ratio*) of the conservation and propagation of things.

Next, there is "particular" physics, which treats of natures, forces, properties and effects, and which includes medical sciences and the *historia animalium & plantarum*. The terminology links up with both the essentialist notions and the idea of innate causal powers which we came across in Voetius' account of species and intrinsic motors. The use of the term *conservatio* moreover, indicates a Christian orientation with regard to divine government. It is however, the definition of the subject of physics itself, which removes all doubt as to the context in which, according to Danæus, we should put the study of physics:

> What is physics?,

the studious pupil asks. Danæus answers:

> It is the true knowledge or treatment of the causes by which the creation and the distinction of this whole world and its parts are brought about and of the effects that follow therefrom, serving the glory of God's works.[19]

---

[17] In fact, Danæus goes even further and, besides pagans, also lists Christian writers among the quarrelling opiniators of natural knowledge. See above, note 8.

[18] Danæus, *idem*, Cap. II, p. 30: "Physice utrum homine Christiano digna."

[19] Danæus, *idem*, p. 27.

It is hence no surprise that, in chapter 2, the first benefit of learning physics for the Christian is said to be "that from [physics] we come to know God." Further, "that we understand the powers and natures of created things". Thirdly, that we know ourselves, that is to say that we come to know our mental and bodily condition. Fourthly, that,

> contemplating these many, diverse and wonderful works of God with our minds and seeing them with our eyes, we shall only with the help of this art [i.e. through physics] be seized [in such a way as] to praise God in the highest and thank Him with zeal and passion (*impetu*).

Finally, through physics we are able to explain many places of Scripture which remain inscrutable to the unknowing.[20]

Danæus adds that knowledge also gives us mental satisfaction and commodities for human life. The value of physics for the Christian however, lies in the five main advantages mentioned. These put God (and man) in a very central place. The value of pagan thought remains controversial.

3.1.3 Moses or Aristotle?

It was, in other words, the question whether to chose Moses or Aristotle, and of the two systematic writers on physico-theology praised by Voetius, together with Voetius himself, Lambertus Danæus was by far the most critical of ancient thought. Where Voetius for instance claims that the Scholastic concept of substantial form is consistent with the Biblical idea of nature, explaining the concept of "nature" both as *species* and as *intrinsic motor*, Danæus was not at all willing to identify the Aristotelian and the Mosaic views on this point.

3.1.3.1 Lambertus Danæus on the Concept of Nature

In the second part of Danæus' *Christian Physics*—a treatise of natural phenomena in hexaemeric form—various chapters are dedicated to the question of how to interpret the concept of nature. Danæus never doubts the existence of *natures* as such and accepts precisely those functions which common sense invests them with, namely the classificatory, metaphysical idea of essence on the one hand and the notion of an innate causal power on the other. *Nature* is thus said to stand

---

[20] Danæus, *idem*, pp. 30-31.

> for the subjects and supporting principles (*pro subiectis &* ὑφισταμένοις), that is, for the bodies and individuals themselves which form some composite whole that depends on an essence (*essentia*) and property (*proprietas*) and on some accidents besides, as quantity, quality and various other such things, which mark off its proper being (*hyposthasis*) and substance (*subsistentia*).

This may seem a circular explanation, but the fact is that it is not an explanation at all: it is simply a definition of nature. The character of the definition only testifies to Danæus' allegiance to the idea of essential natures. And it is clear what he means:

> thus men, horses and particular animals are called natures, which with another word are also called *natural things*.[21]

Besides this aspect, the immanent motor of change is also referred to:

> some internal force or cause common and inherent to all things of the same sort (*genus*), by which they act and move.

The concept of nature may exceptionally be used for other things as well,[22] but when Danæus comes to a conclusion, it is the essence and the internal motor that play the central parts. Nature is

[21] Lambertus Danæus, *Physices Christianæ Pars Altera, sive De Rerum Creatarum Natvra*, Genevæ (Apud hæredes Eustathij Vignon) 1589^3, p. 2 / verso.

[22] *Nature* may thus refer to those affections, inclinations, attachments and qualities which are not brought about by a habit of action (*consuetudo agendi*), nor by any superadded artificial practice or determined effort, but merely "by admission"—that is to say, by admission of sin. These are thus exceptional forms of nature, of which Paulus speaks when he says we are "by nature the children of wrath" and that "men burned in their lust one toward another." Cf. *Ephesians* 2:3 and *Romans* 1:24-27. The idea is, that homosexual lust is not a "real" tendency, but merely a momentary fall: "Quia præter naturalem & insitum à Deo motum hic tam foedus appetitus in quibusdam inest." Yet the example remains awkward, since the Apostle seems to speak only of what is *against* nature, not of a homosexual nature itself. Or does he? John Boswell has given an interesting reading of παρὰ φύσιν in this context, arguing that "nature should [here] be understood as the *personal* nature of the pagans in question." This would indeed lead to the positing of homosexual versus heterosexual natures, a violation of natural tendencies being implied only in cases in which one goes against one's *own* nature. Cf. John Boswell, *Christianity, Social Tolerance and Homosexuality*, Chicago and London (University of Chicago Press) 1980, pp. 107-117. For the quotation, see *idem*, p. 111. Further examples of the use of the concept of nature in Danæus are those in which *nature* stands for "God" as the Supreme Nature, or for genitals as those members by which natures are begotten and which are "most necessary for the

> an organical something implanted and poured into individual things by God at their first creation, such as that it is common to all individuals of the same type (*species*) and by which, [at least when they are] free and with no part of them obstructing them they each ordinarily [fulfil] their duties and go towards the end which God has set for them.[23]

The passage has obvious similarities to one which Voetius gives in his essay on substantial forms:

> a stone falls downwards, stars rise and finally all natural things perform their motions [...], [since] they are thus created by God and thus, according to the faculties impressed upon them, aim at their ends as an arrow at its target.[24]

Again, natures are, through the intermediary of their internal faculties, connected with the ends imposed by God. On the whole, we may safely say that both Danæus' and Voetius' analyses are dependent on the idea of relating natures to ends. Contrary to Voetius, Danæus even makes a point of distinguishing between the technical terms of essence, nature and hypostasis.[25] But if Danæus draws as least as much on Aristotle as does Voetius, he also makes a point of distinguishing more sharply between the Mosaic and the Aristotelian idea of nature.

So what was wrong with Aristotle? To answer this, we must first keep in mind what we said at the end of the former chapter, namely that the *finesses* of technical philosophy were regarded as the "articulation of common sense".[26] Hence there was no problem for Danæus to assert on the one hand that the concept of nature was precisely to be given the properties which were usually bestowed on them by those who followed Aristotle, whilst on the other, Danæus denied that Aristotle himself had an exact idea of such natures. The exact idea could either be known by natural reasoning, or by reading Scripture. In the book of *Genesis*, Moses had made use of the notions of *species* and *genera*. Aristotle had also written on these things, so

---

conservation of nature". Danæus, *Physices Christianæ Pars Altera, sive De Rerum Creatarum Natvra*, Genevæ (Apud hæredes Eustathij Vignon) 1589[3], pp. 2/verso-3.

[23] Danæus, *ibidem*, p. 4.

[24] *Narratio*, p. 45 / *Select. Dispp.* I, p. 877 / *Responsio*, pp. 24-25 / *Querelle*, p. 110.

[25] Danæus, *Physica Christiana* II, Chapter III: "Quid differrant inter se essentia, natura, hypostasis."

[26] Theo Verbeek, *Descartes and the Dutch*, Prologue, p. 7.

for studying Scripture, Aristotle might be of some help. But of course Aristotle could only approximate Moses.

According to Danæus, Aristotle was not altogether successful. What was wrong with his position was exactly what the Fathers of the Church had said to be wrong with all pagan learning. Aristotle had defined nature as

> the principle of motion and rest in those things in which [the principle] exists as first and by itself.[27]

Danæus has problems with talking of a principle of things which is first and by itself, when you are not talking of God. God is in fact the only real cause which is *per se* and does not operate through "any other motor":

> *causa per se* is said of that which sustains (*sistit*) itself and moves when it wills and is neither urged nor held back by something else when it wants to operate.[28]

As the Apostle teaches, God acts and operates in everything.[29] All other things, and hence all other natures, are only instruments with respect to this first cause. In the light of Danæus' sceptical stand towards pagan thought, it is no surprise that the Aristotelian definition in terms of a principle of motion and rest, is found too meagre a description. It only indicates that Aristotle—to quote St. Basil of Cæsarea—lacked a notion "that there is a conscious cause presiding over the origin of things".

3.1.3.2 Hieronymus Zanchius on Moses and Aristotle

How very different was Hieronymus Zanchius' position! Recalling in his *Hexaëmeron* the appearance of living natures on the face of the Earth, he sidesteps the discussion for a moment, in order to reflect some more on *The Method of Moses and Aristotle*. There, it is bluntly stated that the method of Moses which we are here dealing with

> is not so very different from the one that Aristotle followed. It is therefore useful to observe, when you are reading Moses, who

---

[27] Danæus, *Physica Christiana* II, p. 3/verso: "[Ex] Aristotele dicetur [...] naturam esse principium motus & quietis in iis rebus in quibus primò & per se inest."

[28] Danæus, *ibidem*.

[29] Danæus refers to *Ephesians* 1:23, where it is said that Christ's body is "the fullness of him that filleth all in all".

> recounts the works of God in separate days, that, when you are looking for a longer explanation of these matters, you know where to turn to: namely to Aristotle's works, in which these matters are discussed more widely.[30]

The total confidence Zanchius shows in Aristotle's explanation of the works of Nature, clearly sets him apart from the careful Danæus. Behind their different appreciation of pagan thought, there is however at the same time a certain similarity in their ways of weighing Moses against Aristotle. What is at stake is, again, the defence of truth. In Scripture, one finds factual truths which may be explained by philosophy. One may either, like Danæus, distrust the philosopher's capacity for acquiring like truths, or, like Zanchius, take a more positive stand towards the philosophers. The result is more or less the same. Just as Danæus, Zanchius also takes the approximation of *Genesis* to be the only goal for Aristotelian thought. We should always keep to the rule, Zanchius says:

> not to twist Scripture according to [philosophical works], but, on the contrary, to test these in the light of Scripture as their touchstone.[31]

The argument is the same as in Danæus. Yet the consequences are entirely different. Zanchius for instance has no problem at all with Aristotle's concept of nature, which he gives in much the same words as did Danæus: nature is

> the origin (*initium*) of motion and rest in the subject in which it inheres primarily and *per se* and not accidentally.[32]

This, to be precise, is only the definition of "second natures", the first nature being God. Zanchius values this philosophical idea of a second nature as a principle of motion and rest for its attempt to approximate, or even elucidate the Biblical notion of nature. He therefore does not criticize Aristotle's phrasing, as did Danæus. Rather, Zanchius adds to Aristotle's credit by referring to Plato, who is supposed to offer a very similar idea:

---

[30] Hieronymus Zanchius, *Opera Theologica* III, *De Operibus Dei intra Spacium Sex Dierum Creatis* [...] (Excudebat Stephanus Gamonetus) 1613, col. 223.

[31] *idem*, col. 224.

[32] *idem*, col. 219.

> Plato called this nature "world-soul" for the reason that, just like the soul in all animated beings, Nature is the principle of all natural action in natural things.[33]

It may seen somewhat strange to see Plato's concept of a world-soul be compared—and even identified[34]—with Aristotle's concept of nature. The argument however, is clear enough: just as Aristotle held natures to be responsible for the action of natural substances, so Plato too presents a motive principle governing the material world of natural objects. Looking for evidence in pagan thought for Christian ideas central to the Mosaic account of Creation, Zanchius thus also offers the world-soul as a pagan idea in favour of the Biblical idea of nature. However, not only are Aristotle's "second nature" and Plato's world-soul identified with the Mosaic idea of nature; they also offer a fine parallel to the Christian idea of an all-moving *Spiritus Dei*:

> For as the Apostle says of the first Nature, that is, of God, that *it is He in whom we live and move and have our being*, so [it is] likewise this second nature, or world-soul, through which, as the Platonist rightly teach, all things are formed, established, moved and acted upon.[35]

Aristotle's and Plato's conceptions of nature are not only seen as alternatives to the Mosaic idea of *species*: moving all things natural, these "second natures" also parallel the first. In other words, as active principles, they point to the ultimate principle in which all things are moved: they point to the God of *Genesis*.

The conclusion must be, that despite Danæus' and Zanchius' at first sight wholly incompatible ways of appreciating pagan thought, their methods and ideas are in various respects the same. Both hold the view that philosophy should try to approximate Biblical truth. Both moreover, like the Church Fathers before them, evaluate theories of natural philosophy according to their succes in presenting the world as governed by divine administration. According to Danæus, no philosophical theory comes near to it. According to Zanchius, all philosophical attempts are worthwhile and some in fact parallel the Biblical

---

[33] *idem*, col. 219. For Plato's idea of a world-soul, see *Timaios*, 34 B-36 C.

[34] Zanchius, *De Operibus Dei intra Spacium Sex Dierum Creatis* I, I, III, Thesis II, *Opera Theologica* III, cols. 249-250: "Alterum à primo [principio] profectum, illúdque internum verum principium, est ipsa Natura rebus à Deo insita: quæ tum secunda natura à Peripateticis: tum Anima Mundi à Platonicis appelari solet."

[35] Zanchius, *Opera Theologica* III, col. 219. The verse is taken from *The Acts of the Apostels* 17:28 "For in Him we live and move and have our being."

message in various ways. In both cases, the objectives of a Christian Physics are the same. It is more or less a matter of emphasis whether or not these objectives were thought to be met by the Greeks.

3.1.3.3 Conclusions

As we saw, St. Basil pitied the Greek philosophers for their ignorance of divine administration. Centuries after him, Danæus and Zanchius too, and in fact writers on Mosaic physics generally, subjected theories of matter and physical principles to the test of whether or not they supported the idea of divine providence. As the example of Danæus shows, a scepticism as regards the possibility of linking Christian to profane ideas was still very much alive in Protestant works on physico-theology. Zanchius is more sanguine in relating philosophical to theological dogma.

Both "physico-theologians" however, attempted (1) to draw parallels between the Mosaic texts and profane learning and (2) to demonstrate the power of God's Hand not only in the Bible, but in the Book of Nature itself. Divine administration was thus the main issue and profane philosophies such as Aristotle's were valued accordingly. What R. Hooykaas has said in relation to the Copernican issue then, holds just as much for theories of physics generally:

> it depended on the preconceptions of the interpreter whether 'Mosaic philosophy' would turn out to be more or less Aristotelian or, on the contrary, violently anti-Aristotelian.[36]

Contrary to both Danæus and Zanchius—the physico-theologians he so much praises in his reading guide for students[37]—Voetius chose for both Mosaic physics *and* Aristotle. Voetius' defence of Aristotelian notions against the anti-Aristotelian currents of his day shows him at once to be far more confident than Danæus in the possibilities of linking Christian ideas to philosophical standpoints and far stricter than Zanchius in his choice of which philosophical doctrines could and which could not be brought in line with religious ideas.

---

[36] "But all protagonists of a 'Mosaic philosophy' claimed that the Holy Spirit guided the hand of the biblical writers, so that they wrote down scientific truth for all times." R. Hooykaas, *G. J. Rheticus' Treatise on Holy Scripture and the Motion of the Earth*, Amsterdam / Oxford / New York (North-Holland Publishing Company) 1984, p. 30.

[37] See above, note 2.

Thus, Voetius was at once more positive and more critical with regard to the accomplishments of pagan thought than were these fellow Protestant divines. Contrary to his non-Aristotelian colleagues, the Aristotelian theologian had the task of explaining how Aristotle and Moses could be brought in line. We have already seen how the concept of substantial form was thought to match the text of *Genesis* 1. But the same could be done with regard to what Church Fathers and Protestants alike thought to be crucial: the notion of divine government and administration. On this question, Aristotle had been consulted by almost every one of his Christian commentators, and, in the Renaissance, the commentary-tradition was alive as never before.[38] As in the case of the question of the mule, where Voetius used Danæus alongside Aquinas and Pereyra, he made ample use of Catholic Neo-Scholastic sources regarding the question of administration also. It is to this question as presented in the Aristotelian tradition that I shall now turn.

3.2 The Commentary Tradition

If Voetius saw Aristotle's work as the articulation of common sense, common sense alone did not urge him to embrace the great philosopher. Danæus did just as well without him. Nor was saving common sense the primary objective: the objective was to stay in line with Moses. This however, was exactly what the Aristotelian commentary-tradition claimed to do when interpreting Book II of the *Physics*. To see why Aristotle and Moses were thought to agree, let us, again, go back to Aristotle.

3.2.1 Aristotle on Luck and Chance

In *Physics* II 8, Aristotle refutes the idea that Nature might be governed by necessity, without there being an end in Nature. In *Physics* II 9, he explains what kind of necessity (ἀνάγκη) is compatible with an end. The first of these chapters is directed against the view of the materialists—Empedocles in particular—who hold that there is no τέλος to be taken into account. Aristotle denies this on the basis of the constancy of natural processes and the regularity of natural patterns. Such constancy is by definition incompatible with chance and spontaneity (τύχη, αὐτόματον). Secondly, as we have

---

[38] Cf. Charles B. Schmitt, *Aristotle and the Renaissance*, Cambridge, Mass. / London, England (Harvard U. P.) 1983, *passim*.

already seen, the purposefulness of natural processes is defended by comparing Nature with art. Like products of art, natural products develop in stages. The orderly pattern of such stages depends on the result which is desired or the end which is to be achieved. Rather than deviant behaviour, like olives growing from vines, most of the time we observe that identical processes lead to identical results. There is therefore every reason to suppose that natural processes are dependent on the result (or end) to be achieved. Hence, there is no room for chance or spontaneity of results. Also, we may see why (as Aristotle tries to show in *Physics* II 9) there is necessity in Nature without this "necessity" making it superfluous to take "ends" into account. For to achieve some end in art, certain materials are "necessary". The same is true in the case of natural processes. However, no product of art will necessarily come about from the materials themselves. The next step in the argument is that the same counts for natural materials. Though being necessary for some purpose, the purpose itself is what initiates the process and the materials being used for its fulfillment. Hence, Nature is governed by ends, whilst the necessary material for such ends, are necessary in a hypothetical way only. They are *necessary conditions* for the end to be achieved. The material itself does not necessitate any process.

3.2.2 Commentaries on Physics *II 8*

Making God a direct participant in natural action, the Christianised version of Aristotle brought religious ends and meanings into natural life. The famous Jesuit philosophers of Coimbra (or *Conimbricenses*) for instance answered the question whether Nature works for the sake of something by saying that Nature did indeed operate for an end (*finis*). Not only was this the common opinion of philosophers and of Plato and Aristotle: it was also proved by Nature's beauty and grace which in such a variety of things could not have come about without factors of harmony and agreement. The regularity of causal patterns and inherent rarity of chance factors also pointed to an "impulse and intention of nature". "What is this but striving towards an end?", the commentators rhetorically ask. Further, who could say that the parts of animal bodies and "so many organs, so many powers and potencies, so many instruments of multiform coherence", were the result of a "vague and uncertain movement of an unsteady Nature" instead of being "aimed at an unmistakable goal?"[39] One had further to take into account "the discordant harmony of the elements", "the bounty and opulence of the Earth", "the beauty and fecundity of the Ocean" and the "greatness, beneficence and beauty of the heavenly world".

In some of these examples the Scholastic commentators appear to repeat Aristotle's arguments, but in fact they go far beyond Aristotle. Not only do they invest natural teleology with aesthetic and moral force, they also use the concept of final causation in ambiguous ways and at different levels of theory. To see this, let us return to a notion already discussed in the former chapter, the notion of perfection.

3.2.3 Concepts of Perfection

One ingredient of the Scholastic concept of the *causa finalis* is of genuine Aristotelian origin. It is the use of the term in its function of the fourth causal determinant in natural change: the "that for the sake of which" (τὸ οὗ ἕνεκα). In *Physics* II 3, this end, τέλος, or completion, aquires the status of a causal factor. Aristotle presents the τέλος as a possible answer to questions in which we ask for the "why" of things. Giving answers to such questions was considered by Aristotle to be the proper object of science.[40]

To this idea of a τέλος functioning as the "end" of processes that can be explained by natural teleology, the Scholastic concept of a *finis* introduces a new level of causality and a new concept of perfection.

3.2.3.1 First and Second Perfection

In question 73 of the first book of the *Summa Theologica*, Aquinas deals with the Seventh Day of Creation. The first article concerns the completion of divine works. As we have seen, the problem of how to account for entities and even entire species of things which only later arose, was answered to by Aquinas' saying that, on the Seventh Day, all species and things were there in some way or another.

---

[39] *Commentarii Collegii Conimbricensis S. I., In octo libros Physicorum Aristotelis Stagiritæ* [etc.], Ioannes Baptista Bvysson, Lyon, 1594 (= Reprint Georg Olms Verlag, Hildesheim 1984), pp. 323-324. According to the Conimbricenses there are five sorts of ends: (1) the formal end of the growing animal; (2) the more general end of preserving the species; (3) the heavenly bodies as final causes, ordering the sublunar world by their embrace; (4) the end of all corporeal nature, i.e. man; and finally, (5) God as the "author" and "parent" of Nature. See *idem*, pp. 327-328.

[40] Aristotle, *Physics* II 3, 194^{b}17-20; *Posterior Analytics* I 14, 79^{a}23-24. See also our introduction, note 3.

More difficulties were, however, involved. For one thing, the Bible speaks of "the harvest [at] the end of the world"; of "the fullness of time" when "God send forth his Son" and of Jesus, when dying on the cross, saying "it is finished". Should the completion of divine works therefore be thought only to have arrived at a much later point in time? Also, it is written that "God rested on the Seventh Day". If so, then how can He have completed His works at the time?[41]

In order to answer to these possible objections, Aquinas first draws a distinction. According to him, the "perfection" of a thing can be either of two things. Perfection in one sense is the perfection "by which a thing is perfected in its substance". The second type of perfection, however, is the end (*finis*) of a thing, and is causally dependent on the first type. With this in mind, Aquinas goes on to answer the questions concerning the completion of the divine works. According to Christian dogma, the ultimate perfection of the second type, i.e. the end of the whole Universe, is the perfect happiness of the saints. However, before their happy state of glory can be accomplished at the end of time, two other conditions need first to be fulfilled: Nature must be completed and so must grace. In other words, before there can be any sanctification of the elect, they have first to be created and subsequently saved by God's merciful act of grace. Aquinas explains that the sort of completion of divine works which occurred on the Seventh Day bears only on the "completion according to nature". The coming of Christ and the end of the world, which were both mentioned as later completions of God's work in the objection to *Genesis* 2:1-2, are not completions "according to nature", but according to grace and glory. Although these aspects of the divine plan with the Universe were only later fulfilled, as far as Nature is concerned, the work was finished on the Seventh Day. Moreover, although God did not work on the Seventh Day in the sense that He "put together [some] new creature", He did, from the Seventh Day

---

[41] Aquinas, *Summa Theologica* I, qu. 73, art. 1: "Utrum completio divinorum operum debeat septimo diei adscribi. *Sent.* 2 d. 15 q. 3 a. 1 Ad primum sic proceditur. Videturquod completio divinorum operum non debeat septimo diei adscribi. 1. Omnia enim quæ in hoc sæculo aguntur, ad divina opera pertinent. Sed *consummatio sæculi* erit in fine mundi, ut habetur Mt. 13, 37 ss. Tempus etiam incarnationis Christi est cuiusdam completionis tempus: unde dicitur *tempus plenitudinis*, Gal. 4, 4. Et ipse Christus moriens dixit: *Consummatum est*, ut dicitur Io. 19, 30. Non ergo completio divinorum operum competit diei septimo. 2. Præterea, quicumque complet opus suum, aliquid facit. Sed Deus non legitur septimo die aliquid fecisse: quinimo ab omni opere quievisse. Ergo completio operum non competit septimo diei."

onwards, "administer" what He had created and "move it to [perform] its proper operation".

This divine administration of things is said to "somehow already relate to a certain beginning of the second perfection".[42] What Aquinas seems to mean is that, from the Seventh Day onwards—i.e., when God started to administrate what He had created in the six days before—the ultimate aim of the Universe is already in the process of coming into being. This "second perfection" however, could not do without the first perfection, on which—as Aquinas explicitly says—it is causally dependent. Hence, before God started to administer what He had made, He had completed, or "perfected" what He had made in the first sense of perfection. That is, "in the first arrangement of things", He had perfected the things He had made "in their substances". In other words, He had created the species mentioned in *Genesis* 1.

3.2.3.2 Tέλος *versus* Finis

As was said above, according to Aquinas, an object may be called "perfect" because of the "wholeness" (*integritas*) of its parts. From such a complexity, its "form" may arise, which is in fact identified with "first perfection". This description has obvious links with Aristotle's idea of perfection and natural teleology.[43] In our example above, the full-grown individual organism is the τέλος for the embryo, explaining its development as a goal-directed one. This τέλος may therefore be identified with the complete, whole being. Thus, if Aquinas says that perfection lies in "the form of the complete thing which arises out of the wholeness of its parts", he is calling to mind the Aristotelian idea of a natural teleology in which natural processes

---

[42] Aquinas, *Summa Theologica* I, qu. 73, art. 1, ad 2: "quod iam aliqualiter pertinet ad inchoationem quandam secundæ perfectionis."

[43] Apparently, as against the second type of perfection, this type of perfection does not refer to any finalistic scheme. The object under consideration does not have to be looked upon as an object with a certain end, or—to put it yet another way—God does not have to have any plan in mind in order to perfect a thing "in its substance". He only "completes" it "according to nature". Aquinas further explains that this type of perfection "is in fact the form of the complete thing which arises out of the wholeness of its parts." Aquinas, *ibidem*, "Quæ quidem perfectio est forma totius, quæ ex integritate partium consurgit." The description is of a rather cryptic character, but what Aquinas apparently intends to make clear is that without reference to its end—i.e. in a non-final context—a thing cannot be called "perfect" unless the condition be satisfied that it is "whole" as to its parts. Out of this complexity, its "form" may arise. In fact, "first perfection" is identified with the form, or at least with its presence.

are directed at, and may be explained by, the end or completion of the process.[44]

For Aristotle however, the "end" of a thing, as we have seen, lies in its completion. In fact end, completion and perfection coincide. The end, or τέλος, is what natural processes aim at. Scholastic authors traditionally translate the concept of τέλος as *finis*. Yet Aquinas' restrictive use of the concept of *finis* in the passage where he distinguishes two types of perfection, is indicative of the difference between the Scholastic and the Aristotelian accounts. Where Aristotle uses τέλος to indicate the type of perfection identified by Aquinas as "first perfection", Aquinas restricts the use of *finis* to "second perfection" only.[45] This "second perfection" however is wholly lacking in Aristotle.

---

[44] It is an interesting question whether either Aristotle or Thomas went so far as to completely identify the "complete" or "perfect" (τέλειος; *perfectus*) on the one hand and the "whole" (ὅλον; *totum*) on the other. Discussing infinity in *Physics* III, VI, Aristotle says that infinity is not so much "what has nothing outside it", but rather "what always has something outside it", *Physics* III 6, 207ª1-2. For "what has nothing outside it" is rather a definition of the "whole". In this context, Aristotle adds that "whole and complete are closely akin, if not identical", *idem*, 207ª13-14. Moreover, in *Metaphysics* V 16, Aristotle gives a similar definition of τέλειος in the sense of something outside of which nothing can be found that belongs to it; see above, chapter 2, note 39. Aquinas, in his commentary on *Physics* III 6, even says that "that of which there is nothing outside is the definition of the perfect and the whole" (Aquinas, *In Physicorum* III, lectio 11, Maggiòlo, p. 189), thus immediately identifying "whole" and "perfect". He adds, however, that a difference may be found in the fact that the adjective "whole" has only limited application, since it cannot be used for simple entities that have no parts. Apparently, this is the only way in which "whole" and "perfect" differ. Nevertheless, the notion of second perfection, which Aquinas does not seem to have in mind at this point, introduces a different level of perfection entirely.

[45] Not always , however. In the example of Aquinas' commentary on the Seventh Day of Creation, there is a clear dividing line between the teleological (first) perfection of Creatures and substances that were created within the Six Days and the finalistic (second) perfection which they are meant to achieve by fulfilling the ends for which God created them. However, in other sections not only of the vast Thomist oeuvre, but even within the *Summa Theologica* itself, Aquinas draws different distinctions. The concept of perfection is analysed into a variety of different meanings according to the context in which it is used. For instance, in answer to the question "whether by the sanctification the Blessed Virgin obtained a plenitude (or perfection) of grace in her womb?", Aquinas mentions not two, but three types of perfection: "in things natural, there is first the perfection of disposition, as when I say that matter is perfectly disposed for the form. The second perfection however, is the perfection of form, which is more powerful, for that fire which comes forth from the form of fire is more perfect than the one which disposes

Aquinas' second perfection of the objects of Creation, i.e. their *finis*, is an end in the sense of a goal which is to be achieved but which lies outside the object's being completed or "perfected in its substance". It refers to God's plans with the Universe which begin to be developed only after the first perfection has been achieved. While Aristotle's natural teleology is restricted to the completion of substance, the Scholastic account adds another level of finality. To avoid conceptual confusion, I shall use the term "teleology" for the Aristotelian account of the attainment of an end, using "finalism" and its derivatives for any superadded goal-directedness.

3.2.3.3 Ends and Aims: Rational Deliberation and the Τέλος

Aquinas' idea of perfection hinges on two notions of the goal-directedness of natural action. The first corresponds to the naturalistic, Aristotelian type of *perfectio* which consists of the achievement of form—in Thomistic terms, the *perfectio prima*. The second type, *perfectio secunda*, does not correspond to any Aristotelian notion of perfection. Aquinas explains second perfection as *finis*, describing *finis* in its turn in the following way:

> Now *finis* is either an operation, as the *finis* of the cithara-player is to play on the cithara, or it is something which is attained through operation (*per operationem*), as the *finis* of the builder is the house which he makes by building it.[46]

---

something to the form of fire. The third perfection is the end (finis), as when the fire has its qualities in a most perfect way when it arives at its [proper] place." Aquinas, *Summa Theologica* II, qu. 27, art. 5, ad 2. In the Thomistic distinction between "first" an "second" perfection that we presented earlier, the first was also identified as "form", whilst the second was identified as "end", finis. It would be wrong, however, to identify the "form" and "end" types of perfection of the passage just quoted with those of the earlier passage. In fact, all three types of perfection which occur in the passage just quoted, are "naturalistic" types. What I mean by this, is that all would fit into a teleological explanation of natural change without reference to God's plans with the world, or other issues relating to a non-natural goal-directedness. Being naturalistic types of perfection, all three types mentioned here might occur as subclasses of a non-finalistic—though "teleological"—type of perfection. Thus, they all belong under the heading of first perfection in the earlier classification. Since Aquinas there uses *finis* for second perfection only, this would mean that Aquinas makes an ambiguous use of the term *finis*. I argue that, indeed, he does.

[46] *Summa Theologica* I, qu. 73, art. 1.

The two types of *fines* which are here distinguished differ only in respect of the first being an example of an "operation" which is aimed at for its own sake (presumably because it is, for instance, a desireable thing to play the cithara), whilst in the second example it is the *product* of the operation which is the aim (i.e., the house, not the aspect of building it). Both however, are examples of ends being pursued as a result of intelligent deliberation. It is on the basis of this explanation of the concept of *finis* that Aquinas explains second perfection as the achievement of God's aims with the Universe, leading ultimately to the "perfect happiness of the saints."

The analogy between the rational choice of ends by which one's conduct is directed in the case of playing an instrument or building a house and the divine goverment by which the world turns according to God's plans, is a fine example of the incorporation of a finalistic concept of causality into the theological body of Christian dogma. Just as our affairs are governed by our aims and projects, so the world turns according to God's plans. I have argued that this type of finalism should be distinguished from Aristotle's, but it might be objected, that a concept of a goal-directedness based on an analogy with human deliberation occurs just as well in Aristotle as it does in Aquinas. Explaining his theory of teleological causation, Aristotle in fact exclusively offers examples of purposeful human action: walking, taking drugs or operating for the sake of health.[47] If Aristotle himself gives the analogy of rational goal-directed choice for teleological causation, does he not thereby draw a parallel between the course of Nature and an intelligent scheme that lies behind this course and which could be interpreted in terms of a divine plan?

The answer is: *no*. Aristotle does not use his examples in such a way as to suggest that he identifies goal-directedness in Nature with purposeful human behaviour. What Aristotle in fact does in *Physics* II is to sum up the various ways in which we give explanations of phenomena. Since we sometimes explain human behaviour by referring to its purposefulness, or to the goal which is aimed at, it might be useful to investigate whether an analogous type of explanation could be of use in explaining natural processes also. Aristotle's answer is that in explaining Nature it is in fact necessary to give some sort of teleological explanation as well.[48] This is not to say

---

[47] *Physics* II 3, 194^b32-195^a3.

[48] I shall not further embark on the question whether we should interpret Aristotle as saying that nature is therefore governed by teleological *causes*. For a

that natural goal-directedness should be identified with human intentionality—nor does Aristotle ever make such a claim.[49]

It is true that, in a famous passage at the end of *Physics* II 8, Aristotle does draw a comparison between "art" and "Nature", saying that if

> purpose is present in art, it is also present in Nature.[50]

This however, does not mean that Aristotle identifies art—that is, the products of human intention—and Nature in every respect. For one thing, Aristotle does not claim that Nature is purposeful in the animistic sense that it deliberates how to act in a given circumstance. But neither does he hold that purpose has a place in Nature in the sense that there is a plan to be achieved. What Aristotle does at the end of *Physics* II 8 is to counter the objection that there would be no purpose in Nature *because* "Nature does not deliberate." Aristotle agrees: of course Nature does not deliberate. But then neither does art.[51] What he means by this is not that an artist may not have doubts about what he plans to work on. Rather, what he means is that the production of an object of art is governed by the idea the artist has of the product. If the artist is master of his art, his action will without any deliberation lead to the desired result. In building a ship for instance, the goal of the art of ship-building is fixed and aimed at.

> If the ship-buiding art was in the wood, it would produce the same results by nature,

as Aristotle goes on to add.[52] Of course, ship-building is not "in the wood". But growing-into-trees is. What Aristotle intends to make clear is that, in our explanation of natural processes, the τέλος has to be taken into account for Nature in the same "fixed" way as for art. Art and Nature correspond to each other in the sense that

---

discussion of the problem whether or not Aristotle is to be interpreted as a "causal realist", see the introduction, note 3, above.

[49] It therefore seems to me to be somewhat beside the point that David Charles concludes that there are "major differences between the types of teleological causation at work in the cases of agency and natural organisms". David Charles, "Theological Causation in the *Physics*", in Lindsay Judson, ed., *Aristotle's* Physics*: A Collection of Essays*, Clarendon Press, Oxford 1991, p. 108.) Charles seeks to find "a unified account of teleological causation" which Aristotle nowhere intends to offer.

[50] Aristotle, *Physics* II 8, 199^b30. I here follow the Revised Oxford Translation.

[51] *Physics* II 8, 199^b28.

[52] *Physics* II 8, 199^b29-30, Revised Oxford Translation.

> where there is an end, all the preceding steps are for the sake of that [end].[53]

The question of rational deliberation does not occur. As for the connection of processes and ends, it is as fixed in Nature as it is in art.[54] This does not, however, call for any further identification of human goal-directedness and natural teleology.

3.2.3.4 *Universal Cosmic Principles*

As a consequence, one should neither read into Aristotle's teleological explanations of natural processes the finalistic picture of a deliberate goal governing them from an external point of view. Neither does Aristotle at any point posit the idea of an intentional plan inherent in Nature and comparable with human deliberation.[55] Aristotle offers

---

[53] *Physics* II 8, 199^{a}8-9, ROT. See also W. D. Ross, who, in his analysis of the passage summarizes Aristotle's argument as follows: "Where there is a terminus to a course of action, the earlier stages are for the sake of the terminus. Now the course of nature corresponds to the course of action. Therefore the course of nature is also for an end. If a house were a natural object, it would be constructed as it in fact is by art; and if natural objects were produced by art as well, they would be produced as they are by nature. Nature, therefore, like art, is purposive." *Aristotle's Physics*, a revised text with introduction and commentary by W. D. Ross, Oxford, Clarendon Press, 1936, p. 357.

[54] Although the artist may, of course, decide not to build a ship, but, e.g. a raft, or fail in building altogether. See W. Wieland, "The Problem of Teleology", in Jonathan Barnes, Malcolm Schofield, Richard Sorabji (edd.), *Articles on Aristotle*, 1. Science, Dukworth, London 1975, p. 155: "Art is most truly art precisely when the artist is so much in control of the material that he already knows what to do at any particular point without having to give special thought to it. Deliberation only enters the picture when immediate involvement with the material and commitment to it are at any point disturbed. The artist who has to deliberate simply proves that he is not in perfect control of his material. Thus art is most truly itself precisely when it forgets itself. The goal-directedness which in nature is quite divorced from any deliberation is therefore most easily understood in terms of this perfected skill." This interpretation of Aristotle's statement that "art does not deliberate" also makes sense of the concluding sentence of the paragraph: "The best illustration is a doctor doctoring himself: Nature is like that."Aristotle, *Physics* II 8, 199^{b}30-32, translation from ROT. Again, Aristotle presents Nature as analoguous to that artistic action in which the art is perfectly goal-directed. At the same time, the example bridges the gap between purposeful action and Nature (a gap which Aristotle nowhere denies), by presenting an agent that aims at a τέλος inherent in himself, in this case, the doctor and his health.

[55] Accordingly, *Physics* II does not aim to offer the "unified account of teleological causation" that modern commentators seem to demand of Aristotle. See above, note 49.

examples of an artist and of someone who walks for reasons of health, in order to compare these to instances of purpose and ends in Nature. He denies that the fact that Nature does not deliberate would mean that there is no purpose in Nature.

Aquinas' examples occur in quite a different setting from Aristotle's. The intentional *fines* that the cithara-player and the builder have in mind (in the one example, the act of playing the cithara; in the other, a house) are not presented in order to analyse the analogy between such intentional goals and natural goals, but function as examplary instances of finalistic planning. The difference between Aquinas and Aristotle on this point is not merely a question of conceptual analysis versus theoretical application. In fact Aquinas uses the analogy between human goal-directedness and goal-directedness in Nature in a sense in which it does not occur in Aristotle. Where Aristotle draws an analogy between artistic and natural teleology, claiming that in both there is a connection of process and end, Aquinas uses human goal-directedness as examplary for second perfection and interprets the Universe in the light of God's plans and purposes behind it.[56] In an exellent article, W. Wieland has discussed this question as a conflict between Aristotle's notion of natural teleology and the concept of a "universal cosmic principle" which was later formed on the basis of this notion.[57] As far as *Physics*

---

[56] Of course, this is not to say that Aquinas does not also make use of the notion of a final determinant of natural action in the Aristotelian sense. In fact, in his commentary on *Physics* II 8, he explains the passage in which Aristotle says that "art does not deliberate" in quite the same way as we have done above: "Nec artifex deliberat inquantum habet artem, sed inquantum deficit a certitudine artis: unde artis certissimæ non delibererant, sicut scriptor non deliberat quomodo debeat formare litteras. Et illi etiam artifices qui deliberant, postquam invenerunt certum principium artis, in exequendo non deliberant: unde citharædus, si in tangendo quamlibet chordam deliberaret, impertissimus videtur. Ex quo patet non deliberare contingit alicui agenti, non quia non agit propter finem, sed quia habet determinata media per quæ agit. Unde et natura, quia habet determinata media per quæ agit, propter hoc non deliberat. In nullo enim alio natura ab arte videtur differre, nisi quia natura est principium intrinsecum, et ars est principium extrinsecum." S. Thomæ Aquinatis, *In octo libros Physicorum Aristotelis expositio* Liber II, lectio 14, ed. Marietti, Roma 1965, pp. 130-131. The fact remains however, that in his use of the concept of *finis* in secondary perfection, Aquinas himself extrapolates the *principium extrinsecum* of human goal-directedness to Nature as guided by the aims and goals of God.

[57] W. Wieland, "The Problem of Teleology", in Jonathan Barnes, Malcolm Schofield, Richard Sorabji (edd.), *Articles on Aristotle*, 1. Science, Duckworth, London 1975, pp. 141-192.

II is concerned however, natural teleology bears only on individual processes directed at a certain τέλος.

3.2.4 Conclusions

Aquinas' examples comparing rational deliberation with the finality of second perfection add a more anthropomorphic type of "aims" to the "ends" of natural teleology. Of course, in attributing the aspect of goal-directedness to natural processes as much as to rational deliberation, the Aristotelian explanation of natural change is itself, at least to some extent, liable for introducing an anthropomorphic picture of causal relations into scientific explanation. Yet Aristotle's analytical approach of investigating which of the concepts with which we aim to explain human conduct are applicable to natural action as well, differs radically from the Thomistic position by which rational deliberation is supposed to be mirrored on a higher level by the finality of God's plans leading to a second perfection of things natural.

Although it seems impossible to read into Aristotle's teleological explanations the finalistic picture of a deliberate divine goal governing Nature from an external point of view, both Aquinas and later commentators on the *Physics* showed far more interest in the question of divine government, than in the original Aristotelian idea of natural teleology. Accordingly, the Scholastic commentaries which were used in Descartes' day abounded in theological considerations where Book II of the *Physics* was discussed. In the famous commentary of the philosophers of Coimbra, for example, the commentary on *Physics* II, chapter 7, adds to the considerations on the efficacy of "secondary" (i.e. ordinary, everyday, physical) causes, some specific questions concerning divine interference in their action. We shall discuss these theological questions in chapter 9, below.

In any case, theological considerations were discussed in relation to physics as well. As we have noticed, the philosophers of Coimbra answered the question whether Nature works for the sake of something by saying that Nature did indeed operate for an end (*finis*). It was for this reason, that Aristotle and Moses could be combined. But when Aristotle's Renaissance commentators pointed to an inclination and end of Nature, their notion of such an end was equally ambiguous as was that of Aquinas.

In Voetius' day, the discipline of Physics in the Dutch academies was explained not so much in relation to Aristotle's natural teleology as to the Neo-Scholastic picture of a world governed by an hierarchy of substantial "powers" or "agents", according to the divine plan.

3.3 *Physics in the Dutch Academies*

As regards some of the aspects of final causality, the Scholastic commentators resume Aristotle's arguments. Mostly however, these Aristotelians go far beyond Aristotle. To illustrate this, let us take an example from the work of Voetius' one-time student and Descartes' great enemy, the Groningen professor of philosophy Martinus Schoockius (or Martin Schoock, 1614-1669).[58]

3.3.1 *Martin Schoock on the Concept of* Finis

Distinguishing various types of *finis* in his *Physica Generalis*, Schoock first recalls a genuinely Aristotelian type:

> The *finis* [...] must be considered in two ways: absolutely or relatively. The first way coincides with physical perfection. For a natural body is thought to be physically perfected when it is so disposed as is sufficient for [it to perform] all natural operations appropriate to its species.

Thus, in the "absolute" sense, "ends" are the perfections of natural growth. But there are other ends. As Schoock continues:

> Relatively however, the *finis* here bears not only on the absolutely ultimate end (*finis*), or God's glory, to which all natural things, even the very smallest, are subjected, but also on the mutual end (*finis*) of natural bodies themselves, which are ordened by a most wise God in a way that they mutually benefit each other both in their being and in their operation. And here, as a centre of Nature, stands man, whom all other natural things are adapted to serve.[59]

At this point, final causality is introduced not only to account for individual development, but also to explain organic processes on a larger scale. To a late twentieth-century reader, questions of ecology and an intuitive notion of symbiosis may come to mind upon reading this passage. Yet one must realize that the biological phenomena only occur as examples of a general theory of natural change and that

---

[58] As a professor of philosophy in Groningen, Schoock wrote wrote the *Admiranda Methodus*, or *Admirable Method of the New Philosophy of René Descartes*, a violent attack on Cartesian philosophy of which a French translation may be found in the *Querelle*. See further: Theo Verbeek, *Descartes and the Dutch*, pp. 19-33 esp.

[59] Martinus Schoockius, *Physica Generalis*, Groningæ, Typis Johannis Coelleni, Bibliopolæ & Typographi, Anno 1660, pp. 138-139.

neither ecology nor biological theory at large is what is here focussed upon. In the natural philosophy of the Schoockian—and Voetian—type, the biological or organical paradigm which leads to the supposition of criss-cross, interdependent (i.e. "relative") ends, is subservient to the theological world-outlook and never examined for its own sake. Accordingly, God appears at practically every stage. Apart from playing the role of an ultimate, *absolute*, end of Nature itself, it is God who invests Nature with intrinsic ends and arranges things *relative* to the needs of its most noble and eminent inhabitant, man.

Schoock's distinction between "absolute" and "relative" ends does not coincide with our distinction between the *fines* of an observable natural teleology and the *fines* of a deliberate kind imposed upon the system from without. But then, Schoock never intended his distinction between absolute and relative processes to be a distinction between observable ends and non-apparent causes of these ends. One could for instance incorporate a variety of "relative" ends into a teleological biology without making any reference to the rational planning of an intelligent Creator. Suppose that we could in some empirical way determine that the preservation of mankind were to be regarded "the central aim of corporeal nature." This might be argued for without there being any reference to a "hidden" plan of divine origin or to the "goodness" of the universal scheme. Aristotle's Christian interpreters, however, never sought to distinguish between the empirical and the theological domains.

Something like a distinction between empirically observable and non-observable *fines* is implicit in other writers. Commenting on a passage in *Physics* II 4, Aquinas for instance introduces the concept of "immanifest" causes.[60] In the chapter on which he was writing his commentary, Aristotle examines the opinions of his predecessors on the idea of chance. He mentions the view of those who hold chance (τύχη) to be a cause, but one

> inscrutable to human thought, because it is divine or supernatural in character.[61]

---

[60] St. Thomas Aquinas, *In Octo Libros Physicorum Aristotelis Expositio*, Liber II, Lectio VII, "De diversis opinionibus circa fortunam et casum, causas immanifestas" (ed. P. M. Maggiòlo O. P.), Turin / Rome (Marietti) 1965, pp. 99-102.

[61] Aristotle, Physics II 4, 196^{b}6-7. The translation is from William Charlton (ed.), Aristotle, *Physics Books I and II*, Oxford (Clarendon Press) 1992, p. 33.

Where Aristotle may be thinking of a variety of authors or religious cults,[62] Aquinas interprets the passage in a specifically Christian way, saying that there is "a true ground" for this opinion. Translating τύχη with "fortune" (*fortuna*), Aquinas says it is only the use of this *name* which is wrong with this position. Divine government cannot be called Fortune, since, for something to be part of a scheme (*ratio*) or an order (*ordo*) is exactly the same as for it to be *withdrawn* from the scheme of fortune (*ratio fortunæ*). Fortune is an "inferior" cause, the superior causes being those which determine an order.[63]

Having brought order and planning into the natural world, God cannot be said to have brought fortune in as a general cause. However, although the concept of fortune is thus criticised and the word "fortune" rejected as not being fit to denominate the hidden causes of a divine order and government, the idea that such "immanifest causes" exist, is strengthened rather than rejected by Aquinas. In fact it is exactly this type of causes which correspond to the "hidden" area of Providential government in general.

If we use "immanifest causes" in the general sense of causes determining phenomena and processes in a "hidden way", the distinction between manifest and immanifest causes may serve to explain the difference between the Aristotelian idea of final causation and that of his Christian interpreters, but we must be aware of the fact that the distinction was not made by Aristotle's interpreters themselves. Aquinas does not use the concepts of "manifest" and "immanifest" causes in the same way.[64] Schoock distinguishes between ends of individual processes and ends of interrelated systems. Other philosophers might offer different distinctions altogether, applying various logical distinctions to the finality of both individual processes and the supposed organic unity of these processes on a larger scale.[65]

---

[62] Cf. W. D. Ross (ed.), *Aristotle's Physics*, Oxford (At the Clarendon Press) 1936 / 1966^4, Commentary, p. 515.

[63] Aquinas, *In Octo Libros Physicorum* II, Lectio VII, 206, ed. Maggiòlo p. 102.

[64] Aquinas uses the term *causes immanifestas* for luck and chance (*fortuna, casus*) only, reserving the term *causes manifestas* for matter, form, efficient and final causes, in all their various interpretations. See the *Prospectus* or summary, right at the start of his commentary: *In Octo Libros Physicorum*, ed. Maggiòlo, p. vii.

[65] The *Summa Philosophica* of Eustace of St. Paul for instance, a consice course of Scholastic doctrine which Descartes new well (Étienne Gilson, *Index Scolastico-Cartésien*, Paris (Vrin) 1979, pp. iv-v), distinguishes firstly between a *finis cujus* and a *finis cui*. For the *finis cujus*, or "end of which" Eustace takes the example of health, for which end we take medicine. As we have seen, Aristotle uses similar

No Scholastic author, however, seems to make a clear distinction between the manifest, natural realm of teleological processes and the immanifest, finalistic realm of divine providence. Like natural philosophy and Mosaic physics in the educational sphere, both types of final causality were supposed to be to a large extent interchangeable. They equally contributed to the idea of a living Nature governed according to a divine plan.

3.3.2 Schoock's Disputation De Aqua

"Absolute" and "relative", "manifest" and "immanifest" notions of final causality were all put to use in natural philosophy. Specimens

---

examples in *Physics* II, without claiming that such deliberate ends and the natural ones are in any way identical. Offering health as an example of a general concept of final *causality*, Eustace does seem to identify the natural and the deliberative process in some way, but we must not press the point. In fact, the question of natural teleology is not the issue here. Eustace only compares this use of the end of something (for instance, of taking medicine) to another, the *finis cui*, or "end for which". In the example of health, this end is the sick person, "for whom", or, rather, "for whose health", medicine is taken. Aristotle's teleology natural processes does come to the foreground in Eustace's second distinction, that is, in the distinction between the end "of generation" and the end "of the generated thing". Here, an ambiguity is dealt with which we have already dealt with above, namely that Aristotle's τέλος or the natural concept of *finis* may refer both to the natural end of a process and to the well-functioning of that end. The third of Eustace's distinctions is: "between the end preceding the operation and the end subsequent to the operation." The first is "the one at which the agent aims (*in quem tendit agens*) not in order to produce it, but in order to gain it, or to unite with it, or assimilate to it. In this way, we say that God is the end especially of intellectual creatures. The second [is the one] at which the agent aims in order to produce it. In this way, the production of form by a physical motion is thought to be the end of the physical agent." Eustachius à Sancto Paulo, *Summa Philosophica quadripartita, de rebus Dialecticis, Moralibus, Physicis et Metaphysicis*, Parisiis, Carolus Chastellain, 1609. The whole passage is quoted by Gilson in his *Index Scolastico-Cartésien*, p. 39. The third distinction is the most interesting with regard to drawing a dividing-line between Aristotle and his Scholastic followers in that it refers to God as the end of intellectual beings. Eustace's distinctions do not, however, offer a clear-cut division between natural, teleological and finalistic, theological ends. All of his distinctions are logical distinctions supposedly to be applicable to any type of ends. Eustace does not distinguish between natural ends and deliberate planning. Yet in fact his distinctions cannot all be applied to natural teleology. Both the example of God as an end in the last distinction and of the sick person as an end in the first (that is, both the "end for which" and the "end preceding the operation") cannot be applied to a natural teleology without adding a great deal of theory referring to divine planning or to a universal cosmic scheme behind natural phenomena.

may be found in university disputations. Let us take Schoock's disputation *De Aqua*, "On Water" as an example. This *disputatio physica* was held in May 1646, with Schoock's German student Christoph Wittich, who was later to become famous as a Cartesian theologian, acting as *respondens*.[66]

As dialectical custom would have it, the short tract presents the element water in a variety of ways, starting with such generalities as the etymology of its name. *Aqua* is said to derive from *à quâ omnia*: that from which everything is brought about.[67] Not endorsing the presocratic view of single elements as principles, however, Schoock does not elaborate the point and talks of water only as one of the four elementary constituents of composite bodies. The notion of final causation enters in another way. A final cause is given, for example, for the phenomenon that water always appears in the form of drops. The most important reason for this particular shape is said to be deduced from the "sacred field of first philosophy", which teaches that there is a certain

> natural Goodness, by which water preserves itself and holds out against the enemy which it shuns (*quem fugit*). It cannot, however, hold out in a more convenient way than by its spherical form, which it in fact keeps in order that the water does not scatter and would thus be conquered and destroyed.[68]

Nature, in other words, invests water-particles with an "incentive" to stick together in globular form because, presumably, scattered

---

[66] For Wittich, see Theo Verbeek, *Descartes and the Dutch*, pp. 74-75 and p. 130, note 152. According to the title-page of *De Aqua*, Christoph Wittich was in fact the author of the dissertation: "Publicè defendere conabitur Christophorus Wittichius, Bregâ-Silesius Auth. & Resp." According to Magreet Ahsmann, this way of mentioning oneself as *auctor* or *auctor et respondens* was typical for German students, who may thus have wanted to indicate that they themselves had written the text for the disputation, something which was unknown in German universities. See Ahsmann, *Collegia en Colleges*, pp. 315-316 and above, chapter 2, note 9. Schoock will in any case have supervised the preparation of the final text.

[67] Similar attempts at etymological derivation led to an identification of *aqua* with *æquus*, *æqualis* and *æquor*, with regard to the equality of the substance of water and the evenness of its surface. Closer to Schoock's explanation is the one that relates *aqua* to *à qua iuvamur*, i.e. that by which we are assisted. Cf. *Thesavrvs Lingvæ Latinæ*, Lipsiæ [Leipzig] (In ædibvs B. G. Tevbneri) 1900-1906, Vol. II, p. 347. Presently, the word is thought rather to relate to *acer*, "sharp", with reference to violence and swiftness of motion. Cf. T. G. Tucker, *A Concise Etymological Dictionary of Latin*, Halle (Max Niemeyer Verlag), 1931, p. 21.

[68] Schoock, *Disputatio Physica de Aqua*, Thesis XIV.

particles would fall below a certain minimum value for keeping their substantial identity in a hostile (airy) medium.[69]

Another example of final explanation occurs where the *locus*, or "place", of water is discussed. According to the Aristotelian scheme of things, the element water takes up the region above that of the element earth and below the elements air and fire. Leaving it at that, however, would necessarily lead to the conclusion that the face of the Earth must be covered with water—which is contrary to actual fact. It was therefore to be concluded that

> to the advantage of living things, God brought water together in certain cavities, in order to let dry land stand out and provide a suitable dwelling-place for both man and other animals, and [...] "for the good and the better"[70] subdued the power of necessity, by which water would be bound to have enclosed the Earth completely.[71]

Here God's benign providence is appealed to, supplying dry land as a safe haven for the living. It serves as an explanation for other phenomena as well. According to the *De Aqua* text, most rivers find their origin in sources—instead of, for instance, being formed by rainfall. Sources let out water which earth has "drunk" from the sea. In some cases however, "subterranean fires" warm up the water while it is still underground. Thus, water may reemerge in the form of a hot spring, which, it is explained,

> is made by a particular favour of God for man's use and for averting and curing various illnesses.[72]

In this case, the "particular favour" (***singulare beneficium***) is presented more or less as a convenient side-effect of the natural subterranean processes. It is not entirely clear whether the explanation of water being heated by "subterranean fires" would in itself have

---

[69] Note that according to Scholastic theory, a certain equilibrium was always conserved in the constant transformation of elements. This fact alone was reason enough to rule out chance-factors in Nature. Cf. Conimbricenses, *In octo libros Physicorum*, p. 224. The particular shape of water drops was a standard argument in favour of an innate "appetite" (***appetitum***) for conservation. The Conimbricenses refer to it in relation to the famous question of the *horror vacui*. Cf. Conimbricenses, *In octo libros Physicorum* II, p. 62.

[70] Schoock here quotes from the *Exercitiæ* of Scaliger: "τὸ εὖ καὶ τὸ βέλτιον εἶναι."

[71] Schoock, *De Aqua*, Thesis XVI.

[72] Schoock, *De Aqua*, Thesis XXIII.

sufficed as a natural explanation for the phenomenon of springs without reference to the "particular favour" made to man by his Creator. However, such a distinction would probably not have been of much relevance to the Scholastic. To see this, we may compare the example of the springs with another example of finalistic explanation occurring in the *De Aqua* text. The question there is why there is a "non-natural" motion of rivers describing various curves rather than straight lines. This, it is argued, is

> to the end (*finis*) that [such waters] irrigate various lands and that they may lead men to various regions.[73]

In this case of meandering rivers, no mention is made of any divine planning, let alone of a special *beneficium* of God to mankind. Nor is the explanation presented as a convenient side-effect of processes that may be explained without referring to finalistic considerations. And yet the account is not really any different from that of the springs. The bizarre explanation that rivers are formed in such a way as best to serve man as a means of transport is hardly feasible without the idea of deliberate divine planning. Besides, the flow of rivers—despite all evidence to the contrary, which it is curious for an empirically minded writer as Schoock not to have noticed[74]—is explicitly said to be "non-natural", that is, contrary to what the regional order of the elements would have as a "natural" effect. The natural "tendency" is to flow downhill (*quantum in se est per declivia*).

As in the case of providing dry land, the idea seems to be that God in fact acts *against* natural causality and subjects the general law of Nature to His particular decision of supplying certain commodities for man and beast. However, in *De Aqua* such notions are rather muddled. At one point, where the "altitude" of water is discussed, water is said not to be "higher" (*sublimior*) than earth, for the reason that it would then have to cover the Earth or be "violently" kept from

---

[73] Schoock, *De Aqua*, Thesis XXVIII.

[74] In good Baconian fashion, Schoock made studies of matters as diverse as the art of cutting peat, of brewing beer, of making cheese and butter, and of curing herrings. He does not seem to have studied the Dutch waterworks in equal detail. For an assessment of Schoock's teachings and writings in the context of contemporary traditions, see A. J. Vanderjagt, "Filosofie tussen humanisme en eclecticisme. Van Ubbo Emmius (1547-1625) tot Martinus Schoock (1614-1669)", in G. A. van Hemert / J. Schuller tot Peursum-Meijer / A. J. Vanderjagt (edd.), *'Om niet aan onwetendheid en barbarij te bezwijken': Groningse geleerden 1614-1989*, Hilversum (Verloren) 1989, pp. 31-49.

doing so. Violence however, is contrary to the order of Nature. Hence, earth is "higher" than water, despite the fact that it is elsewhere concluded that the "natural place" of water is above earth and that God therefore has to conquer the forces of necessity in order to procure dry land.[75]

It is not clear why the "altitude" of water does not call for non-natural divine interference whilst the regional "place" of water does. Schoock probably saw the question of altitude, which is discussed alongside that of the profundity of the seas, as a minor empirical matter. It would therefore not matter too much to place the altitude of earth above the altitude of water. The regional diversity, on the other hand, was axiomatic to Aristotelian philosophy and may therefore have summoned exceptional interference to account for the obvious counter-example of dry lands. In any case, although the contrast between "natural" and "non-natural" occurs in a wholly intelligible way, it is not a very satisfying distinction within a natural philosophy in which God would, after all, have been present all along.

De Aqua ends with a final consideration entirely related to final causality:

> The uses of water are varied: both to contribute to mixtures, to form a habitat for fish, to serve shipping and promote human commerce and trade, and indeed, to form part of the universe and contribute to the beauty (*decorum*) and harmony of the world.[76]

In the organic, mysterious, but well-ordered world of Renaissance philosophy, finality in its "manifest" sense is completely superseded by final causality of the "immanifest" type: divine providence and mutual harmony replace natural perfection, finalism substitutes teleology. In accordance with the idea of a universal harmony, every step on the scale of nature had its proper function and place in relation to others.[77] Rivers meander in order to irrigate the soil and to be of use for travelling, but they also flow in order to preserve themselves from

---

[75] Schoock, *De Aqua*, Thesis X.

[76] Schoock, *De Aqua*, Thesis XXXVI.

[77] "[Far more broadly disseminated] was the association of Jacob's ladder with the elaborate 'world picture' current during the renaissance, according to which the universe is an imposing system of interdependent levels of existence arranged hierarchically 'from the Mushrome to the Angels.'" C. A. Patrides, *Premises and Motifs in Renaissance Thought and Literature*, Princeton, New Jersey (P. U. P.) 1982, p. 46.

pollution and hence conserve the fish.[78] In such a wonderfully meaningful world, it is hardly appropriate to settle the question of the natural versus the non-natural. In order to save Aristotle, God has sometimes to make an exeption, but in fact it is not clear why the uniform course of Nature should predominate over the particular beneficence with which Nature is invested.

We find various sorts of ends which were all somehow covered by the notion of final causality. God appears at every stage. Being an ultimate end of Nature Himself, God also arranges things according to the needs of Nature's most noble inhabitant, man.

3.3.3 Philosophy and Theology

The account of natural harmony in Schoock's disputation *On Water* may not strike the modern reader as a particularly well-developed scientific argument. More important than its scientific merits, however, is the way in which *De Aqua* points to the harmony not only between the objects of Nature, but, on a higher level, between the disciplines of theology and philosophy in their Neo-Scholastic forms. Theological considerations were never far away when Renaissance commentators or philosophers looked at the world through "Aristotelian" eyes. In fact, the whole body of Aristotelian philosophy functioned primarily as a structured interpretative means for explaining both Scripture and the Book of Nature in a way that lies much closer to common sense and anthropomorphic prejudice than its extensive arsenal of concepts and distinctions would seem to imply.

This again calls for a reconsideration of the idea that, as the New Philosophy of Regius and Descartes gained ground, their denial or negligence of certain ideas of metaphysics and logic did not fail to alarm the members of the Utrecht faculty of theology. The question of final causality for instance, could not possibly be discussed as a matter of philosophical excercise only. Neither could related ideas of substance and accident, substantial and accidental union, the unity of body and soul, the efficacy of quantity, and so on. Once more we may point to the fact that the Senate's verdict against the New Philosophy, in referring to the problems brought about in the higher faculties, is not at all to be read as a failure of conservative scholars to discuss the real issues. The unity of university teaching was paralleled by the intertwined character of the subjects that were taught.

---

[78] Schoock, *De Aqua*, Thesis XXVII.

3.4 Conclusion

When Voetius in his essay on substantial forms tells the followers of the New Philosophy that

> a stone falls downwards, stars rise and finally all natural things perform their motions without the use of reason, [since] they are thus created by God and so, according to the faculties impressed upon them, aim at their ends (*fines*) as an arrow at its target[79],

he comprises in one sentence all ideas of substantial being, immanent powers and final causality that we have discussed in chapters 2 and 3, and links them, as the Church Fathers and Scholastics had done before him, to the God of *Genesis* who created *substantial* objects each with their various functions and ends. It is in this way that, for Voetius, the commentary tradition served to corroborate the views of the Protestant physico-theologians whom he thought so highly of. Aristotelian concepts only strengthened the main issue of Mosaic physics: the notion of divine administration. As we shall see in chapter 9, God administers things very directly, by constantly acting along with them. However, let us for the moment leave Voetius and take a closer look at the philosophy that provoked his defence of Scholastic ideas on causation: Cartesianism.

---

[79] *Narratio*, p. 45 / *Select. Dispp.* I, p. 877 / *Responsio*, pp. 24-25 / *Querelle*, p. 110.

CHAPTER FOUR

THE MECHANICAL ALTERNATIVE: PERCEPTION AND CAUSALITY

Ainsi ie croy que nous auons cy-deuant
confondu la notion de la force dont l'ame
agit dans le corps, auec celle dont vn corps
agit dans vn autre.

Descartes to Princess Elisabeth
21 May 1643[1]

Descartes' philosophy is best typified as a "mechanical", or "mechanistic" philosophy. I shall use both of these terms interchangeably. "Mechanical philosophy", "mechanist philosophy", "mechanicism" and the like may however stand for a much broader type of philosophic or scientific reasoning.[2] It will have to wait to the next chapter to define the Cartesian type of mechanism more precisely. For the time being I shall start with that part of the subject-matter of philosophy through which the mechanical way of thinking is most easily exemplified: perception theory.

---

[1] Descartes, AT III, p. 667.

[2] E. J. Dijksterhuis' classic *Mechanisation of the World-Picture*, opens with a note in which the author explains his terminological preferences: "Het is moeilijk, een geheel bevredigende terminologie vast te stellen. Mechanisch doet te veel aan werktuigelijk in den zin van gedachteloos denken. Mechanistisch ontmoet als zodanig geen bezwaar maar vereist als correspoderend substantief mechanisme, [welk] woord echter ook voor de inwendige samenstelling van een werktuig (mechaniek) in gebruik is. Wij geven daarom de voorkeur aan het substantief mechanicisme ter aanduiding van de denkrichting en begaan dan de inconsequentie, daarnaast het adjectief mechanistisch te gebruiken en van mechanisering van het wereldbeeld te spreken." E. J. Dijksterhuis, *De Mechanisering van het wereldbeeld; De Geschiedenis van het natuurwetenschappelijk denken*, Amsterdam (Meulenhoff) 1989^6, p. 1. For Dijksterhuis however, the term *mechanisation* covers much more than Cartesianism or other seventeenth-century corpuscular philosophies alone. As far as our subject is concerned, the association of "mechanism" with the internal structure of artificial instruments is actually an advantage. The danger moreover that "mechanical" is too easily associated with "thoughtless", is perhaps even greater in Dutch than it is in English.

4.1 *The example of* Les Météores

Descartes' *Météores*—the second, that is, of the essays which, in 1637, were published along with the *Discours de la Méthode*—deals with a large number of subjects, reaching from the corporeal structure of salt to the meteorological phenomena of storms, thunder and rainbows. Besides an interesting variety of topics, however, it also presents a sharp image of Descartes' use of "mechanistic" argumentation.

4.1.1 Suppositions and mechanistic explanations

An important point to notice first, is that Descartes' terminology, his sequence of topics and much of his argumentation is wholly Scholastic. As has been pointed out by Étienne Gilson, Descartes' text closely follows that of the Conimbricenses' commentary on Aristotle's *Meteorology*.[3] In fact, resemblances between Descartes and Aristotle are often so great, that they must have been obvious to any contemporary who had studied Aristotle's views or those of his commentators.[4] It is therefore all the more interesting that Descartes presents his meteorological essay as a sample of the new method which he had claimed to have developed in the preceding *Discours de la Méthode*.

Descartes does not seek complete originality as regards his ideas on meteorological phenomena. What he does claim, is to have found "general principles of Nature" on which a correct understanding of these phenomena is supposed to depend. As it is said in the *discours premier* of the *Météores*:

> It is true that since the knowledge of these matters depends on the general principles of Nature, which, as far as I know, have not yet been well explained, I must, at the start, make use of some suppositions, just as I have done in the *Dioptrics*; but I shall aim to

---

[3] Étienne Gilson, "Météores Cartésiens et Météores Scolastiques", in Étienne Gilson, *Études sur le rôle de la pensée médiévale dans la formation du système cartésien*, Paris (Librairie philosophique J. Vrin) 1984^5, pp. 102-137.

[4] See e.g. the passages from *Les Météores*, AT VI, pp. 231 and 232 in Gilson, *Index*, pp. 180-181. These and other passages are also referred to in Ferdinand Alquié's edition of excerpts from the *Météores*, in René Descartes, *Œuvres philosophiques*, Textes établis, pésentés et annotés par Ferdinand Alquié, Paris (Garnier) 1963, Tôme I, pp. 719-761.

> make them so simple and easy, that you will probably not find it difficult to believe them, even though I have not demonstrated them.[5]

Hence the novelty of Descartes' presentation lies in the "general principles" which are here—at least for the time being—represented by certain *suppositions*.

These suppositions take up most of the rest of the *discours premier*. Descartes first formulates a very general hypothesis concerning the microstructure of the elements. All are supposed to consist of smaller parts, leaving spatial intervals between them which are in turn filled with still smaller parts of matter, forming the so-called subtle matter (*matière subtile*). Next, the supposition is made that liquids, such as water, are made up of long, closely united, smooth and easily separable particles, whilst solids, such as earth, air and "the greater part of other bodies" are made of particles of an "irregular and uneven" shape, which makes it more difficult for them to be taken apart. Microscopic water-particles—that is to say Cartesian water-molecules—can thus be compared with a bunch of eels in the macroscopic world, the particles of solids on the other hand with the entangled branches of shrubs in a hedge.[6]

The third supposition postulates (*il faut penser*) that the subtle matter constantly moves in all directions at great speed through the intervals, or "pores" of bodies which it fills. Moreover, being agitated by the light of the Sun, the particles move more swiftly in the daytime, in summer, near the equator and near the face of the Earth, than at night, in winter, further to the poles and in the clouds.[7]

It should be noted that in this third supposition, the boundary between general principles of Nature and applied theory becomes rather diffuse. The subtle matter-hypothesis can only be validated by its success in offering a consistent mechanical explanation of, for

---

[5] Descartes, *Météores*, AT VI, p. 233.

[6] Descartes himself draws the analogy: *Météores*, AT VI, p. 233.

[7] It is not entirely clear why there should be more agitation of particles of the subtle matter near the Earth's surface. It may be because the subtle matter is supposed to be set free from Earthly matter by sunlight. Within the Earth's "atmosphere", however, there is supposed to be plenty of subtle matter around on higher elevations as well. Might not these particles just as well be agitated by the Sun's rays? Note that Descartes refers to the *Dioptrics*, in which he had explained light as being "a certain movement or action, by which the luminous bodies push this subtle matter rectilinearly to all sides surrounding them." AT VI, p. 234.

instance, the influence of the Sun's rays on earthly phenomena.[8] Let us, for the moment, only notice this and return to Descartes' list of suppositions.

Fourthly, Descartes says, it is to be assumed that, although all subtle matter is made up of very small particles, some are yet bigger than others. Also, it is the latter which have more "force", just as, in general

> all large bodies have more [force] than the lesser ones (*moindres*) when they are moved in the same degree (*quand ils sont autant esbranlés*).[9]

The imprecise use, here, of such terms as *force*, *grand*, *moindre* and *autant* indicates the mathematically unsophisticated way in which Descartes deals with questions of impact. It anticipates the erroneous laws of collision in the *Principles*.[10] What is presently of concern, is that with the supposition that the bulkier parts of subtle matter have more force than the finer ones, Descartes now in fact wholly steps over to offer some specific mechanistic explanations, first of the difference between our experience of heat and cold in distinct substances, and, second, of the formation of ice.

The first offers a rather imaginative picture of the effluences of subtle matter from the pores of bodies. Supposing (*Et il est à remarquer*) that all bodies have pores through which the subtle matter may pass, but that some smooth objects have only very narrow pores, we may see how these can only let very fine particles pass, that is,

---

[8] All of Descartes' remarks here form an introduction to the explanations of more specific meteorological facts. In that sense, they are indeed "general" remarks. But it is important to see that Descartes does not make any logical or epistemological distinction between the "simple and easy" suppositions and more detailed mechanistic hypotheses. The difference between the *suppositions* and particular mechanistic explanations is only a matter of detail. I shall come back to this when discussing the status of the *a priori* and the metaphysical in Cartesian physics.

[9] Descartes, *Météores*, AT VI, p. 235.

[10] Although clearly at odds with macrosopic experience, Descartes' rules of collision can nonetheless be shown to be consistent if one takes into account his axiom that change always arises from a "stronger force". On this matter and on the relation between the macroscopic and microscopic levels in the Cartesian theory of collision, see Martin Kalmar, *Some Collision Theories of the Seventeenth Century: Mathematicism vs. Mathematical Physics*, PhD. dissertation John Hopkins University, Ann Arbor (University Microfilms International) 1981. I shall return to the general question of the status of the mathematical in Cartesian physics in Section 5.4, below.

only those with very small force. It is for this reason, Descartes argues, that such objects feel colder than others—heat and cold being supposed (*Car je suppose ici*) to be nothing but a varying degree of agitations and impressions upon "the small strings of those of our nerves which form the organs of touch". Marbles and metals thus feel particularly cold, and ice even colder than these.[11]

Second, we may imagine how the formation of ice comes about. The difference between liquids and solids being explained as above, it is clear that the subtle matter filling up the intervals between the particles of the macroscopic body will agitate those of liquids and solids differently. The worm-like, smooth particles of water are easily separated because of the gaps between them through which the subtle matter flows. In the case of the interwoven, branch-like particles of solids however, the suble matter has no such effect of separating particles and—consequently—macroscopic parts. It can only make these particles move and tremble in a way not unlike the wind moves the branches of a hedge.

Further, one should suppose (*il faut penser*) that, if particles of subtle matter are very much agitated and/or grosser than average, they may separate water-particles and hence make water liquid. On the other hand, when subtle matter

> is not as much agitated, nor less subtle [and this is the way] it normally is in the higher regions of the air, or sometimes in winter towards the Earth [and at night, and towards the Poles], it has not enough force to bend (*plier*) and agitate [the particles of water] in that manner, which causes them [i.e., the water-molecules] to stand still, joined in a disorderly manner the one above the other, and thus to compose a solid body, namely, ice.[12]

In a cold environment, no particles of subtle matter are strong enough to keep the water-molecules apart. The molecules will therefore stick together and form a solid mass. This is how ice is formed and Descartes imaginatively compares the difference between water and ice on a microscopic level with the macroscopic difference between a shoal of eels in water and the same group of eels when dry and stiffened by the cold.

The *discours premier* offers some more suppostitions, but we may, for the moment, leave it at these. From what we have seen, we may

---

[11] Descartes, *Météores*, AT VI, pp. 235-236.
[12] Descartes, *Météores*, AT VI, pp. 236-237.

already draw some important conclusions. The first is that, as I have said above, the boundary between general principles of Nature and applied theory is rather diffuse. Indeed, there is no discernible borderline between general suppositions of corpuscular philosophy and mechanical interpretations of specific macroscopic phenomena. Accordingly, there is no evidence that by *suppositions* Descartes means anything else than that part of mechanistic explanation which so far has not yet been fully dealt with. There is, in any case, no clear evidence in this work for the necessity of any *a priori* demonstration or other type of argument that has epistemological primacy. What is *supposed*, are only the more general accounts of particle motion. I shall come back to this matter in chapter 7, below.

A second conclusion is that Descartes offers explanations in terms of particle motion *only*. Both with regard to perception-theory (such as in his explanation of the experience of cold surfaces) and with regard to physical causation (as in the case of the formation of ice), Descartes exclusively introduces images of particles of matter influencing other particles of matter. Admittedly, there are some obscurities even in the text I have been quoting from. Descartes for instance makes a somewhat obscure statement concerning the less subtle particles, which are said to be found more often at those times and places at which there is more agitation, such as in daytime during summer at the equator and near the Earth's surface:

> The reason for this being that the larger of these parts, having greater force (*le plus de force*), are better able to go to the places where, because the agitation is greater there, they can more easily continue their movement.[13]

This is a strange supposition indeed. Why should these larger particles go to those places? What brings them there? It may be that Descartes considers that since change always arises from a "stronger force", only the more "forceful" particles can reach other regions. Once brought from places of less to places of greater agitation, their motion is augmented further through the influence of the Sun. Yet the passage clearly mentions a "tendency" of grosser particles to be in places with a higher potential of motion. Such "tendecies" also occur at other places in Descartes' writings—his explanation of gravity in

[13] Descartes, *Météores*, AT VI, p. 235.

particular.[14] It is, on the other hand, his proclaimed aim to ban them from his explanations, to sharply distinguish the mental from the physical.

4.1.2 Primary and Secondary Qualities

In *Le Monde*, the work which, following Galileo's trial, he abandoned for fear of condemnation by Church authorities, Descartes argues that all physical theories should be rejected that attribute to material things any analogy to mental considerations. A famous example is the thesis that Nature abhors a vacuum. Descartes rejects this *crainte du vide*. Taking the example of wine which supposedly does not leak from a hole at the bottom of a barrel for as long as the top is firmly closed, he argues that it is improper to attribute this fact to a *horror vacui*, as was usually done. According to Descartes,

> One knows full well that the wine has no spirit for abhorring anything.[15]

The pun is somewhat silly and even faulty in its ascription of an abhorrence to the wine instead of to Nature itself. Nevertheless, Descartes' unwillingness to accept the notion of a *horror vacui* is the consistent effect of his attempt in the first chapters of *Le Monde* to demolish the common-sensical attitude that

> the ideas of our thought are entirely similar (*semblable*) to the objects from which they come forth.[16]

This emphasis, right at the start of *Le Monde*, on the difference between objects as we see them and objects as they are in themselves, introduces into modern philosophy the idea of primary versus

---

[14] Cf. Descartes, *Principles* IV, § 23 especially. AT VIII-I, pp. 213-214 / CSM I, pp. According to Descartes, heavy objects have lesser "lift" as a result of the relative absence of celestial matter within them. See also P. van den Hoeven, *Metafysica en Fysica bij Descartes*, Gorinchem (J. Noorduyn en Zoon N.V.) 1961, pp. 139-143. Thus, there is a remarkable absence of the notion of impact with regard to the very phenomenon with respect to which the philosophy of impact was so fiercely defended by Huygens and Leibniz at the end of the century.

[15] Descartes, *Le Monde ou Traité de la Lumière*, AT XI, p. 20.

[16] Descartes, *idem*, p. 3: "Car encore que chacun se persuade communément que les idées que nous avons en nostre pensée sont entierement semblables aux objets dont elles procedent, je ne vois point toutesfois de raison, qui nous assure que cela soit; mais je remarque, au contraire, plusieurs experiences qui nous en doivent faire douter."

secondary qualities. The distinction is presented in terms of perception-theory. Its function however, is to make room for a purely mechanist conception of causation. Just as in the *Météores* examples of the experience of cold surfaces and the formation of ice, so here too, the perception theory and the theory of mechanical causation are two sides of the same coin. The perception-theory calls for a sceptical doubt as regards the likeness of mental experiences and the natural process by which they are caused.[17] At the same time, it calls for a restriction of causal factors to such as properly belong to the objective realm. Animistic notions like the *horror vacui* are for example to be excluded.

It is this dismissal of all "mental" notions from natural philosophy which sets Descartes' explanations in *Les Météores* apart from those of the Scholastic tradition. Descartes avoids discussion with the School-philosophy:

> Let it be noted also that, in order not to break the peace with the philosophers, I do not at all wish to deny the things that they imagine to exist in bodies besides those which I have mentioned, such as their *substantial forms*, *real qualities* and the like.[18]

Yet the possibility of such a neglect of internal bodily principles and real (*secondary*) qualities precisely points out why they might as well be rejected:

> it seems to me that my arguments should be only the more approved of when I make them depend on fewer things.[19]

The Scholastic principles are, in other words, superfluous. A comparison between Descartes' argumentation in the *Météores* and the type of argument we encountered in the *De Aqua*-example of the former chapter—and which may be found in other university disputations of Descartes' time—may give us a clearer insight as to

---

[17] See the former footnote: there are various reasons to doubt (*qui nous doivent faire douter*) the identity between the natural objects and our experiences of them. Descartes offers various arguments for such a doubt in the following pages of *Le Monde*, offering (1) an analogy with words, which also do not resemble the things to which they refer, (2) a physical account of sounds, which bear no resemblance to the vibrations causing them, and, (3) the example of the tactile sense, arguing that the itchy feeling bears no resemblance to the mechanics of the feather brushing the lips of a sleeping child. Descartes, *Le Monde*, AT XI, pp. 4-6.

[18] Descartes, *Météores*, AT VI, p. 239. For the original text, see chapter 5, note 86, below.

[19] Descartes, *ibidem*. For the original text, see chapter 5, note 87, below.

the various ways in which the banishment of explanations other than those from particle pressure set Cartesian theory apart from his Scholastic contemporaries.

4.2 Schoock and Descartes on Sensory Perception

According to Descartes, lifeless things have no inclinations.[20] These are the privilege of spiritual beings only. Descartes therefore formulates alternative theories of natural philosophy in which he claims to be able to reduce the number of principles and do without the "imaginary" principles of the philosophers. Let us compare the two positions—Scholastic and Cartesian—using the example of the formation of ice.

4.2.1 The Formation of Ice: De Glacie *and* Les Météores

In the same series as that in which the disputation *On Water* was published—Martin Schoock's *Disputationes physicæ* of 1646-1647—we also find an interesting discussion *On Ice*. In it, Schoock and his pupil Bernhard Vriesen from Zwolle argue that there are basically four conditions for the formation of ice. First, in order to account for the fact that ice can float on water, it is necessary to suppose a warm, earthy exhalation that "tempers" the heaviness (*gravitatem temperet*)

[20] Except, that is, for such fundamental "inclinations" as the tendency of objects to pursue their movement in a straight line, to be moved to the Earth's centre etc. Descartes uses the phrase *inclination naturelle* for instance in *Le Monde* in relation to the phenomenon of the liquid in a glass, rising against its "*inclination naturelle*". Descartes, *Le Monde ou Traité de la Lumière*, AT XI, p. 18. Such "inclinations" are, however, experimentally observed and/or commented upon by the use of analogies. Both in *Le Monde* and, later, in the *Principles*, Descartes for instance illustrates what—in the *Principles*—was to become his second law of Nature ("that all motion is, from itself, rectilinear") using the example of a gravel thrown from a sling. The gravel follows its path always in the direction of the tangent to the point at which it left the sling. This is because it is, in every single instance, "determined to continue its motion in a certain direction along a straight line". Descartes, *Principia Philosophiæ* II 39, AT VIII-I, p. 64. Rectilinear motion is therefore taken to be a natural tendency of particles, for instance in *Principia Philosophiæ* IV 19, AT VIII-I, p. 211, where it is said that celestial globules for instance will always continue their movement "along straight lines in as far as they can" (*semperque quantum possunt secundùm lineas rectas*). In this case, the obvious anthropomorphic way of expression should presumably be taken as rhetorical finery, validated by the empirical example of the sling and the metaphysical argumentation concerning the invariability of God's conservation of motion. For Descartes' argumentations regarding his three laws of motion, see Daniel Garber, *Descartes' Metaphysical Physics*, Chicago (The University of Chicago Press) 1992, pp. 197 ff.

of ice. For, normally, what has become hard, has become heavier. It must be the hot fumes which often steam out when ice is broken which account for its levity. Secondly, a—denser—earthy vapour is needed for congealing the waters, since, according to Schoock, pure water itself never condenses to ice, but water freezes only according to the degree of turpidity.[21] Thirdly, one of course needs a watery fluid, since there is no ice without water. And, finally, an external cold is necessary. This, in fact, is the efficient cause of the *constrictio* (compression, condensation) of water into ice.[22]

If these four conditions are met, ice is formed in the following way:

> The cold brings together (*cogit*) the denser parts of water, that is, the earthy and watery parts, expelling many lighter parts of water. And when they are thus clotted together [the cold] makes them congeal (*concrescere*), so that they even become hard.[23]

This account differs markedly from that in Descartes' *Météores*. But in what way exactly? For there are similarities as well. Descartes also distinguishes various types of elements according to their macroscopic differences. Both Descartes and Schoock, moreover, see the expulsion of smaller intermediate particles from the vacuola between grosser ones, and the consequent congealing of the latter, as an essential step in the formation of the ice. Finally, both hold low temperatures to be the efficient cause of ice-formation.

These similarities however, are as trivial as the various minor disagreements in comparison with the one general difference between both theories, which lies in the scope of what they purport to explain. Where Schoock argues that "the cold brings together" some and

---

[21] This idea was in fact generally accepted. Since water is an extremely humid element, the *concretion* of water must be the result of earthly components. See e.g. the *Commentarium Collegii Conimbricensis S. J., In Libros Meteororum Aristotelis Stagiritæ*, Coloniæ, Impensis Hærendum Lazari Zetneri, 1631, col. 68: "non sola frigoris efficientia concurrat [ad eam congelationem]; sed admistio etiam quædam terrenæ concretionis, quæ congelationem magnopere juvat, eò quod terra ex se densa existit, & crassitatem ultrò appetit."

[22] Martinus Schoockius, Bernhardus Vriesen (respondens), *Disputatio Physica de Glacie*, Groningæ (Typis Augustini Eissens, Typographi) 1646, Theses XVIII-XI.

[23] *Idem*, Theses XII. The text goes on to explain that what is left of the lighter parts of water, is turned into an airy substance by the cold (*eas facit aërescere*). This substance is pushed (*pellit*) to the inside (*ad intus*) together with the warm exhalations. Thus, the internal pores are formed "which are frequently found in ice, like in pumice stone, albeit not as big."

"expels" other parts, he takes the cold to be a "force", or quality, capable of bringing about various motions in the elemental particles. The situation is wholly different in Descartes, who claims that the sensible quality of cold corresponds to a certain degree of motion in the material particles themselves. The cold does not *make them* move in a certain way: cold *is* simply a situation of slow particle movement.

The difference is a difference between a qualitative description and Descartes' "mechanical" one (a term we have yet to define more sharply). Fortunately, the two were once weighed against one another by Schoock.

4.2.2 Two Conceptions of Efficient Causality

With the *Admiranda Methodus*, or *The Admirable Method of the New Philosophy of René Descartes*, Martin Schoock—doing for his Utrecht tutor Voetius what Father Mersenne had refused to do before[24]—wrote a malicious book in which he shows anything but admiration for Cartesianism.[25] Descartes himself refused to take the book seriously, arguing that it contained nothing but slander.[26]

---

[24] Voetius had tried to incite Marin Mersenne to write against Descartes. Cf. *Correspondance du P. Marin Mersenne*, commencée par Mme. Paul Tannery; publiée et annotée par Cornelis de Waard, Vol. XI, Paris (Éditions du C.N.R.F.) 1970, pp. 376-377. Most of what Voetius wrote to Mersenne is now lost. Descartes kept no less than five letters, to which he refers in the *Lettre apologétique aux Magistrat d'Utrecht*. Cf. *Querelle*, pp. 409-410 and notes. Further: *Correspondance du P. Marin Mersenne*, Vol. X, Paris (Éditions du C.N.R.F.) 1967, p. 166 esp. and Vol. XI (1970), pp. 363-377. Voetius' request to Schoock to write a book against Descartes, came at a time when the conflict between Voetius and Descartes had started to become a more personal one, as a result of the unfavourable picture Descartes had given of the Utrecht theologian in his *Lettre à Dinet*. Cf. Theo Verbeek, *Descartes and the Dutch*, pp. 19 ff. and *Querelle*, introduction.

[25] A French translation of the *Admiranda Methodvs Novæ Philosophiæ Renati des Cartes* (Vltraiecti, Ex Officina Joannis van Vvaesberge, 1643) may be found in *Querelle*, pp. 153-320. I shall refer to both editions.

[26] See the *Epistola ad Voetium*, AT VIII-I, p. 5 / *Querelle*, p. 328, where Descartes says of the *Admiranda* that it contains "nullæ [...] rationes priùs scripta probanda vel excusanda, sed novæ tantùm calumniæ prioribus atrociores." Schoock certainly did not shun vicious *ad hominem* attacks on Descartes. Even in the first lines of the Preface, he refers to the French philosopher as "Mr. No-one, unknown, though no doubt the chief of the philosophers", which he later on alternates with epithets as "that mendacious tongue", "that most mendacious biped", "Chief of Cretans" etc. Often enough Schoock pities the Cartesians and prays for their salvation (*Admiranda*, p. 79 / *Querelle*, p. 219), but he never fails to add a sarcastic comment on the illiterateness of the Cartesians and even dedicates an entire chapter to the Cartesian warfare against reading (*Admiranda*, pp. 26-31 / *Querelle*, pp. 193-

Nevertheless, Schoock's *Admiranda* contains various points of interest, one of them being his critique of the Cartesian way of reducing sensible qualities to the primary, "mechanical" properties of insensible (that is, minute, microscopic) particles. Schoock ascribes to the Cartesians the following definition of heat:

> heat is the diverse agitation of insensible particles, whilst cold is their [state of] rest,[27]

a definition which is in fact taken over from Regius.[28]

Although Schoock mentions the supposedly insensible character of the particles, he does not in fact distinguish between insensibly small[29]

---

195). Of course, Descartes' personal claim on authority could be tolerated, says Schoock, if a certain perfection of philosophy were to be expected, but although "pigs prefer dirt to perfume", it could never be tolerated that anyone would despise "the meals of the more solid philosophy" and prefer "Descartes' droppings" (*Admiranda*, p. 34 / *Querelle*, p. 197). In the Preface, Schoock also criticizes Descartes' personal life, referring to the Frenchman's numerous removals throughout the Seven Provinces. This would make him suspect of being one of the Rosicrucians—who tended to seek anonimity—if it were not for the fact that "this most ambitious man wants his name to be sung out everywhere." Schoock writes that Descartes would certainly have followed the example of Ignatius of Loyola and enter the latter's Jesuit sect after an unsuccessful career in the army, if it were not for the fact that "not being able to bear the constraints of the habit", he feared his "indomitable lust" would soon betray him and reveal his hypocrisy" (*Admiranda*, pp. 11-12 / *Querelle*, p. 185)—an allusion to Descartes' illegitimate Deventer daughter Francine (1635-1640), also referred to immediately before at *Admiranda*, p. 10 / *Querelle*, p. 185. See further: Theo Verbeek, *Descartes and the Dutch*, pp. 20-21.

[27] Schoock, *Admiranda Methodus*, p. 224 / *Querelle*, p. 293: "calor est varia agitatio particularum insensibilium: frigus autem est earum quies."

[28] Henricus Regius, *Physiologia sive Cognitio Sanitatis*, Utrecht, Ægidius Roman, 1641, I, I (17), p. 6. The *Physiologia* consists of a series of academic disputations on various questions of natural philosophy, organised and presided over by Regius from 17 April 1641 onwards. Schoock's "Cartesian" sources were indeed basically Regius' publications in medicine and physics, three of which are quoted in the *Admiranda*: First, Regius' unpublished manual to which Descartes refers as the "Physiologia Integra" (Descartes, *Letter to Father Dinet*, AT VII, pp. 582-583 / Querelle, p.) and Schoock as the "Dictata Physica" (*Admiranda*, pp. 132 and 234 / *Querelle*, pp. 246 and 299). Further: Theo Verbeek, "Regius's *Fundamenta Physices*" in *Journal of the History of Ideas* 55 (1994), pp. 533-551. See also the *Physiologia*, and, finally, Regius' *Responsio*. Regius' definition of heat reoccurs in his major work, the *Fundamenta Physices*, Amsterdam, Apud Ludovicum Elzevirium, 1646, p. 98.

and larger, sensible bodies in his counter-argument, which runs as follows. If heat is to be reduced to the various agitations of (small) particles, then why is it that we do not feel any heat when we shake a dry lump of heather and all the tiny particles of earth dart away in every possible direction? Or how can it be that we burn ourselves when we touch a piece of metal which is red-hot, but at the same time solid? Or again, why is it that my mouth and tongue are heated when I am slowly and carefully chewing ginger and can be cooled off by hastily eating an apple and reducing it to particles even totally insensible?[30]

Schoock is obviously playing, in a very non-Cartesian manner, with the notions of the smallness and velocities of particles. Since he fails to distinguish between the sensible and the non-sensible, his Cartesian adversary might argue that, in the case of the red-hot piece of metal for instance, the agitation is to be found in small fiery particles in the pores of the object. It is questionable, however, if that will do. The only thing Schoock does at this point is to argue that experience does not point to a strict connection between the agitation of small particles and the quality of heat. Even granting the existence of a world of insensible particles, he might argue that, if on the macroscopic level movement is not strictly correlated with heat, there is no reason to suppose that on the microscopic level it is.

Schoock further asks whether the insensible particles are themselves the heat, or only its efficient cause.[31] If they are themselves the heat, they must form a sensible substance rather than an insensible one. If, on the other hand, the insensible particles are only the efficient cause of heat, then the Cartesians take the effect for its cause. Most significantly, the first possibility—which need not concern us here, since Descartes would never assert that heat could be a substance of its own[32]—is elaborated by Schoock, whilst the latter is not further commented upon. One might get the impression that Schoock is trying

---

[29] According to Regius, particles can also be insensible as a result of their velocity. See *Physiologia* I, I (5), p.1: "insensibilis est, quæ ob suam exiguitatem, vel motus sui celeritatem sensum fugit."

[30] Schoock, *Admiranda Methodus*, pp. 224-225 / *Querelle*, pp. 293-294.

[31] Schoock, *Admiranda Methodus*, p. 227 / *Querelle*, p. 294.

[32] Note that the alternative was accepted by Pierre Gassendi for instance, according to whom heat was caused by calorific atoms. Cf: François Bernier, *Abregé de la philosophie de Gassendi*, Lyon 1684², II, pp. 95-98, quoted by Marie Boas in "The establishment of the Mechanical Philosophy": *Osiris* 10 (1952), p. 430.

to draw attention away from this type of answer because it comes too close to the reply a Cartesian would in fact have welcomed. As far as Descartes is concerned, there is actually no quality of heat outside of human awareness. Heat is to be regarded as the subjective, sensible, effect of the objective and efficient cause of particle motion acting on our nervous system. According to Descartes, violent movement actually *is* the efficient cause of our sensation.

The discussion is, however, obscured by a confusion of two different notions of *efficient causality*. In the Cartesian sense, the causal chain terminates in the mind, which, as it were, "reads off" the feeling of warmth from a certain brain-state.[33] In the material objects themselves, however, there is nothing that resembles our sensation. There are only movements causing other movements: a chain of movements extended all the way up through our nervous system to the brain. The sensation of heat itself only emerges once we feel the particular pain or tickling proper to this mental experience:

> And as we have already said that there is nothing, outside our thought, which is similar to the ideas that we perceive of tickling or pain, we may believe likewise, that there is nothing similar to that which we perceive of heat, but that everything which can variously stir the small parts of our body, may arouse this feeling in us.[34]

Our ideas of material objects and their qualities are formed by a mechanical process only. The sensation—for instance of heat—is only its subjective effect. No quality of heat is to be found in the mechanical process itself. Or, as Jean-Luc Marion has put it:

> Pour Descartes [...] il suffit qu'une pression s'excerce, pour que mécaniquement naisse l'impression.[35]

In this sense, matter in motion is, for Descartes, the *causa efficiens* of our feeling.

---

[33] In his article on "Descartes' Corporeal Ideas Hypothesis", E. S. Reed argues that the Cartesian idea that "all awarenesses are awarenesses of brain states" was to become a central idea in later scientific psychology, for instance in Johannes Müller's hypothesis of specific energies. See *Review of Metaphysics* 35 (1981/'82), pp. 731-752. For a discussion of the philosophical implications of Müller's theory, see Arthur Liebert, "Johannes Müller, der Physiologie in seinem Verhältnis zur Philosophie und seiner Bedeutung für dieselbe", in *Kantstudien* 10 (1915), pp. 357-375.

[34] Descartes, *Le Monde ou Traité de la Lumière*, AT XI, p. 10.

[35] Jean-Luc Marion, *Sur l'ontologie grise de Descartes*, Paris (Vrin) 1981, p. 120.

Schoock also gives an example of a violent agitation being the efficient cause of heat. One only has to visit the joiner and feel a piece of wood just after it has been planed.[36] Schoock's idea of the efficient cause of heat actually does not involve any theory of perception. The example starts and terminates in the objective realm with motion causing heat in a wholly common-sensical way: rubbing a surface causes heat. Schoock does not go into the physiological point. Discussing Regius' definition of heat, Schoock is not concerned with the question of how a mechanical process in the nervous system can cause the effect of our mind experiencing a certain sensation. Schoock just asks himself how heat may be defined in terms of agitation.

4.2.3 Localizing Real Qualities

The strange thing is, that, five years before, Schoock had written a dissertation under Descartes' first ally in the Netherlands, Henricus Reneri,[37] in which he comes quite close to the mechanical viewpoint. The dissertation, *On the Nature of Sound and of the Echo*—by which Schoock became the first Utrecht student to acquire the degree of *magister* in philosophy—starts off with a disavowal of the "fanatical" position of the Sceptics and an appeal to scientfic autonomy, both of which cannot but remind us of the project of scientific reform that Descartes had only recently presented in the *Discours*.[38] Its main

---

[36] Schoock, *Admiranda Methodus*, p. 226 / *Querelle*, p. 294.

[37] On Reneri and his contacts with Descartes, see Theo Verbeek, "Henricus Reneri (1593-1639)", in H. W. Blom, H. A. Krop and M. R. Wielema (edd.), *Deventer denkers. De geschiedenis van het wijsgerig onderwijs te Deventer*, Hilversum (Verloren) 1993, pp. 123-134.

[38] Martinus Schoockius, *De Natura Soni, & Echus*, in *Lusus Imaginis Iocosæ sive Echus a Variis Poetis, variis Linguis & numeris exculti*, Ultrajecti, Ex Officina Ægidii Roman, Acad. Typog. 1638. Since the pages of the dissertation are not numbered, I shall refer to pagenumbers in brackets. The reference to "Pyrrhonis sectatores, Scepticos vulgò dictos" occurs at p. [6], where we also find the following "Cartesian" considerations: "Quo enim compendio quis facilius, & felicius in quaque re verum & rectum invenire potest, quam si sepositâ omni authoritate humanâ, præconceptâque opinione, ipse ingenium suum intendat, & abhibitis legitimis sciendi adminiculis non quæ alii dixerunt credere, sed quæ ipse videt rationi, sensibus, & experientiæ, consentanea esse, amplecti laboret. Libertatem omnibus utilem, philosophis etiam necessariem esse existimo." See also p. [7]: "Veritatem non ex Præceptoris schola sapiens adfert, sed in ipsis rebus invenit, aut saltem in iis scrutatur", and: "Quid alii senserint fideliter annotabo, quid mihi videatur, ingenuè dicam." I do not doubt that it is under the influence of Descartes' friend Reineri that Schoock formulated his method in this way. Note however, that, as Theo Verbeek has argued, Reineri's own scientific position

theme, however, is a discussion of the nature of the sound. The Aristotelians, it is said, hold that sound is

> a sensible quality that can be heard, [arising] from air or water which has collided or which has been broken by a violent shaking (*concussio*) of solid bodies.[39]

Schoock immediately discusses this "Aristotelian" position, criticizing the notion of "sensible quality":

> Let us consider the following: *Sound* is called a *sensible quality*. Now is this right and is this an accurate description? The Peripatetics seem to think it is. [...] But if [sound] is [a *quality*], then it differs from the movement of the bodies themselves and, [in that case,] air-in-motion is one thing and sound is something else. But do they not in this way multiply things beyond necessity? Wherefore this quality? As if air, when it is moved and prompted in a certain way, does not itself move the eardrum. [...] If air moves the eardrum and the movement of the eardrum is perceived (*sentitur*) by the mind, what else is there to say, but that sound is the air-in-motion itself, or rather, repercussed air?[40]

Much like Descartes in the example of heat, Schoock here questions whether a quality is to be found apart from the mechanical process causing our qualitative sensation. In fact, he straightforwardly concludes—against Otto Casmann and Clemens Timpler as authorities on the subject—that "sound is a repercussion of air".[41]

---

differs from Descartes' in various respects and may rather be thought of as "Baconian". Theo Verbeek, "Henricus Reneri (1593-1639)", in *Deventer Denkers*, pp. 132-134 esp.

[39] Schoock, *De Natura Soni & Echus*, p. [8].

[40] Schoock, *ibidem*.

[41] Schoock, *De Natura Soni & Echus*, p. [9]: "acriter hic reclamat Casmannus in Somotolog. cap. 13. q. 2 quem & sequitur Timplerus in Phys. Gener. cap. 8 q. 1." Otto Casmann (d. 1607) was a German Rektor and voluminous writer. See *Jöcher* I, cols. 1724-1725. In the year of his death, his *Somatologia, Physica Generalis, seu Commentationvm Disceptationumque Physicarum Syndromvs Problemat. I. De Natvralivm Corporum in Genere Essentia, & Qvalitatibvs Physicis, tum manifestis, tum occultis* was published in Frankfurt. As Schoock also noticed ("quem & sequitur", *De Natura Soni & Echus*, p. 9), the book seems to have been the basis for Clemens Timpler's *Physica sev Philosophia Natvralis Systema Methodicvm*, Hanoviæ Apud Hæredes Guilielmi Antonii, 1613. Timpler was Professor at the Steinfurt *Gymnasium* who was converted to the Reformed belief. In the early decades of the seventeenth century, he wrote introductory works on various disciplines, purging them of Catholic aberration. Cf. *Jocher* IV, col. 1751 and Max Wundt, *Die deutsche Schulmetaphysik des 17. Jahrhunderts*, Tübingen (Mohr)

The question remains, however, whether Schoock's criticism of sensible qualities coincides with Descartes'. Let us follow the text of Schoock's dissertation somewhat further:

> if sound is a quality that has arisen apart from the movement of air, it must of necessity be brought about by something other than the motion of air. But what would that be? [...] If you were to say that it is brought about by movement in the same way as heat and cold, you would say nothing. Heat and cold are brought about and pass away successively and by degree. Sound on the other hand, is brought about and dies away in an instant and is not spread (*intenditur*) by increase, like [heat], but at the start it is most powerful and in the first instance at its height. So if this quality were produced through motion and the fraction of air, it would follow that the effect would be prior to the cause. At the same time as the air is fractioned, the sound arises. According to the Aristotelians, however, the air produces it through motion. All motion is successive. What remains, then, but that from some thing which causes by degree and successively, something arises in an instant?

Schoock seems to have some difficulty here with getting the full meaning of his argument across. His intention seems to be, that to argue for the existence of a quality of sound over and above the existence of the motion of air, involves a reversal, or, rather, a chronological incongruity of cause and effect, on account of the impossibility of a successive motion producing an instant effect. The question might be put why the motion should *precede* the quality. Realising the possibility, Schoock comes up with an argument to which, in the *Admiranda*, he might have given more thought:

> It will not do that someone says that air is here the vehicle of sound: the production of sound is one thing, its diffusion another. When, for instance, a canon is fired, its sound is heard at the same time as it is fired, but one hears it because the air is broken asunder. Who would then say that sound, which they see as a quality, is in addition

---

1939, pp. 87-89. Casmann and Timpler both defended the idea of a separate quality of sound. As Casmann says: "ad sonum faciendum duo corporum genera concurrere extrema percutientia, & medium interceptum". Nevertheless, sound is a *quality*, resulting from the crush (*elisionis motu effectam*). Casmann, *Somatologia*, p. 484. For Timpler's discussion of the question whether or not sound is to be seen as a quality, see *Physica*, pp. 116-117.

> produced by a motion and fraction of the air, when it has already been heard?[42]

Diffusion is not the issue. The question, whether a quality *exists* besides the motion of air, is reformulated in terms of whether a quality *is produced* by the motion of air. The difference is important, because it is here that Descartes and Schoock go different ways. Descartes disconnects quality and motion, seeing the quality as a mental effect of a motion of air in the physical realm. The quality is the mental appreciation of a phenomenon in the objective world. It has no "similarity" to this motion. Nevertheless, in Descartes' case, there is every reason to say that a quality is *produced* by a motion and this is exactly what Schoock denies.

The reason is, that qualities are located differently in each case. Descartes accepts the quality, but puts it in the mind of the observer only. Seemingly anticipating the mechanical viewpoint, Schoock also denies the existence of a quality in the physical world. He does not, however, look for the quality in the mental realm. If there is to be a quality, it must be there from whence the sound came. But since it is impossible for a motion of air to produce the quality instantly, the phenomenon of sound must be defined in terms of motion only. There is simply no time for a quality to arise.

4.2.4 The Scholastic Concept of Quality: Squeaking Air

Explaining, in the first pages of *Le Monde*, the difference between physical processes and our sensations of them, Descartes takes the example of sound to be an obvious one in favour of the mechanical view:

> a man opens his mouth, moves his tongue, and exhales: I do not see anything, in all of these actions, which is not quite different from the idea of sound which these cause us to imagine.[43]

If our sensation were to give us an adequate idea of the process causing it, we should, instead of hearing a sound, rather be informed about the movement of air working on our eardrum. It is at this point that Descartes refers to the Scholastics in a remarkably positive way:

> the majority of the Philosophers affirm that sound is nothing but a certain trembling of air.[44]

---

[42] Schoock, *De Natura Soni & Echus*, p. [9].
[43] Descartes, *Le Monde*, AT XI, p. 5.

The reference has not gone unnoticed.[45] Yet in fact Descartes is mistaken. As Étienne Gilson has noted, Descartes wrongly attributes to Aristotle the comparison of sound with the circular undulation of water in which a stone has been thrown.[46] Gilson offers two quotations from the philosophers of Coimbra as proof for this point. The simile of undulation apart, these also confirm that many philosophers did *not* regard sound as simply a *tremblement d'air qui vient frapper nos oreilles.*

According to the Conimbricenses, the motion of air does not necessarily extend to the ears. For in that case, when a bell is heard a mile away (*ad leucæ distantiam*) it would mean that all the intermediate air would have to be stirred—something the Conimbricenses hold to be "ridiculous". Moreover, if it were true that sound is propagated by air all the way to the ears, two people calling out to each other from opposite positions would each find their own cries impeded by those of the other. Finally, the Conimbricenses argue, are not fishes in a closed pond, and we ourselves, when we are shut up in a room which is closed on all sides, still able to hear although no movement of air can pass?[47]

The philosophers of Coimbra had an altogether different idea of sound. Although a certain movement of air does play a role in their theory, it is not wholly constitutive of sound. Certainly, the *generatio soni*, the coming-into-being, or production, of sound, does follow upon a certain bodily movement, and, intermediately, upon a "fraction of air". It is, however, not itself a motion of bodies, for such a motion is

---

[44] Descartes, *ibidem*. Descartes does not keep up this positive stand vis à vis the School philosophy for long. Already in the next sentence, Descartes moves on to another example, since, as he says: "not everybody would perhaps believe what the Philosophers say."

[45] As Ferdinand Alquié notes in his edition of Descartes' *Œuvres Philosophiques*: "Le cas du son est privilégié, ce pourquoi Descartes prend ici le son comme exemple. En ce cas en effet nous avons l'expérience directe de ce qui provoque la sensation, et du fait que ce qui la provoque est de l'ordre du mouvement (ouvrir la bouche, remuer la langue). Bien plus: la scolastique admet que le son est produit par un *tremblement d'air.*" Descartes, *Œuvres Philosophiques*, Édition de Ferdinand Alquié, Paris (Garnier) 1963, Tome I, p. 317, note 1.

[46] Étienne Gilson, *Index Scolastico-Cartésien*, Paris (Vrin) 1979, p. 273, under "SON".

[47] *Commentarii Collegii Conimbricensis S. J.*, *In Tres Libros de Anima Aristotelis*, Coloniæ, Sumptibus Hærendum Lazari Zetneri, 1629 (fourth edition) 2, 8, 1, 1-2, col. 262. Gilson quotes the 1592 Coimbra edition, p. 424, in *Index Scolastico-Cartésien*, p. 274.

merely local motion , and local motion only concerns the "where" and the "place", without being able "out of itself" to "bring upon the movable any new reality."[48] Being a quality, the sound falls under a category which is different from the category of place. Ontologically transcending the sole bodily motion, the quality cannot be identified to it. Therefore,

> the production of sound is formally another action, terminating intrinsically in the sound as being the *terminus* produced by it. This action could be called a "sonation", if it were not for the fact that this word sounds so awkward.[49]

Awkward though the word may seem to these late-Renaissance commentators, the *sonation* itself is thought to be a necessary intermediary between motion and sound. But what is this sonation? What function does it have, given the fact that sound is also said to immediately follow upon a *fractio aëris*—a breaking of air—, and ultimately upon a collision of bodies?

The answer lies in getting the chronological order right. First, we should realise that, according to the Coimbra Commentators, the motion of air does not necessarily have to be propagated all the way up to the ear. This is the same thing as to say that sounds can be heard from afar without any intermediate medium. The reason is, that we hear the sound at the very point where it is produced. The motion of air does not have to be propagated after this production of sound (or *sonation*), because it does not follow the production at all. It precedes it, causing the action of *sonation*. In other words, there is first a collision of bodies. Next there is a fraction of air. This occurs

> because the air, being worn away by the two bodies, is first fractioned, and, when fractioned, sounds.

No motion of air is propagated, because according to the Commentators, the motion of air precedes both sonation and sound. Sound is a quality arising not in the ear or in the mind, but somewhere in the audible field—in the physical realm itself, that is, where bodies collide and air consequently gets stuck between them. A *sonation*

---

[48] Conimbricenses, *In tres libros de Anima Aristotelis* 2, 8, 1, 1-2, col. 259 (= p. 424 in the 1592 edition; Gilson, *Index Scolastico-Cartésien*, p. 273).

[49] Conimbricenses, *ibidem*. The first sentence is the best that I can make of the phrase "[Itaque] generatio soni formaliter est alia actio terminata intrinsece ad sonum ut ad terminum per ipsam genitum."

results: the air, as it were, cries out. It squeaks. And the cry may be heard whatever the medium between the sound and its hearer.

4.2.5 Conclusion: The Mechanics of Perception

Although the Philosophers of Coimbra mention various authors as proponents of theories attributing to air the role of a medium for the propagation of motions and even sound-waves,[50] they clearly do not themselves accept such a view, let alone a Cartesian idea of sound itself as being nothing but a *tremblement d'air*.

Contrary to Descartes' own suppositions, there are significant differences between the Scholastic and Cartesian viewpoints.[51] There are also important distinctions between Schoock's account of sound and that of the Aristotelian commentators. Schoock, and his Utrecht teacher Reneri, do not seem to draw on any earlier "mechanistic" accounts.[52] Schoock only discusses and rejects the arguments in favour of a distinct quality of sound which occur in the textbooks on physics of Otto Casmann and Clemens Timpler.[53] This is important,

---

[50] Conimbricenses, *Commentarium in tres Libros de Anima*, ed. 1629, col. 261, where Alexander, Averroes and Themistius are named as proponents of the theory that reduces sound to a motion of resounding objects and media. See also Gilson, *Index*, p. 274.

[51] Étienne Gilson, *Index*, p. 273: "Le texte des Conimb. ici rapporté [i.e., *Index*, pp. 273-274] [...] rend très vraisemblable l'hypothèse d'une erreur de mémoire de la part de Descartes." Gilson has the comparison of sound and undulation in mind that Descartes erroneously attributes to Aristotle. However, the same may be said as regards the general view that "la plupart des philosophes assurent que le son n'est autre chose qu'un certain tremblement d'air." It is true that the idea of sound being merely a propagation of a movement of air was attributed to Aristotle, who had said that sound is a *motum & ictum aeris*, or κίνησις ἢ πληγὴ ἀέρος. Aristotle, *De Anima* II 8, *passim*. However, Aristotle's phrasing was generally taken as a figurative way of speaking. Timpler, for instance, establishes that "Aristotelem in citato loco aut grauissime errasse, aut tropice loquutum esse." Of course, the latter must be true. (Timpler, *Physica*, p. 117.) See also Otto Casmann's akward formulation of the same point, in *Somatologia*, p. 485: "Aiunt quidam Philosophi, sonum esse motum esse ictum: Sed tropice pro eo, *quod est sonus fit motu, ex ictu procedit*." The Conimbricenses too, when reflecting on what Aristotle may have meant by saying that sound is a motion or collision of air, argue that he merely meant that there sound normally follows upon some bodily impact. (Conimbricenses, *In tres Libros de Anima* II, cap. VIII, qu. 1, art. 1, 1629 edition, col. 259, = 1598, p. 424 / Gilson, *Index*, p. 273.)

[52] That is to say, of Alexander, Averroes and Vitruvius and the like, who occur as authorities of the view denying a distinct quality of sound in the commentary on *De Anima* of the Conimbricenses. See above, note 50.

[53] Schoock, *De Natura Soni & Echus*, pp. [9]-[11].

since it shows that a potential tension between Schoock's two major discussions of the nature of qualities—his dissertation of 1638 and his critique of Cartesian method, the *Admiranda Methodus* of 1643—lies not so much in an antagonism between a Scholastic / anti-Cartesian view and a (proto-)Cartesian view under the influence of Henricus Reneri, but between an Aristotelian and a non-Aristotelian view, both of which are wholly un-Cartesian. This tension then, if indeed there was one, became insignificant in the light of Cartesianism, at a time when Aristotle was fully restored. But is there in fact a tension at all between Schoock's former and latter discussion of the nature of sensible qualities? Our digression into the Scholastic theory of sound points in a different direction.

4.2.5.1 Schoock on Sound and Heat

The Conimbricenses' discussion of sound suggests why, in the 1638 dissertation, Schoock formulates the problem of qualities in terms of a physical sequel in which movement produces quality. Schoock does not accept the squeaking of air as the origin of sound; on the contrary, he accepts the propagatory theory of sound. Arguing from the common experience of a canon-shot in favour of the idea of a motion of air which is propagated all the way to the eardrum, there is no need for a quality to arise in the meantime. The quality Schoock thus rejects is the quality ("of the Aristotelians") resulting from the movement of air. In the propagatory-account of motion, there is no place for any such quality. It would involve a successive motion of air which *returns* to the original location of the quality, in order to produce the sound there. The sound however, is instantaneous. Therefore, the production of sound by a propagatory movement of air leads to a chronological impossibility and a confusion of cause and effect. Not accepting the propagatory function—indeed, thinking that it was "ridiculous"—, the Conimbricenses on the contrary had no trouble with this argument.

The whole discussion does not concern the Cartesian idea of a purely mechanistic account of perception at all. In the dissertation, Schoock only argues that, in the case of sound, the addition of a "quality" is superfluous. It may not be in other cases, such as heat. This is what explains Schoock's standpoint in the *Admiranda*, where he criticizes Descartes' mechanics of perception. Although according to Schoock there is no need, since there is no time or place, for a "sonation", there *is* a time and place for warming action. In fact, in the dissertation, Schoock explicitely denies any correspondence between sound and heat (or cold), for the reason that

> heat and cold are produced succesively and by degree.[54]

Unlike sound, in other words, heat and cold *are* qualities produced by some non-instantaneous motions, by friction for instance, or by rubbing a surface. It is only for this reason that, accepting bodily friction as an "efficient cause" of heat in the *Admiranda*, Schoock in the dissertation rejects the idea of air being the efficient cause of sound:

> The movement of a bell for example, is not the efficient cause of its ringing. [The efficient cause is] the sacristan who keeps it in motion. For when the bell is moved, the ringing is heard. Likewise a bird, cleaving the air, makes a noise with its wings: the air however, which is moved and repelled by the wings, is itself the sound.[55]

Here, the step required to arrive at a completely mechanical theory of perception seems only a very small one, but the fact is, that it was not taken. Nor was a mechanistic theory the aim of authors like Schoock, who argued against a quality of sound. The reasons for its rejection were merely those given: a problem of chronology and of the relation of cause and effect which arises when the idea of a quality of sound is accepted. With qualities that did not give rise to such inconsistencies, Schoock had no problems at all. Hence, with regard to heat, the Scholastic idea of a real quality was still fully accepted, a quality inhering in the objective world itself and causing our sensation of it. In Schoock's *Admiranda Methodus*, the fire still heats as much as the air of the philosophers of Coimbra sqeaks.

4.3 Conclusion: The Mechanical Turn

Accordingly, despite Descartes' reliance on the support of *la plupart des Philosophes*, it was only he himself who, in the first section of *Les Météores*, made the crucial step of stripping the physical world of the real sensible qualities that were supposed to exist independently of our sensation of them. It is this radical turn towards a mechanistic interpretation of perception that marks the difference between Descartes' explanations and those of his contemporaries, whatever the

---

54 Schoock, *De Natura Soni & Echus*, p. [8].
55 Schoock, *De Natura Soni & Echus*, p. [10].

obvious similarities of detail.[56] As we saw before, Descartes literally says he will not discuss the qualities:

> in order not to break the peace with the philosophers, I do not at all wish to deny the things that they imagine to exist in bodies besides those which I have mentioned, such as their substantial forms, real qualities, and the like.

But the reason for his silence is clear:

> it seems to me that my arguments should be all the more approved of, when I make them depend on fewer things.[57]

Though not explicitly denying their existence, Descartes affirms he has no need for real qualities. In later metaphysical works, Descartes would argue that the qualities in fact do not inhere in the world at all, but are mental phenomena only. The theory of perception thus bans real qualities from physics. The result, of course, is a new, mechanical physics, reaching further than perception-theory only, into a general account of causation.

Our quotation from *Les Météores* already spells this out. For not only are real qualities thought redundant: *substantial forms* are rejected also. The result was that the idea of causality in which substantial forms played an all-important role was abandoned. Causality was, however, presented with a new image and a new set of metaphors. Instead of the idea of natural objects having individual centres of causal activity—powers of action "from within", as we said in chapter 2—Descartes' philosophy came up with a different

[56] In his "Météores Cartésiens et Météores Scolastiques", Étienne Gilson offers a fine analysis of the way in which Descartes "interpreted" very differently many Scholastic ideas which at the same time he did not contest. Descartes simply took over what he regarded to be true facts or distinctions. Cf. "Météores Cartésiens et Météores Scolastiques", in Étienne Gilson, *Études sur le rôle de la pensée médiévale dans la formation du système cartésien*, Paris (Vrin) 1984^5, pp. 102-137. Emphasizing the similarities between Cartesian and Scholastic thought however, Gilson seems to be granting Descartes too little where he says that the "spirit" of Descartes' writings is different for being less "finalistic" or "naive" (Gilson, *idem*, p. 113). It is, as Gilson himelf puts it elsewhere (p. 115), the spirit of the *explanations* which is different. Rather than calling to mind the Scholastic ideas, they therefore shed a new and wholly different light on natural processes and empirical facts. It is only when one takes a first and superficial glance at them that they seem to be similarly described by Descartes and contemporary Scholastics.

[57] Descartes, *Les Météores*, AT VI, p. 239. See also below, chapter 5, notes 86 and 87.

perspective: causality as an omnipresent pressure “from without”. In the next chapter, we shall deal with the Cartesian type of explanation in more detail.

CHAPTER FIVE

THE MECHANICAL ALTERNATIVE: FROM ENDS TO IMPULSE

In the previous chapter, I maintained that from Schoock's identification of sound with moving air, the step towards a complete mechanical theory of perception was only a very small one. It did not seem so to Schoock himself, who violently attacked Descartes in the *Admiranda*. Yet it does seem so to the modern reader, who ponders the coincidences between Schoock's representation of sound as a propagated movement all the way up to the eardrum, where it is "sensed" by the mind,[1] and Descartes' account of a mind reading off the mechanical information. Nevertheless, Schoock never makes the metaphysical step of denying the material world its real, sensible, qualities.

A similar set of likenesses and differences arises in relation to Schoock's explication of qualitative phenomena in *causal* terms. Schoock denies that air is the efficient cause of sound. This, in itself, is wholly un-Cartesian. For Descartes, the movement of air is the physical, efficient, cause of the mental effect of hearing a sound. Yet, if we confine our attention to the physical realm for a moment, we may find interesting similarities. According to Schoock for instance, the movement of a bell is not the efficient cause of its ringing. The movement is itself part of the ringing. It is the one who rings the bell, "the sacristan", who is the efficient cause. "For", says Schoock, "when the bell is moved, the ringing is heard." The qualitative phenomenon of sound thus needs a mechanical cause only, an efficient cause that forces motion upon an object.

Hence, the immediate cause of the sound is no longer localised in the "sonative action", or "sonation" of the squeaking air of the Conimbricenses. It is simply the man moving the bell through local motion—a motion, that is, involving no substantial change, but only a change with respect to the *where* of the object. Discarding the quality of sound, Schoock at the same time makes room for a non-qualitative account of the *causes* of things heard. These causes are purely

---

[1] Schoock, *De Natura Soni & Echus*, p. [8]: "aër movet tympanum, & tympani motus ab anima sentitur."

mechanical: the movement of the sacristan's hand causing a movement of the rope; the rope which in its turn causes the bell to move, the bell moving the surrounding air, the air the ear-drum.

It is this type of *causal* explanation that Descartes radicalised as well: the mechanical description of the Universe by way of a theory that sidesteps individual sources of action. And where individual sources of action are rejected, substantial forms are rejected too. The mechanical account of causality thus involves a loss of substantiality. Since in this account, there is no localisation of real qualities in individual natural bodies, there is neither a localistion of the source of these qualities, being the substantial form that characterises each individual type. It is precisely these aspects of Cartesian theory which Schoock's teacher Voetius in 1641 tried to neutralize by writing his essay "On the Natures and Substantial Forms of Things".

5.1 The Loss of Substantiality

To appreciate the impact of the mechanistic loss of individual centres of causation, let us return for a moment to the meteorological example which we analysed above: the formation of ice. Descartes exclusively introduces images of particles of matter influencing other particles of matter in his theory concerning the formation of ice, whilst Schoock regards the cold as a *real quality* having causative powers.

5.1.1 Forms and Conditions

According to Schoock, the qualities that, contrary to sound, *do* inhere in natural objects are caused by the natural object's internal powers. Heat and cold for instance, naturally inhere in different substances. As for the elements, water, for instance, is cold. This is also what we learn from *De Aqua*, Schoock's 1646 disputation on the subject of water. According to that text, coldness is even the only unique quality of water:

> Amongst the tactile qualities of water, Humidity fits [Water] to a lesser degree, since Water is not so easily brought within the boundaries of something else as is Air, although more easily than the other elements. Coldness however, [fits Water] to the highest degree, [first] because, of all the elements, Water works the most through coldness, and thus also withstands Fire the most because of its coldness. [Coldness also fits Water to the highest degree] because, when Water which has been heated is removed from Fire, it is

> restored to its native coldness, not by an external cause, but by its proper Form, from which this property of Water originates.[2]

When the pot is put on the fire, the water it contains will heat. Take it off and the water will regain its previous, "natural", coldness. It is on the basis of the hypothesis that water becomes hot only under the "unnatural" influence of outer conditions, that the cause and the effect of the heating of water are called *external*. Internal causes and effects are those which obtain when external influences are taken away. The object then acts only according to its proper *Form*. This is in fact a very basic function of the substantial form. When Gisbertus Voetius, in the essay *On the Natures and Substantial Forms of Things*, offers his arguments in favour of the forms, the first deals with the specificity of individual actions according to individual qualities:

> The first argument concerns the proper actions of natural things, each of which stems from its distinct perfections and qualities.[3]

The qualities intended here are *real* qualities, for it is only these which exhibit *proper action*. Take the case of heat: heat may well be a property of water. The water, for instance, which I am presently cooking has the quality of heat. But according to Scholastic theory, such a property is not a *real quality*. Real qualities are those which may be observed as belonging to the object when it does not occur in such extraordinary circumstances. Heat is therefore not a real quality of water, but cold is.

The reason why Voetius brings "distinct perfections and qualities" (and the actions resulting therefrom) in relation with the concept of form, is because the substantial form is what accounts for the existence of real qualities. Contrary to the qualities which are brought about under the influence of external circumstances, real qualities are in want of explanation. It is, for example, obvious what causes the water to be hot: it is the fire. Temporary, externally prompted qualities need no further explanation. But when external conditions are taken away and the real qualities (re-)appear, what, then, is *their* cause? The idea is that there must be a hidden cause which is responsible for the real qualities. This "cause" is the substantial form. Instead of working from without, like fire, the proper form of water cools it from within, once outer conditions are removed. We have already seen that the

---

[2] Schoock, *De Aqua*, Thesis XXII.

[3] Voetius, *Narratio*, p. 42 / *Select. Dispp.* I, pp. 874-875 / *Responsio*, p. 21 / *Querelle*, p. 108.

substantial form which Voetius defends against Descartes is not merely the source of real qualities: it accounts for its action and its perfection as well. The argument however, is in all cases the same: the substantial form is the inner cause of properties, actions, and of the development of a natural object considered apart from external influences.

The notions of real quality and substantial form are thus linked to a theory of causation in which individual action and external conditions are sharply distinguished. On the other hand, the external influence upon an object may be described in terms of the proper action of another. Thus, in the case of fire boiling water, the "calefactory action" is ultimately to be attributed to the form of the fire as its cause. Although the water is forced unnaturally to become hot under external influences, the action is, in respect of the fire, the result of an inner principle. In the end, all action is to be ascribed to such individual centres of causation. But what happens to causality, when, with Descartes, substantial forms and real qualities are taken away?

5.1.2 Individual and External Causation

The difference between the Scholastic idea of causation and the Cartesian is accurately described as a difference between the ideas of internal and external causation. In fact, the unique characteristic of Descartes' theory of causation is the idea that *every change is brought about by an external cause*. Descartes does not say that there are no internal causes, but, as we have seen, keeps silent on the question of the forms and the qualities accompanying them. We have already analysed a meteorological example—the formation of ice—illustrating the two conflicting ways of thinking. Since, according to Descartes, we may suppose that (i) everything on Earth and in the air consists of material particles; (ii) the pores of these particles and the intervals between them are filled up with yet other particles; (iii) the particles may be of regular or irregular shape; worm-like parts for watery substances and twig- and branch-like parts for solid bodies; and (iv) heat and cold are nothing but a faster or slower particle-movement, we may imagine how, given certain conditions (for instance, that it is nighttime, in winter, high up in the mountains), ice is formed because the subtle matter is too slow to work its way through the worm-like parts of water.

The important point is, that in such an explanation, natural processes are no longer described in terms of the natural vs. the unnatural. Even the natural, that is, "normal" case is a conditional one. Water for instance, is normally cold, that is, when it is not put on a fire in order to be heated. For the Scholastic thinker, water is therefore cold "of itself", that is, without outer conditions influencing things in an "unnatural" way. According to Descartes however, the water is not cold "by itself", but only in as far as meteorological and other external conditions allow it to have the temperature it has. Everything thus happens according to the surrounding conditions. To put it differently, we might say that, according to the Cartesian view, every event is caused *under pressure of circumstances*. The world is governed by the unnatural, external causes and effects previously reserved for artificial processes only. This is, as far as causality is concerned, the crux of the mechanical theory: natural processes are not explained by attributing forces or powers to individual centres of causation, but by referring only to the outer conditions in which the processes occur.

5.2 A Non-Finalist Physics

A serious consequence is that, by its disregard of the forms, such a description of Nature also skims over the ends these forms are striving for. The mechanistic loss of individual centres of causation thus stands on a par with the loss of individual *ends* of substantial motion. Substituted for these ends, is a non-teleological chain of causes necessitating events without any final objective. The idea of external conditions resulting in effects which in turn function as new conditions enforcing new effects, renders it impossible to ascribe *ends* to individual processes. Finalisms are therefore absent from the mechanical perspective—a difference nicely illustrated in the two conflicting accounts of the formation of ice which we examined earlier. Necessity takes the place of finality.

5.2.1 From Ends to Impulse

The substitution of finalistic considerations by mechanical views occurs at each of the various levels of finalism that we came across in chapter 3. To take only one example: Schoock's *De Aqua* explains the shape of water drops. According to this text, the most important reason for the shape of water drops seems to be the one

> which is inferred from the holy shrines of first Philosophy, [namely that there is a] natural Goodness, by which Water preserves itself and holds out against the enemy which it shuns. It cannot, however, hold out in a more convenient way than by its spherical form, which it in

> fact keeps in order that the water does not scatter and so be conquered and destroyed.[4]

The spherical shape is a form of self-preservation; the parts keep together in order to avoid a qualitative leap by which negligible amounts of water would be transformed into the element in which they are mixed. This act of self-love echoes both sides of the kind of finality we so often find in Scholastic physics: the natural tendency to strive after an individual goal on the one hand and the explanation of this goal in terms of a deliberate choice for what is best on the other—the First Cause imposing the *Bonitas naturalis* on the process.

Dismissing all "mental" notions from natural philosophy, Descartes, both in *Les Météores* and in the *Principia* of 1644, explained the spherical shape of water-drops quite differently, trying to get rid of ends and aims along with the notions of deliberate ratiocination. He does not completely manage to do so, since, in his account, the celestial globules always move along straight lines *as best they can*—which is a pretty mentalistic way of putting things.[5] The idea is, that the celestial globules, which dart through the air this way and that, are least deflected by spherical objects. The explanation still owes much to considerations of a final character: water drops are spherical *in order to* let celestial globules pass. Descartes' account is, however, immediately followed by the following consideration:

> If for instance there is a part on the surface of the droplet that extends beyond the spherical shape, the celestial globules dashing about through the [surrounding] air will fling at it with more force than at the others, and thus push it towards the centre of the droplet.[6]

In this argument, all events are explained according to the laws of particle movement and in terms of outer conditions determining the process by which water particles are lined up in the form of spherical drops. Instead of an inner tendency or endeavour to keep together, it is only the external circumstances by which material particles are forced to act in a certain way.

The idea of an inner tendency being abandoned, there is neither any occasion to interpret the spherical shape in terms of an end directing

---

[4] Martin Schoock / Christoph Wittich, *Disputatio Physica de Aqua*, Thesis XIV.

[5] *Principia Philosophiæ* IV 19, AT VIII-I, p. 211: "[...] isti globuli coelestes [...] [semper] quantum possunt secundùm lineas rectas, vel ad rectas quam-proximè accedentes, moveantur [...]."

[6] Descartes, *Principia Philosophiæ* IV 18, AT VIII-I, p. 211.

the process. The physical aim, or end, in its Aristotelian form of a natural terminus of physical action has thereby become redundant. The example clearly indicates the way in which the road was paved for a type of physics according to which natural causes were not governed by aims and ends. Causality, for Descartes, becomes mechanical causality of circumstance. The consequence of this "mechanical" view is that the world becomes a machine, running according to invariable, material regularities. More importantly, the rejection of teleology led to the rejection of the idea which in Renaissance thought had always accompanied it: the notion of a natural goodness imposed on physical processes by a benevolent God.

5.2.2 Descartes on Final Causality

In principle, the Cartesian account of external causation might still leave room for considerations concerning the benevolence of the Creator who first made the process, but since the natural τέλος of Aristotelian physics was not distinguished in any precise way from the end imposed on physical processes from without, teleology and finalism were regarded as two sides of the same coin by both Aristotelians and their adversaries like Descartes.

At the start of *Principles* III, Descartes writes that we should keep in mind that God is infinitely powerful and good, and therefore

> not be afraid that our imagination may over-estimate the vastness, beauty and perfection of His works.[7]

We must never underestimate Him. At the same time, we should not overestimate ourselves and limit His possibilities, thinking that God has put everything together only because of us, or fancying that we can grasp what goals He laid before Himself when creating the Universe. Since we do not know God's motives, they should play no role in physics. Thus, Descartes' arguments are formulated in terms of human pride. Although God's works cannot but give rise to our highest esteem, we should not have too high an opinion of ourselves and

> suppose that we understand the ends that God has aimed at in creating the world.[8]

---

[7] Descartes, *Principia Philosophiæ* III 1, AT VIII-I, p. 80. The translation is from *The Philosophical Writings of Descartes*, Translated by John Cottingham, Robert Stoothoff and Dugald Murdoch, Cambridge (C. U. P.) 1985 (also to be referred to as "CSM") I, p. 248.

Descartes criticises those who argue that things are as they are because in that way they are best for us:

> It is a common habit of men to suppose that they are the dearest of God's creatures, and that all things are therefore made for their benefit. They think that their own dwelling place, the Earth, is of supreme importance, that it contains everything that exists, and that for its sake everything was created. But what do we know of what God may have created outside the Earth, on the stars and so on?[9]

Descartes' insistence on man's relative unimportance has been interpreted as an illustration of "the most strikingly 'modern' feature of his cosmology"[10], and indeed, both in the quotation just cited and in the *Principles*, where Descartes says that "many things exist or once existed [..] which have never been of use to anyone"[11], his "modern" conception of a vast and evolving Universe seems to involve scientific claims that previous writers might not have claimed.

Descartes' originality, however, does not lie in his formulation of these arguments, some of which are literally taken from the writings of Montaigne, whose defence of Raimond Sebond—in *Essays* II 12—had been centered around the very issue of man's quest to reach faith by the use of natural reason and so, amongst other things, of reading divine providence into natural causation.[12] Montaigne had also linked

---

[8] Descartes, *Principia Philosophiæ* III 2, AT VIII-I, p. 81 / CSM I, p. 248.

[9] Descartes, "Conversation with Burman", AT V, p. 168 / Descartes, René, *Entretien avec Burman*, Texte présenté, traduit et annoté par Ch. Adam, Paris (Librairie philosopique J. Vrin) 1975², pp. 90-91. The translation is taken from John Cottinham, *Descartes*, Oxford(Basil Blackwell) 1986, p. 97. See also: *Descartes' Conversation with Burman*, trans. with introduction and commentary J. Cottingham, Oxford (Clarendon) 1976, p. 36, and: CSM III, p. 349.

[10] John Cottingham, *Descartes*, Basil Blackwell, Oxford 1986, p. 97.

[11] Descartes, *Principia* III, 3, AT VIII-I, p. 81 / CSM I, p. 249.

[12] In his critical edition of the *Discours de la Méthode*, Étienne Gilson has pointed out a host of passages in which Descartes quotes or paraphrases Montaigne's text. See René Descartes / Étienne Gilson, *Discours de la Méthode*, Vrin, Paris 1976⁵. Basing his study on Gilson's, Léon Brunschvicg weighed the similarities between the *Essais* and the Cartesian oeuvre in his *Descartes et Pascal: Lecteurs de Montaigne*, Brentano's, New York 1944, which has deservedly become a classic. Brunschvicg does not so much offer a detailed analysis of quotations, as describe the coincidences of the subject-matter and the general background of intellectual issues, on which all further studies on the influence of Montaigne on Descartes have been based. Richard H. Popkin, in *The History of Scepticism from Erasmus to Descartes*, Van Gorcum, Assen 1960, goes even further and sees Descartes' philosophical system as a deliberate reaction to the

this way of thinking to the idle habit of making man's interest the centre of divine government.[13] The idea, finally, of a plurality of worlds may be found in Montaigne's *Apology* as well, where it is offered as a probable opinion common to the most worthy of ancient thinkers and even accepted by some Christians.[14]

It was his exceptional usage of this type of argumentation which marks out Descartes' originality. Instead of epistemological pessimism, it was optimism that led him to abandon the idea of natural goals: the optimism that he would be able to describe the physical

---

"sceptical crisis" instigated by writers such as Montaigne and Charron. In the famous *Apologie de Raimond Sebond*, Montaigne does not stop criticizing the presumptuous habit that men have of "reducing God to their measure." See Montaigne, *Apologie de Raimond Sebond*; *Essais* II, 12, édition Villey, Paris (P.U.F.) 1978^3, Tome I, p. 528 (A).

[13] See e.g. *Essais* II, 12, Villey pp. 528-529: "(B) Et est cause de ce qui nous advient tous les jours d'attribuer à Dieu les evenemens d'importance, d'une particuliere assignation. Parce qu'ils nous poisent, il semble qu'ils luy poisent aussi, et qu'il y regarde plus entier et plus attentif qu'aux evenemens qui nous sont legiers ou d'une suite ordinaire." What Montaigne rejects is our habit of thinking that experiences of harms and benefits are of importance to God, which is closely linked to the habit of regarding mankind to be the centre of the Universe and his well-being to be the aim of physical processes. It was in this form that Spinoza was later to develop the argument against final causation. Cf. Spinoza, *Ethica*, I, *Appendix*. and we find it in Descartes also. In *Principia* III 2 it is said that "[It would be the height of presumption] to imagine that all things were created by God for our benefit alone." Descartes, *Principia* III 2, AT VIII-I, p. 81. The translation, of which the first part is a paraphrase of the original, is taken from CSM I, p. 248. Obviously, the argument is directed against the Scholastic idea of accepting *causes immanifestes* within explanatory theory such as we came across in the various examples from Schoock's *De Aqua*. Just as Montaigne had criticized the idea of a natural theology on the basis of a purposeful Nature, so Descartes criticises those who argue that things are as the are because in that way they are best for us.

[14] "Ta raison n'a en aucune autre chose plus de verisimilitude et de fondement qu'en ce qu'elle te persuade la pluralité des mondes. [...] Les plus fameux esprits du temps passé l'ont creue, et aucuns des nostres mesmes, forcez par l'apparence de la raison humaine." Montaigne, *Essais*, Villey I, p. 524 / Screech, p. 586. Although the context is rather different from Descartes'—none of the Scholastic forms of final causation being referred to—the notion of a plurality of worlds yet again forms part of a critique of human knowledge-claims and of reducing God to human measures. The privilege accorded to our particular planet is (in Léon Brunschvicg's paraphrase of Montaigne's viewpoint) "un premier témoignage de cette incurable 'présomption, notre maladie naturelle et originelle'". Léon Brunschvicg, *Descartes et Pascal: Lecteurs de Montaigne*, New York (Brentano's) 1944, p. 70. Descartes' reference to the plurality of worlds may thus be traced back to a long tradition of authors denying the possibility of inquiring into God's purposes and motivations.

world in mechanical terms alone. Descartes employs the sceptical arguments against cognitive presumption in a rather opportunistic way. The Scholastic identification of different levels of final explanation—and the Renaissance preoccupation with the benignly and well-ordered design of the Universe—obscured the fact that more than one concept of "final causes" was at stake. Final explanations referring to God's Goodness and Wisdom and man's central place in the Universe was one type. Descartes however used the sceptical critique on such knowledge claims in order to exclude any type of final explanation.[15] It was no concern of his to limit the use of reason. The sceptical critique served him in a very different way, namely to clear himself of the Aristotelian set of causal determinants and their Scholastic derivatives such as real qualities and the substantial form.

Thus, Descartes made room for a physics independent of final considerations: a physics entirely based on new model of causation: the mechanical efficacy of impulse.

5.3 The Mechanical Theory of Causation

Descartes' optimism was shared by his Dutch ally Henricus Regius. In Utrecht, Regius had impressed on his students that "mechanics alone is our [type of] physics".[16] Regius even drilled his pupils into citing in verse:

---

[15] In the *Meditations*, Descartes sums up this position in the following terms: "Cum enim jam sciam naturam meam esse valde infirmam & limitatam, Dei autem naturam esse immensam, incomprehensibilem, infinitam, ex hoc satis etiam scio innumerabilia illum posse quorum causas ignorem; atque ob hanc unicam rationem totum illud causarum genus, quod a fine peti solet, in rebus Physicis nullum usum habere existimo; non enim absque temeritate me puto posse investigare fines Dei." Descartes, *Meditationes* IV, AT VII, p. 55 / CSM II, p. 35. See also the *Fifth Replies*, AT VII, p. 375 / CSM II, p. 258, where Descartes restates the point, and: John Cottingham, *Descartes*, p. 98, who refers to both passages. There were not only sceptical authorities for such a viewpoint, but in fact a much vaster group of philosophers and theologians limiting the power of natural reason. An author often cited by Montaigne in relation to the question of human presumption regarding knowledge of divine things, is St. Augustine. Montaigne, *Essais* II, 12, Villey I, pp. 449, 499, 531, 535, 539 and 554 / Screech, pp. 501, 556, 595, 600, 605 and 623. It nevertheless seems likely that it was the sceptical revival of the late sixteenth century which for Descartes supplied a useful body of argumentation on this point. Descartes in any case bases his objections to final causality exclusively on arguments similar to those of Montaigne against human presumption.

[16] "Mechanica sola est Physica nostra": Henricus Regius, *Fundamenta Physices*, Amstelodami, Apud L. Elzevier, 1646, p. 67. A similar remark was quoted in Schoock's *Admiranda Methodus* of 1643, p. 133 / *Querelle*, p. 246, and appears to

Mind, measure, rest, movement, position and shape,
Are together with matter the source of all things.[17]

But what exactly did this type of "mechanics" involve?

5.3.1 *The Machine Metaphor*

The idea of a "mechanical" physics was partly based on the confidence which philosophers such as Descartes and Regius had in using the metaphor of man-made machines for describing natural objects. As Descartes had proposed in the *Discours de la Méthode*, the animal body might well be compared to a "machine", even if, being made by the divine Hand, it is

> incomparably better organised and has in itself more admirable movements than any of those which may be invented by humans.[18]

The unmistakable differences between natural objects created by God and man-made automata or machines were, in the seventeenth century, accepted by all philosophers alike. The irony however is, that these differences were just as commonly put forward in defence of the mechanical views as they were used in opposition to it. For those who were convinced of the advantages of the mechanical viewpoint, the obvious differences between God's creations and those of mankind might well be straightforwardly accepted, but explained in terms of gradation only. In the *Responsio*, Regius for instance admits that "artificial things certainly differ from those called natural". Prompted by Descartes, he immediately adds, however, that they differ "according to the more and less, or to the degrees of perfection." From that point of view, it was easy to maintain the mechanical position whilst at the same time conceding to Voetius that the works of God far exceed those of mankind:

> For the creatures which are begotten every day, even the most insignificant, are made according to the mechanical laws with such admirable workmanship, that even the most ingenious automata constructed by man can in no way approximate their precision

be taken from one of Regius' earlier works, probably the so-called *Dictata Physica* which Regius wrote for his students.

[17] "Mens, mensura, quies, motus, positura, figura, / Sunt cum materia cunctarum exordia rerum." Henricus Regius, *Physiologia sive cognitio sanitatis*, I, I, p. 5 and *Responsio*, p. 19. See also Schoock, *Admiranda Methodus*, p. 198 / *Querelle*, p. 282.

[18] Descartes, *Discours de la Méthode*, AT VI, p. 56 / CSM I, p. 139.

> (ἀκρίβεια). This is obvious in a clock, for instance, the very few cog-wheels of which can in no way be compared to the innumerable bones, nerves, arteries or even the blood and spirit of even the most insignificant little animal.[19]

It is the paradox of mechanicism that its main thesis was justified on the very grounds on which the adversaries of the mechanical theory sought to reject it. The argument, which was dictated to Regius by Descartes in January 1642,[20] may be found even more sharply in Descartes' *Discours de la Méthode* of 1637, where the difference between things natural and artificial had already been used as an argument in favour of mechanicism. Descartes explains that, if mankind could make such admirable automata with the use of such a small number of parts, then what could not God contrive mechanically by "the multitude of bones, nerves, arteries, veins and all the other parts" of which the body of each animal is made?[21] The material incongruity of the parts involved and the incomparable excellence of the Artist of natural things actually made the mechanical view of Nature only the more likelier.

For Voetius and his Groningen ally Martin Schoock on the other hand, who were quick to renounce the mechanical viewpoint, the idea of regarding natural objects as machines was simply blasphemous. As Voetius wrote in his essay on the forms:

> It would follow that the proper and intrinsic faculties and the principles of these in animals would be of no other type than those in automata, or Dedalian statues,[22] and that, as a consequence, the works of God and of Nature, which are brought about through creation and generation, would be essentially and univocally the same as works of art. I must say that I do not yet see how this could quite harmonize

---

[19] Regius, *Responsio*, p. 18.

[20] Having written a considerable amount of what appeared as Regius' answer to Voetius in the *Responsio*, Descartes is also responsible for most of the passage quoted above. Descartes to Regius, January 1642, AT III, p. 504 / *Rodis-Lewis*, p. 88: "quia paucæ illæ rotæ in horologio cum innumeris assibus, neruis, venis, arterijs &c. vilissimi animalculi nullomodo sunt comparandæ."

[21] Descartes, *Discours de la Méthode*, AT VI, pp. 55-56.

[22] Dedalus is mentioned in both Plato and Aristotle as the inventor of various extremely realistic robots. See Theo Verbeek, *Querelle*, p. 469, note 59.

with Psalms 104:[7, 29, and 14-15][23], Numbers 16:22 and and 27:16 and Hebrews 11:9-10.[24]

In all of these verses God is presented as the governor and administrator of animal life and mentioned as its craftsman (*artifex*) and author (*conditor*)[25]. Voetius apparently regards the wonders of Creation and generation as being indicative of the abyss which lies between God's creativity and human mechanical contrivance. As his pupil Martin Schoock would argue, there is no need to think that God, almighty and all-wise as we know Him to be, would submit Himself to the few laws of mechanics that enable us to produce some useful instrument.[26]

Nevertheless, having acknowledged the great—albeit gradual—differences between the works of art and of Nature, Regius sees no fundamental objection in these Biblical verses against the view that regards God as the ultimate mechanic. Regius self-assuredly quotes Voetius' verses in full.[27] The fact however, was that whether, intuitively speaking, the idea of a God Who follows the laws of mechanics was the likelier theory or not, the applicability of the machine-metaphor still had to be proven:

> Just let them take a pigeon and compare it to a clock!,[28]

the Groningen professor of philosophy Martin Schoock urges the Cartesians. Emphasizing that he is quite serious this time and not just making a rhetorical remark,[29] Schoock puts his finger on the spot and demands that it be demonstrated in detail that mechanical principles do indeed govern the workings of natural objects—something which, for that matter, Schoock holds to be quite impossible for practical reasons alone.[30]

However, notwithstanding the charm and appropriateness of Schoock's demand, it was Descartes himself who had already made it

---

[23] Reading *Psalms 104:7, 14-15 and 29* for "Ps. 104 29. & 7. 14. 15." See below, chapter 10, note 7.

[24] Voetius, *De Rerum Naturis & Formis Substantialibus*, in *Narratio*, p. 41 / *Select. Dispp.* I, p. 874 / *Responsio*, p. 13 / *Querelle*, p. 107.

[25] I shall discuss all of these verses in section 10.1, below.

[26] Schoock, *Admiranda Methodus*, p. 133 / *Querelle*, p. 246.

[27] Regius, *Responsio*, pp. 18-19. See also section 10.1, below.

[28] Schoock, *Admiranda Methodus*, p. 134 / *Querelle*, p. 247.

[29] Schoock, *ibidem*: "Non calumnior, aut calumniari didici."

[30] Schoock, *ibidem*: "maximus adhuc supererit labor, quo nunquam poterunt defungi."

his task to offer precisely this type of demonstration in the *Discours*, where he elaborately deals with the mechanics of the heart. It will be necessary to examine the way in which his account of the heart and of blood circulation makes use of the machine metaphor.

5.3.2 The Movement of the Heart

Although it is now mostly read as an independent philosophical classic, Descartes' *Discourse on Method* was originally written as an introduction to the three essays, or "samples" of the new method, which, in 1637, Descartes published along with the *Discours* itself.[31] One important consequence of present-day neglect of this fact is that the elaborate and rather technical description of the heart and blood-vessels in Part V of the *Discours* may appear to be rather misplaced, or at least tedious in the light of the deeper philosophical subject-matter of the fourth part or the methodology of parts two and six. Once the fact is acknowledged that the *Discours* is meant as a prelude to the treatises on geometry, optics and metereology that follow, various interpretational difficulties may be solved. One of these is Descartes' seemingly rather too detailed description of the workings of the heart.

5.3.2.1 The Mechanical Heart

The description of the human heart serves as an example of the type of research Descartes promises his readers as an application of his method. Were it not for his extreme caution following Galileo's trial, Descartes would by this time have published many more samples of his way of doing natural philosophy, that is, in the unpublished work *Le Monde*. As things stand in 1637, he offers three essays and, in the preceding *Discours*, indicates what his method may yield if it is to be applied to a broader area of inquiry.

In this first outline, the example of the heart plays a central role for its clarity in demonstrating the force of a "mechanical argument". Part V of the *Discours de la Méthode* discusses in some detail the

[31] Descartes, *Discours de la Méthode*, AT VI, p. 74. Having informed the reader of his decision to postpone the project of *Le Monde* three years earlier, Descartes goes on to say that he now has new reasons "qui m'ont obligé à mettre icy quelques essais particuliers, & a rendre au public quelque copte de mes actions & de mes desseins." Cf. CSM I, p. 149.

workings of the heart of any "large animal which has lungs".[32] It is at this point that Descartes explicitly draws the mechanical comparison between the heart and a clock:

> the movement which I have just explained follows from the mere arrangement (*diposition*) of the parts of the heart (which can be seen with the naked eye), from the heat in the heart (which can be felt with the fingers), and from the nature of the blood (which can be known through observation (*qu'on peut connoistre par experience*)). This movement follows just as necessarily as the movement of a clock [follows] from the force, the position (*situation*) and the shape of its counterweights and wheels.[33]

What is at first sight a disturbing element in this comparison is the fact that Descartes introduces various qualities. Does he accept a "quality" of heat in the blood and a "nature" of the blood itself apart from the supposedly mechanical process? And in what way can these be compared to a clockwork mechanics? In order to answer these questions, let us first analyse the Cartesian idea of the heart's functioning.

5.3.2.2 Descartes and Harvey

Both present-day and seventeenth-century physiology distinguish two distinct periods in the beat of a mammal heart: the systole and the diastole. In order to explain these in present-day terms, it is necessary to see that a human heart has four major compartments: two auricles and two ventricles. What happens during the systole, is that both the left and the right auricles contract. They thereby propel the blood which they have received from the connecting veins into the respective ventricles. Being filled with blood the left and right ventricles in turn contract almost immediately, propelling the blood into the aorta and the pulmonary artery respectively. In the subsequent diastole, the heart is at rest, whilst the auricles and ventricles are again filled with blood from the veins.

In the fifth part of the *Discours de la Méthode*, Descartes rather elegantly describes this whole process. He rightly criticizes some

---

[32] Descartes, AT VI, p. 47. Descartes' descriptions of the workings of the heart may be found in the *Discours*, AT VI, pp. 46-55; in *L'Homme*, AT XI, pp. 123-126; and in *Description du corps humain*, AT XI, pp. 228-244.

[33] Descartes, *Discours* V, AT VI, p. 50. Translation from CSM I, p. 136.

inconsistencies in contemporary terminology[34] and pays tribute to William Harvey, who had been the first to discover the circulation of the blood.[35] However, Descartes refers to Harvey—the "English doctor" who "broke the ice on these matters"—only with regard to the idea of the circulation of blood in animal bodies and not with regard to the motion of the heart itself. He does so with good reason, for on the point of explaining the heart's motions, Harvey and Descartes disagreed in fundamental ways.

In accordance with present-day theory, William Harvey saw the heart as a muscle—an idea rejected by Scholastic authors for the reason that our will does not have the same influence on the motions of the heart as it has to move muscles.[36] According to Harvey however, the active phase of the heart's movement, the systole, is to be seen as a muscular contraction forcing blood into the arteries. In the

---

[34] Descartes argues that, being an artery, the pulmonary artery is unjustly called "arterial vein". Also, the "venal artery" should not be so called, since it is in fact a vein, which we today call the pulmonary vein. See *Discours* V, AT VI, p. 47. For an explanation of the old terminology, see Étienne Gilson, "Descartes, Harvey et la scolastique", in *Études sur le rôle de la pensée médiévale dans la formation du système cartésien*, Paris (Vrin) 1984[5], p. 56. William Harvey had arrived at the same conclusion before Descartes, yet both Descartes' hypothesis of a circulation of the blood throughout the body and his ideas concerning the nature of the pulmonary artery and pulmonary vein seem to have been developed independently of Harvey. Cf. Gilson, *idem*, pp. 81-82.

[35] Descartes, *Discours* V, AT VI, p. 50. For other positive references by Descartes to Harvey, see Gilson, "Descartes, Harvey et la scolastique", in *Études*, pp. 75-76. Gilson concludes (p. 80) that Descartes "a immédiatement compris la valeur immense de la découverte de Harvey; il a tout fait pour qu'elle fût connue et pour que tout mérite en revînt à son véritable auteur." William Harvey (1578-1657) had published his findings in his 1628 *De motu cordis et sanguinis in animalibus, anatomica excercitatio.*

[36] William Harvey, *Exercitatio Anatomica de Motv Cordis et Sanguinis in Animalibus*, Francofvrti, Svmptibus Gvilielmi Fitzori, 1628, p. 22: "Ex his mihi videbatur manifestum; Motum cordis esse tentionem quadam ex omni parte, & secundum ductum omnium fibrarum, & constrictionem vndique, quoniam erigi, vigorari, minorari, & durescere in omni motu videtur, ipsiusque motum esse, qualem musculorum, dum contractio sit secundum ductum partium neruosarum, & fibrarum, musculi enim cum mouentur, & in actu sunt vigorantur, tenduntur, ex mollibus duri fiunt, attolluntur, incrassantur, & similiter Cor." See also: Gilson, "Descartes, Harvey et la scolastique", in *Études*, p. 55, where the Scholastic objection is referred to.

diastole, a dilation occurs as the result of a muscular relaxation, allowing blood to re-enter the heart from the veins.[37]

Descartes held a wholly different view. Taking up an idea which was at least as old as Aristotle and Galen[38], he argued that the heart is the source and seat of bodily warmth. Indeed, "one can feel with one's fingers", Descartes writes, that

> there is always more heat in the heart than in any other part of the body.[39]

This erroneous idea was coupled to others equally mistaken, in order to arrive at the notorious Cartesian theory in which the heart is presented as something not unlike a combustion engine.[40] What happens, according to Descartes, is that blood enters the auricles and ventricles from the *vena cava* and the pulmonary vein. The valves between these veins and the ventricles easily let the blood through

---

[37] Harvey, *De Motu Cordis*, p. 23: "Vnde qui motus vulgo cordis Diastole existimatur, reuera Systole est." In the systole, the blood is propelled into the arteries as the result of cordial contraction. See *idem*, p. 30: "Motus itaque cordis omnino ad hunc se habet modum, & vna actio cordis est ipsa sanguinis transfusio, & in extrema vsque, mediantibus arteriis propulsio, vt pulsum, quam nos sentimus in arteriis, nil nisi sanguinis à corde impulsus est." See also: Gilson, "Descartes, Harvey et la scolastique", in *Études*, pp. 64-73.

[38] Annie Bitbol-Hespériès rightly argues that neither Aristotle nor Galen accepted a mechanical explanation of heat. Annie Bitbol-Hespériès, *Le principe de vie chez Descartes*, p. 71. Yet to say, as she does, that Descartes' originality lies in the fact that he saw this heat as being a *natural* type of heat, whilst (1) Galen introduced a specific type of heat in the case of the heart, since he calls this heat "innate" and (2) Aristotle distinguishes it from "normal fire", seems to me to be exaggerating the point. Descartes himself in fact introduces a specific kind of heat for the heart: the heat of fermentation: a "fire without light". See *Discours* V, AT VI, p. 46 / CSM I, p. 134. Although this heat does not differ substantially from natural heat, nor do those mentioned by Aristotle and Galen. In any case, the discussion is beside the point: reference to the supposed phenomenon of the heart being the siege of a natural heat is at least as old as Aristotle and Galen, whatever the interpretation, mechanical or otherwise.

[39] Descartes, *Discours* V, AT VI, pp. 50 and 48 respectively / CSM I, pp. 136 and 135.

[40] The similarity between the Cartesian idea of the heartbeat and the workings of athe internal-combustion motor has often been noticed. Étienne Gilson writes that Descartes "[conçoit] le coeur comme une sorte de moteur à explosions." See Étienne Gilson, "Descartes, Harvey et la scolastique", in *Études*, p. 83. G. A. Lindeboom compares the Cartesian heart with an "expansion motor". See Lindeboom, *Descartes and Medicine*, Amsterdam (Rodopi) 1978, p. 72. Both are cited by Annie Bitbol-Hespériès, *Le principe de vie chez Descartes*, p. 94, note 3.

because they open to the inside of the heart. Once the blood has entered the ventricles, however, it starts to evaporate as a result of the heat inside the heart. Taking up more space, the expanded blood pushes back the valves through which it entered the heart, whilst at the same time it pushes open those valves which are at the base of the arteries. The blood is thus propelled into the arteries and the heart shrinks back. The pressure now being on the other side, the arterial valves are forced back again while the venal valves are opened by new, incoming, blood. Hence the process is repeated.

Descartes, in other words, makes the diastole the active phase instead of the systole. In the diastole, a "rarefaction" occurs as a result of heat. Not only is this theory at odds with observable facts, it is, as Étienne Gilson has pointed out, made up of various "*survivances anciennes*" from traditional sources.[41] Accordingly, it is not very fruitful to dwell on Descartes' account in view of his scientific achievement. The Cartesian theory is worth examining for another reason, however, namely the one Descartes himself gives in the *Discours*, that his theory of the heart's movement offers a pre-eminent example of what he purported a "mechanistic" explanation to be like.[42]

---

[41] Étienne Gilson, "Descartes, Harvey et la scolastique", in *Études*, p. 99. Gilson discusses the transformation of the blood in the heart, where it receives "spirits", as another Scholastic fossil in the rock of Cartesian theory. However, although Gilson's analysis of the sources for the Cartesian redescriptions of traditional views has been—and stillis—of great value, I agree with Annie Bitbol-Hespériès this should not lead to the conclusion that Descartes' conception of the heart's movements is simply a combination or amalgamation of traditional elements. Annie Bitbol-Hespériès, *Le principe de vie chez Descartes*, pp. 64-65. As in many other examples of natural philosophy, Descartes here uses traditional theories and traditionally accepted scientific data for a mechanistic re-interpretation which in itself is radically new. For a clear résumé of the various differences between, and similarities of, the scholastic, Cartesian and Harvean accounts of the phases of systole, diastole and pulsation, see Gary Hatfield, "Descartes' physiology and its relation to his psychology", in John Cottingham (ed.), *The Cambridge Companion to Descartes*, Cambridge (C. U. P.) 1992, especially pp. 340-344: "Descartes' physiological program and its relation to previous physiology".

[42] It was, accordingly, not Descartes' theory that lasted, but his method. See Annie Bitbol-Hespériès, *Le principe de vie chez Descartes*, p. 64: "Nous nous trouvons donc devant la situation paradoxale d'une physiologie, celle de Descartes, dont l'élément central en particulier, à savoir la conception du mouvement du coeur, n'a pas résisté aux travaux médicaux ultérieurs, mais a introduit tout un mode de pensée dans ces travaux eux-mêmes", and, p. 65: "Or la théorie cartésienne du mouvement du coeur n'a pas eu les mêmes prolongements dans le domaine médical, que dans l'histoire des idées."

5.3.2.3 The Economy of Explanation

What exactly are the mechanistic aspects of Descartes' theory? The first point to notice is that there are three basic elements in the Cartesian explanation of the workings of the heart. The first concerns Descartes' emphasis on the construction of the heart, the veins and the arteries and especially of the valves, or, as Descartes writes in the *Discours*:

> the eleven little membranes which, like so many little doors, open and close the four openings.[43]

Besides the element of construction, there is the important feature of the supposed heat existing within the heart. It is this heat that leads to a "boiling" of the blood. Thirdly, the "nature of the blood" itself is of importance, since it is on account of its natural properties that blood supposedly "boils", or "dilates" when heated.

Descartes was attacked by contemporaries especially on the second and third of these points: the possibility of heat leading to an *ébullition* of the blood. Having read the *Discours de la Méthode*, Descartes' inventive critic Plempius wrote to the French philosopher asking him how it could be that the considerable amount of blood in a human heart could be heated and rarefied in such a short period. Of course, oil and pitch are also very inflammable, but only when lighted by fire, which itself is far hotter than the inside of the heart. Moreover, fishes have only a small amount of heat, in fact they are cold rather than hot, and yet their hearts beat as regularly as ours.[44]

In his answer to Plempius, Descartes emphasizes that there are various types of rarefaction and that one must distinguish (1) the transmutation of a liquid into a vapor, (2) the gradual increase in volume, and (3) the immediate increase in volume of a liquid, of which three possibilities only the third applies to the rarefaction of blood in the heart.[45] Nature knows many instances of instant rarefaction, and the boiling of water and the "inflation" of milk at a certain temperature are only two of a great multitude of examples that

---

[43] *Discours*, AT. VI, pp. 47-48. The translation is from CSM I, p. 135.

[44] Plempius also offers arguments against the supposed connection between pulsation and presence of blood and denies that pressure as a result of the heart's dilation could cause an instant pulsation throughout the arteries. See Plempius to Descartes, 25 January 1638, AT I, pp. 497-499.

[45] Descartes to Plempius, 15 Februari 1638, AT I, pp. 528-529 / CSM III, pp. 82-82.

Chymia could offer. So why should not the blood rarefy when suddenly put in a warmer environment—as Descartes testifies having seen several times with his own eyes. Of course the temperature in the heart is not particularly high, and yet the quick ebullition may be brought about by a small remaining part of blood which was heated just before and which now works as a ferment upon the blood entering the ventricles anew.[46]

Thus, both the condition of heat and the nature of the blood remained two vital aspects of the Cartesian explanation. The third element is the one which forms its most obviously mechanistic part. Descartes meticulously describes all significant constitutive parts of the admirable mechanism of the mammal heart, especially the valves. Emphasis on construction or on the disposition of parts is of course a fundamental character of any theory comparing the workings of natural objects with those of machines. And yet with regard to the heart, it was precisely the aspect of its marvellous construction which at any time in the history of anatomical studies had roused the awe and wonder of the learned. A certain interest in the mechanical aspect, or the disposition of parts was therefore just as much present in contemporary Scholastic texts as it was in the *Discours*.[47] So what unique aspect remains that makes Descares' description decidedly mechanistic if there is no ban on qualitive description and if an interest in mechanical detail does not distinguish Descartes' views from alternative explanations?

The answer lies in the argumentative restrictions with which Descartes invests his explanations. When he says—in a passage already quoted above—that the movement of the heart follows from the mere arrangements of its parts, from the heat inside the heart and from the nature of the blood

---

[46] *Idem*, AT I, p. 530-531. The idea that part of the heated blood remains behind, "intimos [cordis] ventriculorum recessus replens" and works upon the entering blood as a catalyst, or a kind of yeast, "[vt] panis fermentum fieri solet ex parte farinæ iam fermentæ, vini fermentum ex vuarum reliquijs, & cereuisiæ (= *cervisia*, beer) fermentum ex quâdam eius fæce", would prevail in later versions of Descartes' theory of blood-circulation. See Annie Bitbol-Hespériès, *Le principe de vie chez Descartes*, pp. 84-86.

[47] For Descartes' scholastic contemporary Jean Fernel for instance, it was the *position* of the valves that testified to an "amazing secret of Nature". Jean Fernel, *De partibus corpus humanis*, quoted in Gilson, "Descartes, Harvey et la scolastique", *Études*, p. 57.

> just as necessarily as the movement of a clock follows from the force, position and shape of its counterweights and wheels,[48]

the emphasis should be placed entirely on the small but very significant additions *mere* and *just as*.[49] What Descartes aims to bring across to the reader in this passage is not merely a mechanistic interest in the clockwork contrivances God has invested Nature with. His mechanicism is theoretically set out much more precisely. It is not Descartes' aim to show that organic processes are like mechanical ones: it is that the elements in an explanation of organic processes should be restricted to a minimum in the same way as is done in the explanation of aritificial devices.

The crux of the argument is therefore that the workings of the heart are *just as sufficiently explained* by the three factors mentioned (construction, heat and blood) as are the workings of a clock by the "force, position, and shape of its counterweights and wheels". In other words, although the argument is at once a piece of mechanist propaganda, with its use of mechanist phraseology and of the ultimate seventeenth-century model of mechanicism—the clock—its core is really an "Occamist" one. The idea of restricting the number of explanatory factors is the same as in *Les Météores*, to which the *Discours* serves as a preamble:

> it seems to me that my arguments should be only the more approved of when I make them depend on fewer things.[50]

And what is left out in the description of the heart is again the same as that which Descartes was happy to leave out elsewhere: the substantial forms of the philosophers. As we know, Descartes wanted to avoid discussing them as much as he wanted to avoid discussing real qualities. It is ultimately on the basis of the argument from sufficient reason that Descartes at once establishes the mechanical model and avoids discussing the forms.

---

[48] Descartes, *Discours* V, AT VI, p. 50 / CSM I, p. 136.

[49] Which is to say, on words like *aussi* and *seule* in the French. The original passage reads: "ce mouuement, que ie vien d'expliquer, suit aussy necessairement de la seule disposition des organes qu'on peut voir a l'oeil dans le cœur, & de la chaleur qu'on y peut sentir avec les doigts, et de la nature du sang qu'on peut connoistre par experience, que fait celuy d'vn horloge, de la force, de la situation, & de la figure de ses contrepois et de ses rouës." Descartes, *Discours* V, AT VI, p. 50.

[50] Descartes, *Les Météores*, AT VI, p. 239. See also below, note 87.

The mechanicism is thus established by way of a description of natural processes which, given certain presuppositions, suffices to render an account of the natural process without taking recourse to the idea of internal motors, such as the forms. That is why the example of the heart was of so much importance to Descartes. It is an imaginative and very attractive example of Descartes' way of restricting the number of explanatory factors: given the fact that there is heat and that blood dilates as a result of this, the heart, being constructed in a certain way, *must* propel the blood in the way it does. The process *cannot but* take place. It must take place *just as necessarily as* a clock must perform its movements as a result of "the force, position, and shape of its counterweights and wheels." No further "form" is needed.

Explanatory factors may be restricted by analysing the workings of the natural objects into as few small steps as you need.[51] The likeness of natural objects and machines thus lies primarily in the way in which one may offer an elegant explanation of their workings. This means, first, that for Descartes the machine metaphor is subordinate to the actual mechanistic explanations: there need be no natural objects exactly like machines in order for the mechanistic account of natural change to hold true. The paradox of mechanicism, i.e. the fact that the mechanical model was justified by its adherents on the very grounds on which the adversaries of the mechanical theory sought to reject it, might, in the case of Descartes, not be so paradoxical after all. Whether or not God really fabricated machines and automata would become relevant only on an epistemological, or metaphysical level. Descartes was at pains to vindicate veracity for his mechanical models in his later writings, most explicitly in the *Principia*.[52] We shall not elaborate on these much discussed metaphysical questions here. What may in any case be concluded, is that in the *Discours* of 1637, the argumentation in favor of the mechanical model is based solely on the consideration of sufficiency of explanation, imaginatively illustrated by the description of the heart as an expansion motor needing no further elements to explain its workings than those of construction, heat and blood.

---

[51] "Descartes estimait en effet que son explication purement géométrique et méchanique d'un phénomène aussi important [i.e., du mouvement du coeur et des artères], était de nature à rendre sensible à tous les esprits le caractère profondément original de la science nouvelle qu'il était en train de consituer." Étienne Gilson, "Descartes, Harvey et la scolastique", in *Études*, p. 51.

[52] Descartes, *Prinicpia Philosophiæ* IV 203-206, AT VIII-I, pp. 325-329 / CSM I, pp. 288-291.

Second, the non-qualitative character of the notions occurring in the explanation is also made subordinate to the basic requirement that the explanation be stripped of superfluous arguments. Whatever Descartes may in fact think of the nature of heat and of the nature of blood—which, ultimately, he will analyse in a mechanistic manner also[53]—does not necessarily have to be part of the explanation to make it a mechanistic explanation in the Cartesian sense. The only requirement for mechanicism, at this stage, is a certain restriction of argumentation.

Machine and clockwork models are the result of this way of analysing natural processes, and hence a natural process such as that of blood-circulation may be further mechanicized at a later stage. Still, rather than the machine comparison or the restriction to non-qualitative elements in the explanation of things physical, it is the principle of restrictive description that first defines the mechanistic way of reasoning. In the next section, we shall see that it is this aspect of mechanicism, which also defines Descartes' famous, but equally misunderstood "mathematization" of physics.

5.4 Mathematics and Science in Cartesian Philosophy

Descartes points out the analogy between his explanation of the workings of the heart and the explanation of the movements of a clock in order to convince those readers who might be "tempted to reject this explanation without examining it." Especially prone to such misunderstandings are those

> who are ignorant of the force of mathematical demonstrations and unaccustomed to distinguish true reasons from probable.[54]

Descartes thus links efficient explanation in physics to the demonstrative proofs of the mathematical sciences. This use of *demonstrations mathématiques* is significant in that it indicates what function Descartes appointed to the mathematical in his scientific reasoning.

---

[53] Annie Bitbol-Hespériès refers to *Principia* IV 92-93, AT VIII-I, pp. 259-260, where Descartes gives his mechanistic analysis of the phenomenon of fire. Although I do not think that contrary to Aristotle and Galen, Descartes was the first to accept a "natural heat" inside the heart (see above, note 38), I agree that it was Descartes who alone interpreted this heat in a purely mechanical way. See also section 7.2.3, below.

[54] Descartes, *Discours*, AT VI, p. 50. Translation from CSM I, p. 136.

5.4.1 *In Search of a* Mathesis Universalis

Much embarrassed that "nothing more elevated" had been built on the foundations of mathematics, which were themselves *si fermes et si solides*,[55] Descartes had sought to obtain a mathematical foundation of science in his *Rules for the Direction of the Mind*.[56] There it is argued that, although mathematics is not to supplant all scientific reasoning, meditating upon the mathematical sciences might somehow yield the proper "objects" fit to play a role in science—objects concerning which one might have a certainty comparable to that of arithmetical or geometrical demonstrations.[57]

Which objects are meant here? The answer is given in the fourth of the *Rules*, where "order and measure" are introduced as the objects of all the "mathematical sciences" which together form the subdisciplines of the *Mathesis Universalis*.[58] But then, what exactly is this "mathesis universalis"? The notion is one of the most obscure in the whole of Cartesian philosophy.[59] It clearly does not refer to mathematics in the strict sense. Commentators often interpret it, if at all, in a very broad sense, in relation to a certain method of orderly reasoning.[60] However,

---

[55] Descartes, *Discours*, AT VI, p. 7 / CSM I, p. 114.

[56] The *Regulæ ad directionem ingenii*, AT X, pp. 349-469, have been dated by the editors of Descartes' works as being "de 1628 environ" (See AT X, p. 488). They were, however, only published posthumously.

[57] Descartes, *Regulæ*, AT X, p. 366: "Jam verò [...] est concluendum, non quidem solas Arithmeticam et Geometriam esse addiscendas, sed tantummodo rectum veritatis iter quærentes circa nullum objectum debere occupari, de quo non possint habere certitudinem Arithmeticis et Geometricis demonstrationibus æqualem." See also CSM I, pp. 12-13.

[58] *Regulæ*, AT X, pp. 377-378 / CSM I, p. 19.

[59] "It is a curious matter that an essential aspect of Descartes' method should be the subject of (or play a significant role in) a number of important commentaries without ever being fully defined, or even properly characterized. Indeed, with respect to Descartes' mathesis universalis, even such a simple opening statement will be subject of criticism." F. P. Van de Pitte, "Descartes' Mathesis Universalis", in *Archiv für Geschichte der Philosophie* 61 (1979), p. 154. For a discussion of Van de Pitte's own interpretation of the Cartesian use of this notion, see below, note 80.

[60] See, e.g. E. W. Strong's *Procedures and Metaphysics*, Hildesheim (Georg Olms Verlag) 1966, p. 215: "In Descartes, method is seized upon as the basis of success of mathematical inquiry, and the universalisation of the method (universal mathematics) is conceived to be the way to bring certainty into metaphysics comparable to that achieved in mathematics." Nicolas Grimaldi gives an even vaguer definition of the general science of mathesis universalis, arguing that it is the "système formel de toute connaissance selon un ordre." See N. Grimaldi,

to take the *mathesis universalis* in such a general sense does not clarify anything in respect of the Cartesian use of mathematics within the physical sciences.

Descartes says that all intelligible aspects of music, optics, mechanics and the like, should be put in such propositions as deal with "order and measure"—or, to use the terminology of the *Discours*: with "proportion"—only.[61] Now if he means that to bring "order and measure" into science, only mathematical questions of astronomy, optics, music and the like are to be dealt with, his New Philosophy would be wholly mathematical. In that case, the Cartesian would avoid giving physical interpretations—in optics, for instance, of the nature of light, or, in astronomy, of the causes of the revolutions of heavenly bodies. Instead he should stick to the mathematical description of refraction, or of the revolutions of heavenly bodies, etc.

We do not know what Descartes' thoughts on this matter were at the time when he wrote the *Regulæ*, that is, late in the 1620s. Descartes planned to show how to reduce the *questiones imperfectæ* of physics to "perfect" ones in the third part of this work. Regrettably, it was never written. However, the second part of the *Discours de la Méthode* takes up the question once more.[62] Mathematics providing, as in the *Regulæ*, the order of steps to be followed in scientific explanation, it again seems also to determine their objects. And as in the *Regulæ*, Descartes has those sciences in mind which "do not discuss anything other than the various relations or proportions (*rapports ou proportions*)." This time however, Descartes explains that he will examine "only those proportions in general", imagining them sometimes as lines, sometimes however as figures."[63]

If this passage may be taken to elucidate Descartes' conception of *mathesis universalis*, then it is obviously the descriptive and purely mathematical interpretation Descartes is giving priority to. Indeed,

---

L'expérience de la pensée dans la philosophie de Descartes, Paris (Vrin) 1978, p. 112.

[61] Descartes, *Regulæ*, AT X, pp. 377-378 / CSM I, p. 19, and *Discours*, AT VI, pp. 19-20 / CSM I, pp. 120-121. See also the next footnote.

[62] Jean-Luc Marion even makes the following identification between the *Discours*-text and that of the *Regulæ*: "[AT VI, p.] 20, 2-4 = [AT X, p.] 377, 12 - 378, 1." Jean-Luc Marion, *Sur l'ontologie grise de Descartes*, Paris (Vrin) 1981, p. 67. Marion does not, however, take into account the purely mathematical content of what follows, thereby avoiding the interpretation that a purely mathematical representation of "proportions" would be the subject-matter of Descartes' *mathesis universalis*. See also below, note 80.

[63] Descartes, *Discours*, AT VI, p. 20 / CSM I, p. 121.

"imagining certain proportions as lines or as figures" is exactly what Descartes does, for instance in the *Dioptrique,* where he formulates his law of refraction. There, the trigonometric "proportions" of refraction are geometrically presented without being put into trigonometrical formulae.[64]

The problem is, however, that there is actually no unique scientific method to be found in the *Discours de la Méthode*. As Gilbert Gadoffre has pointed out, the second part of the *Discours* may only have had a very limited scope. According to Gadoffre, it was originally to figure as a preface to the *Geométrie* in which Descartes presented the core of his inventions in the field of analytic geometry, and, amongst other things, refined his mathematical representation of the refraction of light.[65] The *Géométrie* was not finished until after the other two essays—the *Dioptrique* and the *Météores*—and Descartes seems to have taken much pleasure in writing it.[66] Nevertheless, its publication together with the other essays, which are not of a purely mathematical content, and the references made to this essay in the preceding *Discours* itself are rather misleading.

It is clear to anyone acquainted with the Cartesian oeuvre that Descartes' philosophy—even his *natural* philosophy—is not generally concerned with mathematical representations of natural phenomena. Accordingly, Descartes' remarks in the *Discours* concerning the mathematics of proportion may be of very limited significance. They are said to pertain to the sciences "which are usually called mathematical",[67] that is, to optics, astronomy, music, and the like, inasmuch as these represent natural phenomena in an arithmetical or geometrical way. But then, what about the four rules of reasoning

---

[64] Descartes' law would read: "sin i = n sin r", in which "i" is the angle of incidence, "r" the angle of refraction and "n" a constant varying from one body to another according to their relative resistance, or, as Descartes would put it, inversely: according to the "force ou facilité, dont elle [a ball, here used in analogy to a ray of light] entre dans le cors." Descartes, *La Dioptrique*, AT VI, p. 100 / CSM I, p. 161.

[65] Gilbert Gadoffre, "La chronologie des six parties", in N. Grimaldi and J.-L. Marion (edd..), *Le Discours et sa Méthode*, Paris (P. U. F.) 1987, p. 36: "Le chapitre [...] se présente [...] comme une introduction à la Géométrie cartésienne."

[66] Descartes to Père Deriennes, 22 February 1638: "C'est vn traitté que ie n'ay quasi composé que pendant qu'on imprimoit mes Meteores, & mesme i'en ay inuenté vne partie pendant ce temps-là; mais ie n'ay pas laissé de m'y satisfaire autant ou plus que ie ne me satisfais d'ordinaire de ce que i'écris." AT I, p. 458.

[67] "qu'on nomme communement Mathematiques", *Discours*, AT VI, p. 19 / CSM I, p. 120.

presented in the *Discours* as a paradigm for all reasoning? Are these only to play a role in the mathematical sciences? According to Gilbert Gadoffre, they indeed are. To take into account the limited range of the methodology as presented in *Discours* II, is, according to Gilbert Gadoffre, the only way to avoid further

> dialectical acrobatics designed to explain the breach in the development of a scientific method between parts II and IV [of the *Discours*].[68]

There is, however, a way in which, for Descartes, mathematics remains the paradigm for scientific investigation even outside of the "mathematical sciences". No dialectical acrobatics have to be undertaken.

5.4.2 Efficient Explanation and Mathematics

What is important to notice first, is that Descartes himself offers support for Gadoffre's view of the limited scope of what is said in Discours II concerning the method of proportion. He does so in his *Conversation with Burman*, which, although it is an interview, so to speak, is very clear as regards the role of mathematics in the sciences. Hence it is highly improbable that Frans Burman might have mistaken the philosopher's intention, where, reviewing the text of the *Discours*, Descartes says:

> In order to understand what [I have] written in Philosophy, there is no need for Mathesis, except perhaps for a small number of things in the DIOPTRICS, which are mathematical.[69]

Philosophy has no need for "Mathesis". It is clear that Descartes here acknowledges that mathematical considerations hardly affect his philosophical theories. The exception made for the *Dioptrics* is very interesting. Descartes must be referring to those parts of the essay which deal with the mathematical analysis of refraction. Accordingly, Gadoffre is right to disentangle the content of the greater part of Descartes' scientific writings from the mathematical method of *Discours* II, where it is proposed to "imagine certain proportions as

---

[68] Gadoffre, "La chronologie des six parties", in Grimaldi / Marion, *Le Discours et sa Méthode*, p. 40.

[69] Descartes, "Conversation with Burman", AT V, p. 177 / *Entretien avec Burman*, ed. Ch. Adam, Paris (Vrin) 1975², "EX DISSERTATIONE DE METHODO", p. 122 / CSM III, p. 351.

lines or as figures" in those sciences which are "usually mathematical".

Secondly, the exception that Descartes makes for "a small number of things" (*pauca quædam*) indicates that "philosophy" must here refer to *natural philosophy*, since apart from the few mathematical statements occurring in the *Dioptrics*, all the rest of this essay, as well as the complete essay of *The Meteors* deals with natural philosophy only.

This solves a question which has often been put in the face of Descartes' apparent preoccupation with mathematics: why do we find so little mathematical detail even in the purely physical parts of the Cartesian œuvre? As John Cottingham has put it:

> it is precisely the paucity of such mathematical workings that is the striking feature of Descartes' physics, when compared with the work of Galileo, for example.[70]

The paucity of mathematics in Descartes may disappoint the modern reader.[71] Yet as the passage from Descartes' *Conversation with Burman* shows, it cannot have been Descartes' intention to invest his natural philosophy with much mathematical reasoning in the first place. In fact, it is only modern bias to expect that Descartes' rejection of qualitative notions should imply their being supplanted by "precisely measurable *quantities*".[72] Apparently, Descartes saw different alternatives for the use of mathematics in physics.[73]

The mechanical theories which, as alternatives to qualitative reasoning, we do find in essays like the *Dioptrics* and the *Meteors*, make use of mathematics in a wholly different sense. To see this, we

---

[70] Cottingham adds: "The comparison with Newton, whose *Principia* were published some 40 years after those of Descartes, is even more striking in this regard." John Cottingham, *Descartes*, Oxford (Basil Blackwell) 1986, p. 88.

[71] John Cottingham, *ibidem*: "readers who approach Descartes' scientific writings expecting anything detailed in the way of geometrical measurements, arithmetical calculation or algebraic formulae will to a great extent be disappointed."

[72] John Cottingham, *ibidem*.

[73] It is therefore inappropriate to argue, as Cottingham does, that "all that [Descartes] seems to require is that the properties involved be in principle capable of being quantified." The idea is that Descartes only laid the groundwork for a science of "calculation and measurement", without being able himself to—in Cottingham's words—"make much of a contribution to mathematical code-cracking." In fact, for Descartes, "mathematical code-cracking" is just not what physics is about. See John Cottingham, *Descartes*, p. 89.

need only recall the way in which Descartes pointed at the analogy between "physical" and mathematical reasoning in his explanation of the workings of the heart. When Descartes compares the working of the heart to the mechanics of a clock, he says that he does so in order to persuade those "who are ignorant with the force of mathematical demonstrations."[74] Such a reference to mathematics (which, purely methodological discussions apart, is in fact exceptional in Descartes) has nothing whatsoever in common with the project of quantification in science. What is "mathematical" here, is the efficiency of explanation on the basis of which Descartes' theory of the heart and of blood circulation are judged. It is this aspect which, for Descartes, counts as the *mathematical* aspect of scientific reasoning. Natural philosophy may be invested with a "mathematical" way of reasoning by reducing the number of argumentative steps to a minimum. What science has to learn from mathematics is, to quote from the late Dutch scholar P. van der Hoeven, a "reduction to a minimum of effective reasoning."[75]

By pointing to the analogy with "mathematical demonstrations" in his comments on the example of the heart, Descartes emphasizes the significance of the aspect of efficient and "reductive" reasoning, even though the example does not deal with proportion or quantity. Descartes' point is, that the "force" of mathematical demonstrations should be met in explanations of natural philosophy as well. It is in this way that mathematical method is a central feature of Descartes' scientific work, even though "mathesis" is, in the end, restricted to some appendices of Cartesian science alone.[76]

---

[74] *Discours*, AT VI, p. 50 / CSM I, p. 136.

[75] P. van der Hoeven, *Metafysica en Fysica bij Descartes*, Gorinchem (J. Noorduyn en Zoon N.V.) 1961, p. 31. It should be noted that Van der Hoeven uses the expression to characterise the method of Descartes' *Meditations* in particular: "het wiskundig inzicht ligt in de Meditaties veel dieper verworteld [dan in de Regulae]: in de vanzelfsprekendheden die er een rol in spelen; in de dingen welke Descartes, tot ieders verwondering, geen bewijs nodig acht; in de reductie tot een minimum aan doeltreffende redenering."

[76] As a consequence, one could argue that Descartes rightfully claims that the method of *Discours* II is relevant for all sciences despite the fact that, as Gadoffre has argued, these rules were originally meant only to serve as an introduction to his mathematics. However, while the "mathematical method" as a whole stands for a certain way of reasoning, the order and efficiency of which is held to be examplary for all sciences, the famous four rules pertain, as has often been pointed out, primarily only to the *order* of exact reasoning. The first of Descartes' rules of reasoning as presented in the *Discours* is "never to accept anything as true which I did not evidently know to be so, which is to say to carefully avoid any precipitation

This partly explains the emphasis Descartes puts on mathematical reasoning despite his notorious disregard of using mathematics in physics. There is however more to say concerning Descartes' idea of a "mathematical" science.

5.4.3 Efficient Explanation and Mechanics: Cartesian Matter

Besides the purely formal aspect, there is also the aspect of the "mathematical content" of Cartesian scientific argumentation. Yet again, this mathematical content is wholly different from quantification or precise measurement. When, at the end of the second part of the *Principia*, Descartes says that he will not admit any principles in physics which are not also accepted "in geometry, or in abstract mathesis", he is primarily concerned with accepting only the "geometrical" notion of quantity. However, far from thereby making applied geometry the core of physical enquiry, Descartes argues only that, as opposed to the Scholastic explanations in physics, which introduce the concepts of form, unformed matter and qualities, he will introduce no other "matter of corporeal object" than that which can be "divided, shaped and moved in all sorts of ways".[77] The emphasis, in other words, on the mathematical *content* of physical explanations is meant to introduce the Cartesian notion of matter as the proper object of physics. It is, moreover, this notion of matter by which physics is defined in mechanistic terms as the study of "matter in motion".

In the *Regulæ*, the simple notions of "shape, extension, motion etc."[78] together with other *choses communes*, such as the fundamental

---

or prejudice and to include in my judgements only what presented itself so clearly and distinctively to my mind, that I had no occasion to doubt it." Descartes, *Discours*, AT VI, p. 18. The second rule demands that "every difficulty under investigation [be divided] in as many parts as possible and as are required for solving them in the best way." Together, these rules echo the mathematical aspect of dividing problems into smaller ones that can be solved independently of each other. They thus form the basis for what mathematics may teach as far as methodology goes. The proposed "division into smaller parts" may also imply the *reduction* to a minimum of the explanatory elements required. This aspect of mathematical reasoning is said (in *Discours* V) to exhibit "the force of mathematical demonstrations". However, though it seems to be implied, Descartes does not explicitely mention any "Occamist" maxim at this point, and focusses upon the aspect of "order" only.

[77] "In Geometriâ, vel in Mathesi abstractâ"; Descartes, *Principia* II 64, AT VIII-I, p. 78 / CSM I, p. 247.

[78] Descartes, *Regulæ*, AT X, p. 419 / CSM I, p. 45: "figura, extensio, motus &c.".

laws of logic, determine the realm of what is knowable in the world.[79] *Per se notæ*, "known by themselves" the notions of shape, extension, movement, (dis)position etc. define scientific propositions. In fact, all propositions should be reducible to them, or else rejected.

Thus, Descartes had already made clear in the *Regulæ* that he would not admit of any notions save those which refer to place, movement, shape, position and motion of matter, that is, the notions that define his science as a science of quantity. Accordingly, any *qualitative* notions occurring in Cartesian physics should be regarded as temporary. As we have said above concerning the notion of heat and of the "nature" of the blood, such qualitative features of the physiological explanation will ultimately have to be analysed in a "mechanistic" manner also. They are useful in order to render an account of blood-circulation that meets the requirement of efficient explanation, but ultimately this will not do. The form of the "mathematical argument" can, in the end, only have the effect of demonstrative proof if the content of the argument is entirely made of those simple elements which determine strict knowability. Such elements are the "geometrical" elements of shape, position and the like. Therefore, unclear qualitative notions such as "heat", "nature" etc, will ultimately have to be supplanted by notions referring to the purely *material* factors which are responsible for the phenomenon of heat in the heart and for the properties of and reactions of the human blood.

Introducing only explanations which are reducible to the simple (and, epistemologically speaking, *innate*) "geometrical" concepts of movement, shape, position etc., is in fact, according to Descartes, another way of attaining, in physics, the clarity and distinctness proper

[79] In *Sur l'ontologie grise de Descartes*, Jean-Luc Marion has stressed the ontological consequences of this epistemological position: "Il serait dangereux, d'ailleurs, d'occulter en un traité d'épistémologie l'enjeu ontologique [...] qu'encourent les *Regulæ* [...] en élaborant leur épistémologie. [...] En fait, les deux types de "choses commmunes" [...] valent pour toute chose proposée, comme futur objet = x, à la connaissance, puisqu'elles déterminent les transcendantaux (unité, Etre) [...]." Jean-Luc Marion, *op. cit.*, pp. 138-139. The point is further elaborated in Marion's later work *Sur la théologie blanche de Descartes*, Paris (P. U. F.) 1981, especially in Livre II, Section I, "De la science à son fondement", in which the process of sensation is presented as a "défiguration" of the "mathematical" information which our senses recieve from worldly objects, themselves ordered, as it is said in the apocryphal text of The Wisdom of Solomon 11:20 "by measure and number and weight". Marion quotes Descartes referring to this text, *Sur la théologie blanche*, p. 266.

to mathematical demonstrations. Although it seems mistaken to interpret Descartes' references to a *mathesis universalis* as references to this way of supplying physics with a mathematical content—as some commentators have recently argued[80]—it is true that the

[80] Jean-Luc Marion for instance, identifies "le domain de la quantité [...] et de la relation" with that of the object of the *mathesis universalis*, viz. of "ordo et mensura". Jean-Luc Marion, *Sur l'ontologie grise de Descartes*, p. 139. He defends this wider, non-mathematical use of *mathesis universalis* by referring to Giovanni Crapulli thorough examination of the sixteenth-century use of the concept. See Marion, *op. cit.*, p. 67, note 77 and *Sur la théologie blanche de Descartes*, p. 354, note 7. However, as Crapulli argues in his *Mathesis universalis; Genesi di una idea nel XVI secolo*, Roma (Ateneo) 1969, p. 146: "in qualcuna almeno delle espressioni o descrizioni di Diasipodo [for instance], la scienza commune alle discipline matematiche sembra addirittura perdere la sua specifica caratterizzazione matematica", there is yet no reason to attribute the same non-mathematical use of the notion to Descartes, who, as we have seen, specified what he had to say about the general science of proportions in a purely mathematical way. Arguing that "it is clear that [Descartes] employs the word "mathesis" primarily in its Greek sense, i.e. the act or process of learning, the acquisition of knowledge, F. P. Van de Pitte has restated the question, defending Marion's position. See F. P. Van de Pitte, "Descartes' Mathesis Universalis", in *Archiv für Geschichte der Philosophie* 61 (1979), pp. 161. Still, there is hardly a convincing argument to be found for the thesis that Descartes actually did use the concept in a general, non-mathematical way. To argue that "the Mathesis is the master discipline [...] of which various mathematical disciplines (mathematicæ) are the parts" (*op. cit.*, p. 163) is surely right, but trivial. One is still left with the question whether or not these *mathematicæ* in the Cartesian sense are of a genuinely mathematical content. There are, however, various arguments against the Marion-interpretation. Firstly, although each of the notions of "order" and "measure" does reoccur now and again in the Cartesian oeuvre, they are, in the works after the *Regulæ*, never again connected with the notion of *mathesis universalis*, or with any other general science of mathematics for that matter. Secondly, the concept of *mathesis universalis* itself does not reoccur where, in the later works such as the *Principia*, Descartes puts forward the idea of using only "mathematical principles" in science. Thirdly, the second part of the *Discours*, which both summarizes the fourth of the *Regulæ* and serves as a preface to the *Géométrie*, does not leave any doubt as to the purely mathematical content of the "sciences qu'on nommme communément mathématiques". Consequently "darf [es] als sicher gelten, daß das Konzept einer Mathesis universalis, die als eine allgemeine Theorie der Größen und Größenverhältnisse bestimmt wird, enger gefaßt ist als das einer Universalwissenschaft." J. Mittelstraß, "Die Idee einer Mathesis Universalis bei Descartes", in *Perspektiven der Philosophie* 4 (1978), p. 181. Or, as Wolfgang Röd puts it, in his review of Marion's *Sur l'ontologie grise*, *Zeitschrift für Philosophische Forschung* 32-1 (1978), p. 147: "Bei Descartes mag es Ansätze einer Metamathematik geben, aber die Mathesis Universalis ist selbst nicht Metamathematik, so wie sie keine Methodenlehre ist, sondern eine Mathematik, wie es ihre Name ausdrückt."

substitution of all notions in physics by purely "geometrical" ones is one of the most central aims of Cartesian physics. As Descartes wrote to Princess Elizabeth of Bohemia, we have, regarding corporeal objects,

> only the notion of extension, which entails the notion of shape and motion.[81]

Keeping this in mind and distinguishing sharply between physical notions, mental notions, and the intermediate notions of sensations and the passions, we cannot be led astray.[82] In explaining natural processes, the first step is therefore to reformulate them into "geometrical", that is, "mechanical", terms.[83] Doing so, every connection of causes and effects will, provided that the reduction to a minimum of propositions is sufficiently carried through, be as simple and evident as the demonstrations of mathematics. By accepting only "mathematical" principles as for example movement and shape, explanations of natural processes may acquire the certainty of mathematical demonstrations.

Thus, the Cartesian "mathematization" of natural philosophy, instead of calling for quantification and precise measurement, postulates only that physical explanations should (1) meet the formal requirement of reducing effective reasoning to a minimum, and (2)

---

[81] Descartes to Elisabeth, 21 May 1643, AT III, p. 665. The translation is from CSM III, p. 218.

[82] Descartes, *ibidem*: "ie considere qu'il y a en nous certaines notions primitiues, qui sont comme des originaux, sur le patron desquels nous formons toutes nos autres connoissances. Et il n'y a que fort peu de telles notions; car, apres les plus generales, de l'estre, du nombre, de la durée &c., qui conuiennent à tout ce que nous pouuons conceuoir, nous n'auons pour le corps en particulier, que la notion de l'extension, de laquelle suiuent celles de la figure & du mouuement; & pour l'ame seule, nous n'auons que celle de la pensée, en laquelle sont comprises les perceptions de l'entendement & les inclinations de la volonté; enfin, pour l'ame & le corps ensemble, nous n'auons que celle de leur vnion, de laquelle depend celle de la force qu'a l'ame de mouuoir le corps, & le corps d'agir sur l'ame, en causant ses sentiments & ses passions." Sharply distinguishing between these is what science is about: "Ie considere aussi que toute la science des hommes ne consiste qu'à bien distinguer ces notions, & à n'attribuer chacune d'elles qu'aux choses ausquelles elles appartiennent." Cf. CSM III, p. 218.

[83] Note that, according to Descartes, the "principles" of geometry and mechanics are identical. As it is said in the French edition of the *Principles*, part IV, § 203, "[les] regles suiuant lesquelles ces trois choses [i.e., shapes, sizes and motions] peuuent estre diuersifiées l'vne par l'autre [...] sont les principes de la Geometrie et des Mechaniques." See AT IX-II, p. 321.

meet the material requirement of introducing only "geometrical", "mechanical" concepts. However, both requirements lead to exactly the same result: the exclusion of forms and qualities from physics.[84]

5.5 Conclusion: A World Without Forms

Efficiency of explanation is, for Descartes, an essential part of the mechanical re-description of observable phenomena. Yet this is not evidently so. Natural or physiological processes might and in fact were interpreted in a mechanical way by other authors who acknowledged or admired the perplexingly intricate clockworks of Nature.[85] Yet they might still refer to forms or qualities responsible for their functioning. In *Les Météores*, Descartes says that he will not discuss the forms and qualities of the philosophers.[86] The more important point is that he claims to have no need for them.[87] It is this latter aspect of Cartesian theory which was attacked, and indeed ridiculed by adversaries such as Voetius and Schoock. In the next chapter, we shall see why.

---

[84] It might even be asked whether there is really any need for distinguishing between both types of requirements just given. Indeed, as far as the content of physical explanations is concerned, qualities and forms would in any case be excluded by the second condition, since they are not supposed to be "geometrical" or "mechanical" principles. To satisfy the mechanical requirement would therefore be enough in order to exclude all to forms and other non-mechanical principles. However, the significance of the first requirement, well illustrated in the famous *Discours*-explanation of the functioning of the heart, is that forms and qualities are not only excluded as elements of physical description, they are deemed superfluous in terms of causality. It is not only Descartes' objective to *compare* the mechanical aspects of the heart with those of a clock: it is no less his aim to *explain* the heart's functioning as that of a clockwork. His way of doing so is to argue that once the natural process is sufficiently interpreted in mechanical terms, one has no further need of causal principles such as the Scholastic forms. Indeed, not only should we describe natural processes as if they were mechanical ones. As Descartes explains in the *Discours* to "those who are ignorant of the force of mathematical demonstrations", we should also accept the fact that a restriction to the mechanical description alone suffices as an explanation of its workings.

[85] See e.g., above, note 47.

[86] Descartes, *Météores*, AT VI, p. 239: "Puis, sçachés aussy que, pour ne point rompre la paix auec les Philosophes, ie ne veux rien du tout nier de ce qu'ils imaginent dans les cors de plus que ie n'ay dit, comme leurs *formes substantielles*, leurs *qualités reelles*, & choses samblables."

[87] Descartes, *ibidem*. "il me semble que mes raisons deuront estre d'autant plus approuuées, que ie les feray dependre de moins de choses."

CHAPTER SIX

THE COMPOSITE CHARACTER OF SUBSTANTIAL UNITY

Relinquitur quærendus aliquis motor externus, qui potentiam deducat in actum. Quem autem hic dabunt? animam mundi Platonico-Vergilianam, aut intelligentias, aut Deum, aut atomos, aut globulos coelestes. Aliquid hic dicendum erit.

Gisbertus Voetius, "De Rerum Naturis et Formis Substantialibus"[1]

By his "mathematization" of natural philosophy, Descartes excluded forms and qualities. Reacting on this development in his essay *On the Natures and Substantial Forms of Things*, Gisbertus Voetius indicated that the loss of substantial forms would imply the loss of independent sources of action. If there are no forms, there are no causes. Indeed, the forms are as much the centres of causality as the real qualities are the centres of sonation, heating, and so on. How do the Cartesians explain that anything at all may happen in the world when there are no causes, or sources of action?

6.1 The Critique of Mechanicism

Voetius warns that from the New Philosophy

> it would follow that the proper faculties and their intrinsic principles in animals would be of no other kind than those of automata or Dedalian statues; and that, as a consequence, the works of God and Nature, brought forth by creation or generation, would be essentially and univocally the same as works of art.[2]

Neither Descartes nor Regius seem to have been very much impressed by this objection. At the instigation of Descartes, Regius was willing to admit that, the works of Nature being so much more intelligently

---

[1] Voetius, *Narratio*, p. 40 / *Select. Dispp.* 873 / *Responsio*, p. 12 / *Querelle*, p. 106.

[2] Voetius, *Narratio*, p. 41 / *Select. Dispp.* I, p. 874 / *Responsio*, p. 13 / *Querelle*, p. 107.

constructed, there is at least a gradual difference between the works of Nature and of art.[3]

In the *Discours*, however, Descartes had already shown that he had no scruples himself when comparing a "machine [...] made by the hands of God" with "automata" constructed by "the skill of man".[4] Indeed, the whole discussion of the human heart in the *Discours* was meant to illustrate the project of explaining the complete animal physiology in mechanistic terms—a project which, by 1637, Descartes had already worked out in the unpublished treatise on human physiology, *L'Homme*. There, he had made

> special efforts to show that if any such machines had the organs and outward shape of a monkey or of some other animal that lacks reason, we should have no means of knowing that they did not possess entirely the same nature as these animals.[5]

As was said before, such statements were regarded as blasphemous by those opposing the mechanical viewpoint. But there were other objections as well. An important difference between artificial and natural objects, between robots and organisms, is that artificial contrivances lack internal principles of action. Both Voetius and his pupil Martin Schoock were to press this point in a variety of ways. At the heart of their objection lies the consideration that mechanical properties are, in themselves, inert, and cannot be the *causes* of motion. The animal heart, for instance, despite its marvellous mechanical design, is yet in need of a substantial form that moves it.

6.1.1 Forms and Souls: The Example of the Heart

Arguing that mechanics is, in itself "of great benefit to human life", Schoock to a certain degree shares Descartes' opinion that science could help us become "the lords and masters of Nature".[6] However, according to the Groningen philosopher, this concerns technical inventions only. Mechanics being the science of artifacts, it does not provide any key for understanding the operations of natural objects. The idea of regarding all physics—and, more in particular,

---

[3] Regius, *Responsio*, p. 18 and above, section 5.3.1.

[4] Descartes, *Discours*, AT VI, pp. 55-56 / CSM I, p. 139.

[5] Descartes, *Discours*, AT VI, p. 56 / CSM I, p. 139.

[6] Schoock, *Admiranda Methodus*, p. 130 / *Querelle*, p. 254 and Descartes, *Discours* VI, AT VI, p. 62 / CSM I, pp. 142-143.

physiology[7]—as mere parts of mechanics, is simply based on too simple an account of Nature.

Of course, the Cartesian may offer an elegant mechanical explanation of the way in which the blood enters and leaves the heart and compare the disposition of the various organical parts with the disposition of the wheels and the weights of a clock. Yet such a demonstration convinces neither Voetius nor Schoock of the supposed fact that there is any similarity between the natural and artificial realms, or between the living object and the robot. The central question for these authors is: how can the principles of Cartesian mechanics—such as, for instance, the situation and the disposition of the constituent parts of either a clock or the human heart—ever yield any "effective principles"[8] that may give an account of the physical causes of their movements?

Criticizing the Cartesian overestimation of mechanics, neither Voetius nor Schoock directly refer to the passage in the *Discours* where Descartes analyses the heart and the circulation of the blood. Yet both discuss the example of the heart: Voetius presumably in reaction to Regius' teachings and Schoock in reaction to Regius' *Responsio*. Regius had defended the aptness of the mechanical viewpoint by referring to an experiment in which he took out the heart of a living eel and observed that, even outside the living body, it kept

---

[7] Both Descartes and Regius (a man of medicine) were very much attracted to the science of physiology proper. In the seventeenth century however, the term "physiologia" might refer to the science of natural processes in general. See K. Rothschuh, *Physiologie: der Wandel ihrer Konzepte, Probleme und Methoden vom 16. bis 19. Jahrhundert*, Freiburg (Alber) 1968. Schoock and Regius typically used the term in a broad sense, Regius by studying organic functions alongside general theories of physics in his physiological disputations, which were published as *Physiologia, sive Cognitio Sanitatis* in 1641; Schoock referring to "physiologia" in a broad sense as that for which the cartesians offer their mechanics as a substitute: "Quod tamen quis miretur, Cartesianæ philosophiæ præter Geometriam & Algebram regula quoque Mechanica est, cui tantum deferunt novorum deleriorum alumni, ut non dubitent eam cum ipsa Physiologia confundere." Schoock, *Admiranda*, p. 132 / *Querelle*, p. 246. Another example is that of Sebastian Basso, whose book against Aristotelian natural philosophy is at once an attempt to substitute it with the "Physiology" of the Ancients. Cf. Sebastian Basso, *Philosophia Natvralis adversus Aristotelem* [...] *In quibus abstrusa Veterum Physiologia restauratur, & Aristotelis errores solidis rationibus refelluntur*, Amsterodami (Apud Ludovicum Elzevirium) 1649. The first edition of this work appeared in Genève, and was published by P. de la Rovière in 1621.

[8] Schoock's himself uses the term "principium effectivum" for instance at *Admiranda*, p. 206 / *Querelle*, p. 285.

beating regularly for a few hours.[9] This was thought to be a clear indication of the fact that the heart works purely on account of its mechanical construction:

> Indeed, the heart performs its actions merely as a result of the convenient disposition of its parts, in such a way that there is no need to summon the help of the human soul (*anima*) or any other substantial form. This is already clear from the pulse of the little heart of an eel, which beats for many hours after it has been removed from the body and which appears to regain its pulse and to somehow revive by adding blood and by heating it slightly after it has already lost all of its motion.[10]

Relying on his experiments,[11] Regius feels sure he can convince anyone that the heart is a mere machine, working independently from any form or soul. Yet he does not convince Voetius, who, in the essay on forms, had attacked the view that "the animal heart is moved by the sole disposition of its parts". According to Voetius, the heart

> is moved by the soul, or informing form, through the mediation of qualities [operating] as principles and other instruments which are necessary for animal motion.[12]

Of course, Voetius knows that the advocates of the New Philosophy will deny that the heart is moved by the soul, and say that

> the movement of the heart can be diminished or even totally stopped and put to hold even when the animal is alive.

Though equally cruel, this is not exactly the experiment Regius is referring to. Nevertheless, the question in both cases is whether or not a movement of the heart can be observed independently of the heart's

---

[9] Voetius, *Narratio*, pp. 45-46 / *Select. Dispp.* I, pp. 876-877 / *Responsio*, p. 24-25 / *Querelle*, pp. 110-111; Schoock, *Admiranda*, pp. 111-112 / *Querelle*, pp. 235-236; Regius, *Responsio*, p. 27.

[10] Regius, *Responsio*, p. 27.

[11] Note that, in the *Responsio*, Regius most of the time literally follows what Descartes had proposed as an answer to Voetius. In this particular case however, Regius relies on his own theories and his own investigations. It may well be that Descartes accepted Regius' authority with respect to the movement of the heart, suggesting to his Utrecht friend: "And here you could add what you have concerning the movement of the heart." Descartes to Regius, January 1642, AT III, p. 506 / *Rodis-Lewis*, p. 94 / CSM III, p. 209.

[12] Voetius, *Narratio*, p. 45 / *Select. Dispp.* I, p. 876 / *Responsio*, p. 24 / *Querelle*, p. 110.

being attached to a living animal. In reply to Voetius, Regius says that it is no use to simply state that the heart's operations derive from the activity of the soul, or the "informing form". It should first be *proved* that "qualities and instruments" are not themselves sufficient, and that recourse has therefore to be made to "that fictitious principal cause". As long as it is not proven that the soul adds anything to the explanation of the activity of the heart, the adherents of the New Philosophy have all the reason to maintain that

> if it were the soul which does these things, it would be conscious of its operations and would make use of the power of reason.[13]

Since we are not conscious of the operations of the heart, there is, according to Regius, no reason to refer to the soul in order to explain its action.

Voetius however, does not accept this argument. He says that the "informing form"—in this case, the soul—does not have to "make use of reason". Nor does it "[actually] reason", since "forms operate through natural faculties without reasoning." The fact that, in the case of the heart, we are dealing with the human soul which acts as the "informing form", does not alter things at all: the soul may direct bodily movements in the same way as plants and animals act on "natural instinct", bodily organs digest food, stones fall downward and stars rise and set. They all do so without the use of reason.[14]

As for the argument that the movement of the heart may be stopped even though the soul remains present, Voetius says it still does not follow that the soul or form would not be the principle of motion in this case. Likewise,

> [the power of] reasoning stems from the rational soul and yet it is absent in a newly born baby, even though the rational soul is not.[15]

A baby cannot think reasonably, yet the power of reason is dependent only on the rational soul, which it already posesses. Apparently, there is no incongruity involved in the fact that the soul may be responsible for a phenomenon which does not occur at all times. The same may be true of the movements of the heart.

---

[13] Regius, *Responsio*, p. 27.

[14] Voetius, *Narratio*, p. 45 / Select. Dispp. I, p. 877 / *Responsio*, pp. 24-25 / *Querelle*, p. 110.

[15] Voetius, *Narratio*, p. 45 / Select. Dispp. I, p. 876 / *Responsio*, pp. 24 / *Querelle*, p. 110.

Still, Voetius' argument does not seem wholly satisfactory with respect to Regius' experiment with the eel. In Voetius' example, a process is stopped, or a phenomenon held back, even though the source of action remains present. In the case of the eel, on the other hand, the supposed source of the action—namely the eel's animal soul—is in fact no longer present, whilst the phenomenon—the beating of the heart—continues to occur.

So again Regius might ask what need there is to introduce the soul as that which is responsible for the heart's pulsation. Does not the experiment show that the soul has nothing to do with the pulsation? Voetius does not discuss Regius' experiment, but in the 1643 *Admiranda Methodus*, his pupil Martin Schoock did return to the question of the eel's heart. Schoock is not at all disconcerted by the result of the experiment. On the contrary, he regards it as a fine example of an erroneous "experience of reason".[16] Regius holds that the heart beats merely as a result of the disposition of its parts. However, on the basis of the experiment, it should rather be concluded that the disposition of the constitutive parts of the eel's heart is of no significance whatsoever to the continuation of its pulse. This is easily seen, says Schoock, when, after a few hours, the movement stops, while the disposition of the parts remains exactly the same![17]

The "mechanical principle" of the disposition of parts is thus irrelevant to the continuation of the pulse. According to both Voetius and Schoock, the same counts for all mechanical principles: they cannot explain motion.

6.1.2 The Ineffectiveness of Mechanical Conditions

Martin Schoock was willing to agree that, of course, knowledge of the precise disposition of cog-wheels is of utmost importance to the watchmaker, as is the knowledge of the anatomy of the heart to the physician. Yet neither the disposition nor any other mechanical principles effectively explain motion.

Take the principle of "situation", for example.[18] According to Schoock, situation is a property of things already constituted rather

---

[16] "Experientia rationis", *Admiranda*, p. 110 / *Querelle*, p. 235.

[17] Schoock, *Admiranda*, p. 112 / *Querelle*, p. 235.

[18] Schoock, *Admiranda*, p. 209 / *Querelle*, p. 286: "positura". In section III, chapter 5 of the *Admiranda* (pp. 195-211 / *Querelle*, pp. 281-287), Schoock discusses all five of Regius' mechanical principles: "Mens, mensura, quies, motus, positura, figura / Sunt cum materia cunctarum exordia rerum." See Schoock,

than a constitutive principle. When taking *positura* (situation) as a principle of physiology, the Cartesians argue that situation may sufficiently explain an activity of organs. What they actually do, however, is to confuse "circumstances and conditions for action" with "principles of actions".[19] For if "situation" be in itself an effective principle of action, a clock would not need any weights in order to move its hands. A clock does need weights however, since, unlike natural objects, it has no "first and natural principle of movement."

The objection had already been made by Voetius, to whom Regius had answered that he regarded the weights and springs of a clock to be its *causæ moventes internæ*, its internal moving causes. However, the flaw in this argument, according to Schoock, is that weights and springs are not essential to the mechanism of the clock itself. To argue that they are, is like saying that the wind forms an integral part of the windmill, or that

> a roasting-spit would not be a roasting-spit when the dogs that normally rotate it are set loose on the street, or are out hunting.[20]

Just like the wind and the dogs, the weights and springs of a clockwork are also "violent" principles of action only. And even if they be considered internal parts of the mechanism, there are still other violent principles by which their action is to be explained. As Voetius put it:

> a well-disposed and well-equipped clock is still not actually moved without something external moving it, namely a person's hand or the hanging or pulling of a weight in its place; just as a harpsichord, a cyther and a lyre, [though] perfectly disposed and equipped to produce music, do not actually produce it without the addition of an external motor, namely the hand of the performer. Likewise, perfectly disposed mill-stones do not actually grind without an external impulse or pulling of water, wind, horses or mules etc. Let us add that an hydraulic organ or harpsichord does not actually produce sound

Admiranda, p. 198 / *Querelle*, p. 282; Regius, *Physiologia sive cognitio sanitatis*, I, I, p. 5, and above, chapter 5, note 17.

[19] Schoock, *Admiranda*, p. 210 / *Querelle*, p. 286.

[20] Schoock, *Admiranda*, p. 138 / *Querelle*, p. 248. In the *Responsio*, Regius had indeed said: "pondus appensum & spira contorta sunt causæ moventes internæ, cum sint horologij partes essentiales seu integrales" (*Responsio*, p. 26). In the Physiologia, Regius had been more careful, although even there, he does seem to play down the importance of the external agent, when comparing natural respiration with action *in automatis*. Regius, *Physiologia* II, II (3), p. 27.

> without a movement or pulse of water in its underground tubes; yet this water cannot be called a "part" of the organ, much less its "internal motor".[21]

The argument is thus, that none of the mechanical principles under discussion are of a dynamic nature. As far as Voetius and Schoock are concerned, in the mechanical interpretation of the heart, none of the mechanical principles of position and shape can account for the cause of the heart's beatings. Although they are all conditions of action, they cannot be *causes* of action.

The champion of mechanicism might reply that, although this may be true for static principles such as shape, dimension and position, it is not true for the principle of motion. Once motion is admitted as a principle, there no longer remains any incongruity between the description of circumstances and conditions on the one hand, and the action itself on the other. As in the case of a clock, the mechanism need only be operated—the motor need only be started. Yet again, such an argument would not have impressed the Scholastic opponent. According to Schoock, it is futile to try to complete the clockwork-comparison by introducing movement. Indeed, how could movement, which has "come forth from some other principle,"[22] figure itself as an explanatory principle? Instead of explaining anything, movement would rather itself need explanation. The question is not, according to Schoock, whether there is any movement in the heart, the question is: *what causes that movement?* Even granting that there is a second movement by which pulsation may be explained, this movement would again have to be self-explanatory in order to serve as a principle.[23]

---

[21] Voetius, *Narratio*, pp. 44-45 / *Select. Dispp.* I, p. 876 / *Responsio*, p. 24 / *Querelle*, p. 110.

[22] Schoock, *Admiranda*, p. 208 / *Querelle*, pp. 285-286.

[23] Against the principle of movement, Schoock argues moreover that in Nature things grow, die and propagate, thus exhibiting the kinds of movement the Cartesians all too easily forget, for instance *motus generationis*, *corruptionis*, *accretionis*, *decretionis*. (*Admiranda*, p. 209 / *Querelle*, p. 285.) How could local motion account for all these? Surely, a clock does not come into being because of its moving parts? Why is it then, that we should believe natural processes to be governed by this principle? According to Schoock, the accident of motion is no more adequate as an effective principle than are the "static" accidents of shape, size, position and the like.

6.1.3 Conclusion

The Scholastic adversaries of Cartesianism thus argue that if the active principles of natural objects were indeed comparable to those of mechanical devices, an external motor would be needed, since an external motor is necessary in the case of artificial movements, such as that of a clock, of a windmill, or of robots.[24] The question is, in one sense, a question of *inertia*. Since experience tells us that no cart will move without something pulling it, it is very plausible to conclude that every motion needs a force that moves it—which is exactly what all pre-Galilean mechanics taught.

Yet to put the argument in such modern terms would not be fruitful to our understanding of what Schoock and Voetius were trying to say. Some of the arguments which Voetius and Schoock present to the Cartesians with respect to the question of mechanical principles, seem to imply that Voetius and Schoock merely wish to point out that the mechanistic explanation is not complete. Voetius for instance says that the mechanistic principles offer *conditions* for movement only:

> the disposition of the moveable to move in virtue of its quantity, shape, position, is neither an activity nor a causality of an efficient cause, but only a necessary condition and a *causa sine quâ non*.[25]

Therefore, the moveable stands in need of an external mover:

> As a consequence, since no potentially moveable thing can actually move itself or determine itself to move, some external motor remains to be sought which turns the potency into act.[26]

It would seem that according to Voetius and Schoock, the Cartesians must introduce some sort of external motor. Yet the the introduction of such an external motor activating the mechanical world is in fact thought to be no solution at all. When arguing that an external motor would be needed, Voetius and Schoock do not offer the Cartesian a

---

[24] Voetius mentions various automata, or "Dedalian statues", such as "the bronze ox of Phalaridis, the flying dove of Archytas, the bronze head of Pope Sylvester and the speaking statue of Albert the Great." Voetius, *Narratio*, p. 40 / *Select. Dispp.* I, p. 872 / *Responsio*, p. 11 / *Querelle*, p. 106. Schoock also mentions Archytas' wooden pigeon and the talking statue of Albert the Great, *Admiranda*, p. 133 / *Querelle*, p. 246. For further details about these legendary robots, see Theo Verbeek's notes, in *Querelle*, p. 470, note 64. See also above, chapter 5, note 22.

[25] Voetius, *Narratio*, p. 40 / *Select. Dispp.* I, p. 873 / *Responsio*, p. 12 / *Querelle*, p. 106.

[26] Voetius, *ibidem*.

simple way out of their problems. According to them, the option of introducing an external motor is in fact a recognition of one's failure to explain natural motion. In the next section, we shall see why.

Let us, however, for the moment return to the example of the heart. Martin Schoock argued that the experiment with the little heart of an eel was no proof for the mechanistic explanation. The prolonged movement could not be explained by the mere disposition of parts, since the movement ceased to be at a time when the disposition of parts was still in order. However, there was more to say with regard to the experiment. Regius had added in the *Responsio* that the heart appeared

> to regain its pulse and to somehow revive by adding blood and by heating it slightly after it has already lost all of its motion.[27]

Schoock argues that it is no use that the Cartesians seek recourse in the fact that even after having ceased beating, the heart may be reanimated by injecting a few drops of tepid blood.[28] According to him, the reanimation could well be caused by some remaining "animal spirits" which are the "instruments of the substantial form".[29]

Schoock's alternative is neither original nor profound. Yet the argument is an interesting one in that it shows that it would not help the Cartesian to argue that it is for mechanical reasons that a pulsation reoccurs by adding blood. Even when it would be granted that the human hand which adds the blood would be the *external motor*—that is, the "second movement" by which the first, pulsation, might be explained—this second movement would meet the same objections: it could not serve as a principle. Indeed, the only type of cause that could serve as a principle, is one that refers to the substantial form. This is the cause Schoock refers to. The blood, for instance, might contain some animal spirits, but these are only intermediate factors. They are *instruments of the form*. Seeing a heart that beats, the Scholastic philosopher will argue in the same way as the natural philosopher does with respect to all other examples of natural operations, namely that the heart has "something specific (*peculiare*)"

---

[27] Regius, *Responsio*, p. 27.

[28] Schoock, *Admiranda*, p. 112 / *Querelle*, p. 235.

[29] Schoock, *ibidem*. The argument may be by Voetius, since Schoock borrows it from the *Prodromus*, which was published by Voetius' student Lambertus van den Waterlaet in reaction to Regius' *Responsio*.

which "must be its principle." This specific something is nothing else but the substantial form.[30]

Thus, when Voetius and Schoock argue that the mechanical principle of motion is not self-explanatory, they are not simply saying that, in the mechanical explanation, external motors are wanting. In fact, they are not really interested in external motors at all.

6.2 The Composite Individual

Voetius objected to Regius and Descartes that instead of offering principles of motion, they only offer conditions for motion. Therefore, in the mechanical philosophy

> some external motor remains to be sought which turns the potency into act.

Voetius presents the Cartesians with a list of possible alternatives: "What will they have to offer?", he asks,

> The Platonic-Vergilian world-soul, or intelligences, or God, atoms, or heavenly globules? Something ought at least to be named here.[31]

The Cartesians should, in other words, make their choice and pick any one of these alternatives as the external motor forcing the potency of their mechanical conditions into act. In fact however, Voetius does not take the option of introducing an external motor as a serious one. To contemporary readers, his list of alternatives cannot have failed to reveal his scorn. According to Voetius, the mere prospect of having to introduce "external motors" such as a world-soul, God or atoms, should be enough to abandon the mechanical project entirely.

---

[30] The formulation is borrowed from Schoock's remarks with respect to the specific principle of air: "peculiare quod in aëre sit principium, forma scil. substantialis." Cf. *Admiranda*, p. 215 / *Querelle*, p. 289.

[31] Voetius, *Narratio*, p. 40 / *Select. Dispp.* I, p. 873 / *Responsio*, p. 12 / *Querelle*, p. 106. A somewhat similar passage may be found in Schoock's *Admiranda Methodus*: "At quis ille est? num anima mundi Platonica? an vero ad portum locumque communem omnis ignorantiæ confugiendum erit; et quoniam in natura nihil occurrit, Deus sollicitandus? Nisi forte fingere oporteat cum Academiæ principe in Timæo Genios [...]." However, this argument occurs in a slightly different context. Schoock asks how the insensible particles of the corpuscular philosophy can "combine in order to complete natural objects". According to Schoock, the Cartesians will have to introduce some sort of cosmic architect to explain this. Schoock, *Admiranda*, p. 219 / *Querelle*, p. 291.

6.2.1 External Motors Rejected

In formulating his objection concerning external motors, Voetius draws upon a specific body of Scholastic arguments in which the philosophies of those who introduced "external motors" were univocally condemmed.[32] Important examples of these arguments may be found for instance in both *Summæ* and in the *Quæstiones Disputatæ De Potentia* of Thomas Aquinas.[33]

6.2.1.1 Aquinas Against the Platonists

Aquinas defends the inherent activity of natural objects primarily against the Islamic Kalâm, against Plato, against the "Platonists" Avicebron and Avicenna, and, finally, against the atomists, all of whom are thought in some way or another to err with regard to the question of the action of natural bodies.

The Islamic theologians, or Mutakallimûm[34] go furthest in doing so, arguing that there is actually no action of corporeal substances at all, but that every action stems directly from God. It is their view that, since all "natural forms" are accidental and since an accident cannot "step over" (*transire*) into another subject, it is impossible that any natural thing could "induce" its form into another.[35] The argument

---

[32] I have previously dealt with this matter in my article "New Philosophy to Old Standards: Voetius' Vindication of world Concurrence and Secondary Causality", *Nederlands Archief voor Kerkgeschiedenis / Dutch Review of Church History* 71 (1991), pp. 58-91. The following section will recapitulate parts of what I have written there.

[33] Aquinas, *Summa Theologica* I, qu. 115; *Summa contra Gentiles* III, Ch. 69 and *De Potentia* qu. 3, art. 7.

[34] Aquinas speaks of the *loquentes in lege Maurorum* ("Kalâm" being Arabic for "speech") and refers to Maimonides for their views. See Aquinas *De Potentia*, qu. 3, art. 7; ed. Marietti Vol. V-II, p. 56. For Aquinas' discussion of not only the Kalâm, but also Avicebron and Avicenna: Étienne Gilson, "Pourquoi saint Thomas a critiqué saint Augustin", in *Archives d'Histoire doctrinale et littéraire du Moyen Age* 1 (1926), pp. 6-127.

[35] The Mutakallimûm also taught a kind of atomism of non-extended elementary particles. Some of the Mutakallimûm were not atomists, but all believed that accidents have to be created perpetually by God from each moment to the next. Cf. William Lane Craig, *The Kalâm Cosmological Argument*, London (The Macmillan Press) 1979, p. 5. A discussion of Mutakallimûm atomism may also be found in Kurd Lasswitz, *Geschichte der Atomistik vom Mittelalter bis Newton*, Hamburg und Leipzig (Leopold Voss) 1890 / reprint Hildesheim (Georg Olms) 1963 and 1984, Ier Band, "Die Erneuerung der Korpuskulartheorie", pp. 134-150. In the work to which Aquinas refers when commenting on these Islamic writers, viz. Maimonides' Guide for the Perplexed (Book I, Ch. 73; see S. Munk's French edition: *Guide des Égarés*,

seems to be based upon the consideration that accidents cannot change subject. If, for instance, fire—which is a subject—is hot, this heat is an accident, or property of the fire. The idea is, that this property cannot leave the fire in order to be an accident of something else. Hence, if something is heated, there must be a force that creates the heat in whatever object comes in the presence of fire. This creating force can only be God Himself.

Aquinas' objections to this ancient type of "occasionalism"[36] are threefold. First, he says, the theory is manifestly at odds with what experience teaches us. For if the species of heat was transmitted to our sense organs by a different agent from the fire, then the heat we feel would not be the heat of the fire, nor would we feel that the fire is hot, although the judgement of our senses, which is infallible, nevertheless teaches us these things. Thus, if God directly causes all action, we should be deceived by our judgement of sense. Secondly, Aquinas objects, if God were to produce all natural action, forms and virtues would be assigned to Nature in vain, nor would the 'apposition' of fire and wood be needed, if God burned the wood without the fire. Finally, the view that Creation is considered similar to its Creator with respect to being only, but not with respect to acting, is an insult to divine goodness. Hence the idea of God operating in every natural act is to be rejected. But in any case, it is based on unsound suppositions. For although one could say that an accident is not transferable to another

---

Paris 1856 / reprint (Maisonneuve) 1960, pp. 375-419), the atomism of the Mutakallimûm is offered a lengthy discussion in which the aspect of God's continuous creation of accidents is also mentioned. See also: Étienne Gilson, "Pourquoi saint Thomas a critiqué saint Augustin", referred to in the previous note.

[36] Kurd Lasswitz (*Geschichte der Atomistik*, p. 145), Étienne Gilson ("Pourquoi saint Thomas a critiqué saint Augustin", in *Archives d'Histoire doctrinale et littéraire du Moyen Age* 1 (1926), p. 12) and R. Garrigou-Lagrange (*Dictionnaire de Théologie Catholique* Tôme XIII, col. 33) all compare the theory of the Mutakallimûm with later occasionalism. Note however that Martial Gueroult, in his *Malebranche*, Tôme II, "Les Cinq Abimes de la Providence", argues against confusing the ideas of the Mutakallimûm, or of other thinkers preceding the era of Cartesianism, with those of Malebranche. According to Gueroult: "à l'irrationalisme naïf de ces doctrines rudimentaires Malebranche échappe par deux notions essentielles qui leur font défaut: celle d'occasion et celle de loi." Gueroult, *op. cit.*, p. 109. To decide whether or not the Mutakallimûm do in fact accept the notions of occasion and law would involve such a number of theological subtleties, that the question cannot be resolved here. We may, however, without any difficulty refer to both types of theories as "occasionalist" theories in as far as both accept the idea that God acts *at the occasion* of natural phenomenon which, without His action, would not have any natural effect.

subject, this is only numerically so. The *same* heat cannot be at once in the fire and in the heated object. However, by virtue of its accident, the fire can and does produce its specific like in other subjects, which is in fact what all natural action amounts to. Hence may Aquinas conclude that although God concurs in all natural operations, this is no reason to deny the efficacy of secondary causality and to say, as the Mutakallimûm do, that it is God who acts in the presence of what we regard as natural causes.

Close to the view of the Mutakallimûm comes that of Avicebron, who, in his *Fons Vitæ*, taught that all bodies act by force of a spiritual power that penetrates them.[37] Aquinas presents three reasons in favor of this view. First, that every agent but God needs some material subject in which to act. But since nothing material is subjected to corporeal substances, these cannot perform any action. Second, quantity hinders movement. But corporeal substance is bedded (*implicita*) in quantity. Hence it cannot act. Third, corporeal substance stands last among created things, most remote from the First Agent. But the First Agent is pure act. Hence corporeal substance is purely patient and does not act at all.

Aquinas discusses all three of these arguments in *De Potentia*, but what he generally objects to in Avicebron's view, is that the latter overlooks the fact that created substances, or natural objects, are *composites* of matter and form, and that this prevents them from being purely passive. The substantial form being active, and matter being passive, all composites of matter and form are both.

Thus, Aquinas sees in Avicebron's view a misjudgement of the composite character of natural bodies, as accepted in all Aristotelian philosophy. Moreover, he identifies Avicebron's theory as Platonic, since it depends on the supposition that "incorporeal substances are participated, determined and limited (*contractas*) by matter"—an imprisonment that seems to preclude the individual form from being active. Aquinas rejects this particular argument in favor of a universal force on the ground that it

> does not prove that the corporeal form is not an agent, but [only] that it is not a universal agent.[38]

---

[37] Aquinas, *De Potentia*, qu. 3, art. 7, ed. Maggiòlo, p. 57. See also: *Summa contra Gentiles* III, Ch. 69, where the Mutakallimûm are also mentioned.

[38] Aquinas, *Summa Theologica* I, qu. 115, art. 1.

A third account of bodily action rejected by Aquinas is Plato's. As against Avicebron's type of Platonism, Aquinas says that Plato himself only held that the substantial forms were separated from the material world and that they can therefore not be regarded as causing material bodies to act.

Now instead of introducing a spiritual power to account for the obvious fact that corporeal agents do act—in other words, that there *is* motion in the material world, Plato, and, following him, Avicenna, attribute this motion to the *accidental* forms of corporeal creatures, such as "the great and the small", which "dispose matter to the substantial form." The perfection of created objects however, is, according to these thinkers, caused by "an immaterial principle", viz. the separated forms, or "species *sive* [ideæ]".[39] Aquinas' account of the Platonic view is rather vague, but what he seems to have in mind is that Plato, like Avicenna, regarded the action of natural bodies to lie in a material re-arrangement of the accidents (i.e. figure and shape) of the patient, which thereby becomes fit to receive the influence of the separated forms, or Platonic Ideas.[40]

However, according to Aquinas, the idea of separate forms is to be rejected on quite the same grounds as was Avincenna's universal intelligence. Indeed, against all "Platonist" accounts, Aquinas generally alleges that they exhibit a pre-Aristotelian oversight of (1) the notion of potentiality and (2) the composite character of natural objects. Once the view is accepted that substantial forms may be

---

[39] Aquinas, *Summa Theologica* I, qu. 79, art. 3.

[40] Aquinas' phrasing ("an immaterial principle") may recall Aristotle's account of the Platonic doctrine. Cf. *Physica* I 4, 187^{a}12-20. Yet the interpretation seems primarily based on Avicenna, who, in Book IX, Ch. 5 of his *Metaphysics*, elaborated the view that the heavenly spheres influence corporeal bodies in such a way that they become materially disposed to receive a certain form *de intelligentis seperatis*. As the Latin translation reads: "Evenit igitur quod, cum hæc res appropriaverit aliquam de impressionibus cælestibus, absque mediante corpore elementari, vel mediante ita ut ponat illud secundum aptitudinem propriam post communionem quæ erit in sua substantia, tunc ab hoc separato fluet forma propria et describetur in illa materia." Cf. *Avicenna Latinus, Liber de Philosophia Prima sive Sciencia Divina*, Édition critique de la traduction latine médiévale par S. van Riet, Louvain / Leiden (Peeters / Brill) 1980, Tôme II, p. 489 (A 411). Note that Avicenna's description occurs as part of his theory of emanation, and not specifically of causation. Yet, as Étienne Gilson has rightly argued, "le problème posé par les opérations des causes secondes en général [...] n'est qu'un cas du problème universel de la production des êtres." Cf. Gilson, "Pourquoi saint Thomas a critiqué saint Augustin", in *Archives d'Histoire doctrinale et littéraire du Moyen Age* 1 (1926), p. 38.

potentially existent in matter, various problems are solved. Since substantial forms are "concreated" with matter, there is no need of supposing a creation of forms in every single instance of natural action (which, according to Aquinas, the Platonic view amounts to); nor of supposing the forms of material bodies to be dependent on separate intelligences; nor, finally, of reducing natural action to an activity of the accidental forms.[41] Against the latter idea in particular, it may be said that, since in any action the cause must be at least as powerful (*potis*) as the effect, accidents cannot by themselves be the cause of anything substantial. Now be it true, Aquinas says, that an active quality, say heat, is an accident, it nevertheless acts "in virtue of the substantial form" (*sc.* of the fire), whence there is no contradiction in saying that it is able to, as it were, "awaken" a (second) substantial form which was potentially existent in the body acted upon. There is therefore no need to regard accidents as being themselves the cause of action.

Finally, the fourth account of bodily action contested by Aquinas is that of the atomists. His interpretation is based upon that of Aristotle, and amounts to the view that the atomists define corporeal action as an emission of atoms. Passivity would, on the other hand, be regarded as consisting of the reception of atoms in the pores of material bodies.[42] Aquinas' refutation is brief. A passive body would not be passive as a whole (since only its pores would suffer action) and the quantity of an active body would diminish as a result of its action (viz. the emission of atoms). Both of these consequences are, according to Aquinas, manifestly false.[43]

---

[41] Aquinas, *Summa Theologica* I, qu. 45, art. 8; qu. 65, art. 4; and qu. 115, art. 1, respectively.

[42] Aquinas, *Summa Theologica* I, qu. 115, art. 1.

[43] Despite the apparent lucidity of the argument, it is not entirely clear what is meant here. Aquinas refers to Aristotle as a source for his refutation of atomism: "Quam opinionem improbat Aristoteles in I *de Generat.* Sequeretur enim [etc.]" (Aquinas, *ibidem*). In Aristotle's account (*De Generatione* I 8, 326^b6-326^b28), the question is stated in the following way: if the agent does not work upon the patient through contact, then neither will it produce an effect by passing through its pores. On the other hand, if action is made by contact, then the pores are superfluous, since they might as well be filled. For what it is worth, the argument would imply that the pores-hypothesis is unnecessary for explaining the passivity of the whole body. This is not the same as saying that a whole cannot suffer action by way of an influence on its pores alone. The second of Aquinas' arguments, viz. that the quantity of the agent would diminish as a result of the emission of atoms, is a rather unexpected one. At 327^b22-25, Aristotle in fact denies the possibility of growth and diminution by way of "apposition" alone. As Albertus Magnus interpreted the

Aquinas' account of non-Aristotelian explanations of corporeal action either depends on Aristotle's own criticism of his philosophical predecessors, or, in case of later writers, conveys a tendency to identify new theories as reformulations of those already criticized by Aristotle. Accordingly, St. Thomas Aquinas is of the opinion that all problematic aspects of theories which, in one way or another, favour the idea that natural objects are inactive, may be eluded by accepting individual substantial forms. The alternative 'dead matter' conception of natural objects would lead to (1) the attribution of all corporeal action to God as its unique and direct cause; or (2) the introduction of separate intelligences, either as direct causes of corporeal action (Avicebron's view), or as causing the perfection of material bodies (Plato, Avicenna); or, finally, to (3) the acceptance of atomism.

6.2.1.2 Later Sources

In Voetius day, listing alternative explanations of bodily action in the manner of Aquinas was very common, especially in commentaries on Aristotle's Physics. The *loquentes in lege Maurorum* which Aquinas had spoken of were no longer recognized as the Mutakallimûm. Instead, in later years, reference was mostly made to the German nominalist Gabriel Biel (c. 1425-1495), who, regarding the efficacy of the sacrements, had said that

> just as God has ordained that fire produces heat in a proximate patient, so could He ordain that fire produced heat [but] not in the patient or subject. For there is no contradiction involved. That is to say, the former decree was purely contingent; thus it could have been and can be changed.[44]

---

passage, this means that when "active parts" enter the "passive pores" of bodies, the bodies themselves will not in any way have become greater. See Albert's commentary on *De Generatione* in the *Opera Omnia*, ed. Hossfeld, Münster (Aschendorff) 1980, vol. V-II, p. 169. Presumably, the same would hold for the emission of atoms. However, Aquinas does not explain his position.

[44] Gabriel Biel, *Collectorium circa quattuor libros Sententiarum* ed. Wilfried Werbeck / Udo Hofmann, Tübingen (J. C. B. Mohr / Paul Siebeck) 1975, L. IV, dist. 1, qu. 1, art. 3, p. 30. Note that Biel rejects the argument from sense-perception which Aquinas put forward against the Mutakallimûm, arguing that one cannot judge from the senses whether heat arises from the action of the fire or directly from God: "Præterea: Ad sensum apparet quod ad præsentiam ignis incipit calor. Et hoc non negatur. Sed an ille calor incipiat per actionem ignis vel solius Dei, non apparet ad sensum. Et per hoc negans actionem ignis nihil negat apparens ad sensum." Biel, *idem*, p. 31.

It was this argument which replaced the Mutakallimûm doctrine to become the prime example of an "occasionalist" theory holding that God acted without any intervention of subsidiary causes. But apart from that, the examples were still the same and Avicebron, Avicenna and the atomists remained to be cited.

There can be little doubt that Voetius' remark concerning the introduction of external motors is to be placed in the context of the same tradition. Indeed, when Voetius enumerates the alternatives which the adherents of the New Philosophy are expected to offer in order to make good their rejection of the substantial form, his enumeration is identical to that of Aquinas. Arguing that accidental properties—in Voetius' case, the mechanical principles of Regius and Descartes—are simply not enough for explaining any activity, he presumes that the followers of the New Philosophy will come up with either God, spiritual forces, or atoms. Hence his reference to

> the Platonic-Virgilian world-soul, or intelligences, or God, or atoms, or heavenly globules,[45]

---

[45] Voetius, *Narratio*, p. 40 / *Select. Dispp.* I, p. 873 / *Responsio*, p. 12 / *Querelle*, p. 106. Note that Voetius' reference to the "Platonic-Vergilian world-soul" is not taken directly from Aquinas. Voetius is here thinking of the spirit "moving matter" and the "fire of heavenly origin", which, according to the sixth book of Virgil's *Æneidos*, bring life "as long as harmful bodies do not hinder, and earthly limbs and mortal members blunt it." P. Vergili Maronis *Æneidos* Liber Sextus, lines 724-752, pp. 24-25 of R.G. Austins edition, Oxford (Clarendon) 1977, with a commentary on pp. 220-232. See also: Eduard Norden, *P. Vergilius Maro Æneis Buch VI*, Stuttgart (B.G. Teubner) 1957^{4}, pp. 92-95 (Latin text and German translation) and 310-316 (commentary). In Voetius' *Dissertatio Epistolica de Termino Vitæ* of 1634 moreover, we find both the *anima mundi Platonica* (but, this time, also *Paracelsica*) and the *Spiritus (...) universi & communis motor Stoicorum*, mentioned and rejected by Voetius amongst a host of other world-governing principles in the *Dissertatio*. Voetius, *Select. Dispp.* V, p. 47, and section 9.5.3, below. These references however, do not really bring into question the Thomistic origin of Voetius' argument against the New Philosophy. On the contrary, the very fact that reference is made only to the *Platonic* (be it also a "Vergilian") world-soul in the essay against the New Philosophy, prompts us to believe that Voetius was here only thinking of an anti-Platonist argument, such as we find in the works of St. Thomas and not of the various other chemical, astrological and Hermetic theories which occur in the *Dissertatio*. Plato introduces the idea of a world-soul, presented as a mixture of Same, Difference and Being, in the *Timaeus*, 34 B ff. As Desmond Lee, in his Penguin edition of the *Timaeus* rightly says: "Plato (like all Greeks) believed that all motion must have a cause [and] that the soul, as a self-mover, is the ultimate cause of motion." Plato, *Timæus and Critias*, Translated with an introduction and an appendix on *Atlantis* by

all of which are external principles supposedly required to activate the universe of natural objects deprived of their substantial forms.

In attacking the New Philosophy, Voetius does not refer to Aquinas, nor any other writer for that matter. But then, Aquinas' arguments were probably too well known to be explicitly referred to. Two sources which elaborate on the same ideas and which were both known to Voetius, are Antonio Rubio's commentary on Aristotle's *Physics* (1605) and the famous commentary on the *Physics* by the Conimbricenses. Rubio (or Ruvius, 1548-1615) was a Jesuit theologian whom Voetius praises together with the Conimbircenses as the best sources discussing the *controversiæ* of physics.[46] Ruvius' discussion of efficient causality includes a step by step exegesis of St. Thomas' texts on the Arabic Platonists. He refers to the "three positions against an efficacy of created things" and names Avicebron's view as the second and Avicenna's as the third. Instead of referring to the Mutakallimûm, he mentions Gabriel Biel as an example of the first type of theorists rejecting corporeal action.[47] The Conimbricenses likewise comment on those who "deprive secondary causes of their actions", discussing the position of Biel (amongst others) as an example of the Mutakallimûm-argument and further of Avincebron and the "Democritians", i.e. the atomists.[48]

Thus, when Voetius refers to "the Scholastics and the modern metaphysicians and theologians" who have written on these matters,[49] he is referring to a long tradition of thinkers following Aquinas on this point—a tradition moreover, which was not limited to Catholic authors. Voetius' Utrecht colleague Arnoldus Senguerdius (1610-1668) for example, mentions in his course of Metaphysics, the "old

---

Desmond Lee, London (Penguin Books) 1983, pp. 46-47. For an enumeration of later interpretations of the Platonic idea of a world-soul, see A. E. Taylor, *A Commentary on Plato's Timæus*, Oxford (At the Clarendon Press) 1928 / reprint New York & London (Garland) 1987, pp. 109-136.

[46] Voetius, *Excercitia et Bibliotheca Studiosi Theologiæ*, Utrecht 1644, p. 382: "II. Controversiarum scriptores; quales I. *ex recentioribus Scholasticis Theologis* imprimis commendandi *Conimbricenses*, *Ruvius*, Toletus, Pererius; quorum duo priores subtilitate posteriores facilitate & perspicuitate præcellunt." The italics are taken from the original.

[47] "Gabriel in 4. d. 1. quæst. 1. art. 3"; Ruvius, *Commentarii in octo libros Aristotelis de Physico*, Madrid 1605, p. 294.

[48] *Commentarium Colegii Conimbricensis S.J., in octo libros physicorum Aristotelis Stagiritæ* II, Ch. VII, qu. 9, ed. Köln 1625, cols. 346-350.

[49] Voetius, *Narratio*, p. 40 / Select. Dispp. I, p. 873 / *Responsio*, p. 12 / *Querelle*, p. 106.

opinion (...) that created things do not operate at all" and refers directly to Aquinas' texts.[50] In Utrecht itself, therefore, these issues had been discussed in public disputation.

6.2.2 *The Debate About Composite Natures*

It is therefore no surprise that when, in 1641, Regius debated theses which seemed to undermine the idea of individual substances possessing an individual principle of operation, Voetius and his colleagues came to the rescue of the forms.

Regius at various times aroused the suspicion of his fellow professors. The New Philosophy drew more and more attention ever since, on 10 June 1640, Regius had presided over a public disputation on the circulation of the blood.[51] The trouble really started when, on 8 December 1641, Regius had put forward the thesis that "man is an accidental being", an *ens per accidens*. According to the official report, the public disputation was disturbed by Regius' students. Although the master himself was forced to silence, since his opinions were shown to lead to some "often very dangerous absurdities", his students nevertheless made such a noise that further discussion was impossible.[52]

Reactions were soon to follow, both from the senior professor of medicine, Guilielmus Stratenus and from Voetius.[53] It was not just because academic rules were broken. It was also the ideas themselves which troubled Voetius and his colleagues. In the *Narratio*, they quoted students who agreed that

---

[50] Arnoldus Senguerdius, *Collegium Metaphysicum*, Ultraiecti, Ex Officinâ Ægidii Roman, 1640, p. 160: "Quare rejicienda est vetus opinio quæ, ut ex Thoma. *3. Contra Gentes c. 69 & l. p. 105 art. 5.* habemus, statuit, res creatas nihil omnino operari, sed Deum omnia ad earum tantum præsentiam efficere, ita ut non ignem, sed Deum calefacere ad ignis præsentiam; ut & sententia Avicembron qui ponebat, nullum corpus esse activum, sed virtutem substantiæ spiritualis pertranseuntem per corpora exercere operationes, quæ à corporibus provenire videntur; improbandus etiam est Avicennæ error, qui volebat, res corporeas posse efficere accidentia solum, non autem substantias." Senguerd had been teaching metaphysics in Utrecht since 1635.

[51] *Narratio*, p. 14 / *Querelle*, p. 87. See section 1.3, above.

[52] *Narratio*, pp. 22-23 / *Querelle*, pp. 93-94.

[53] *Narratio*, p. 35 / *Querelle*, p. 103. See section 1.3, above.

> it is unbearable that paradoxes concerning man, that he would be a being only by accident, [and] concerning the efficacy of quantity etc., are debated in the university.[54]

It was, according to these students, especially unbearable when such controversial issues were so poorly defended. But they were unbearable anyway. The students who were interviewed must have had a clear idea that both the paradox concerning man and that concerning the efficacy of quantity were nothing but Platonism in disguise, that is to say, samples of a philosophy re-introducing the idea of external motors.

6.2.2.1 The Question of the Ens per Accidens

External motors as such were not unknown to Scholastic philosophy, even to the most orthodox "Aristotelian" types. There are in fact forms which might act as external motors instead of immanent ones. The internal motor of natural objects, the substantial form, is the *forma informans*, or "informing" form. Such a form is internal: it *informs* the matter in which it is inherent, so that the object, consisting of matter and form, is not merely a material object. Besides being material, the object is also of a certain type. Moreover, it is not inactive, since it is not made of dead matter only. The object operates through the form infoming its matter.

However, forms may also act from outside. This is the case with "assisting" forms, the clearest example of which is that of an angel, incarnated in human flesh. The angel guides the body in such a case as a *forma assistens*, overruling, as it were, the substantial form of the human body, or (in case it is a living individual which the angel has taken possession of) of the human being as a whole. Being possessed by an angel or a demon is thus a way of being assisted by a form from without. The angel or demon is the external motor of bodily action. This is also what Voetius had in mind, when he argued in the essay on forms, that from Regius' thesis that man is an *ens per accidens*, it would follow

> that an angel, or a demon in the body of someone possessed [...] is neither more nor less a unity than the soul which is in the body.[55]

[54] *Narratio*, p. 23 / *Querelle*, p. 94.
[55] *Narratio*, p. 48 / *Select. Dispp.* I, p. 879 / *Responsio*, p. 33 / *Querelle*, p. 113.

Contrary to the soul, which is an informing form, an assisting form does not form a unity and a "being by itself".[56] But if man were an *ens per accidens*, then the unity of body and soul would be no different from the unity between a person possessed and the demon taking hold of him!

The reason Regius had for arguing that man is an *ens per accidens*, was that the soul and body "can subsist separately one by one", and

> what can be present or absent without the subject passing away, is called an accident.[57]

Thus, the soul and body being able to exist the one without the other, their union may be regarded as accidental. In his excellent book on causal images in Cartesianism, Rainer Specht has drawn attention to the resemblance between the Cartesian idea of the relation of body and soul and the Scholastic image of the incarnated angel.[58] Voetius had made this conclusion for himself and both he and his students were aware of the dangerous consequences.

The thesis that man is an accidental being, was, in the first instance, attacked on theological grounds. According to Voetius, the Bible shows that

> man is a single kind of substance and animal, created as one essence or nature out of soul and body.[59]

---

[56] Theo Verbeek, "'Ens per Accidens': Le origini della *querelle* di Utrecht", in *Giornale Critico della Filosofia Italiana* 1992 (VI-12), p. 288: "La forma sostanziale 'informa' la materia, cioè forma con essa un tutt'uno individuale e in sé. La forma 'assistente' al contrario non crea un 'essere in sé', ma un 'essere per accidente', che ha con il corpo un rapporto estrinseco e strumentale."

[57] Regius, *Responsio*, p. 35. The argument is literally taken over from Descartes, *Rodis-Lewis*, p. 96 / AT III, p. 508.

[58] Rainer Specht, *Commercium Mentis et Corporis: über Causalvorstellungen im Cartesianismus*, Stuttgart-Bad Canstatt (Fromann) 1966. Note that although Regius quoted Descartes in the *Responsio*, his rejection of Descartes' metaphysical views concerning the nature of the union of mind and body would lead to a separation between Descartes and Regius in later years. In fact, Regius and Descartes were to engage in a very unpleasant polemic in the late 1640s. See Theo Verbeek (ed.), *Descartes et Regius: Autours de l'Explication de l'esprit humain*, Amsterdam / Atlant, GA (Rodopi) 1993 and "Regius's *Fundamenta Physices*", in *Journal of the History of Ideas* 55 (1994), pp. 533-551.

[59] *Narratio*, p. 47 / *Select. Dispp.* I, p. 878 / *Responsio*, p. 32 / *Querelle*, p. 112. Voetius argues that this is shown by "Gen. 2. vers. 7. I Cor. 15. 45. coll. cum. Genes. 22. 27."

Moreover, Christ is both God and man. Yet He is *one* by himself and not by accident. If Christ is one, then so is man:

> Much more a "substance" and "in itself" is therefore the unity of soul and body, since these are far from being as widely separated, and neither are they to the same degree complete substances, as are divinity and humanity.[60]

Voetius offers various other more or less theological arguments against seeing man as an accidental being, among them the danger of Socinianism.[61] However, as Theo Verbeek has argued in connection with the discussions on the *ens per accidens*, Voetius' concern may have been with Averroism rather than Socinianism. According to Averroist sources, man is made of three, rather than two constituent parts—an "intellect", or "spirit" being added as a third element besides body and soul. This intellect, moreover, would escape individuation. Accordingly, the union of soul and body to this Averroistic supra-human *intellectum separatum* would be accidental only.[62]

It may be that Voetius saw Regius' thesis on man as an *ens per accidens* as a return to the Averroist standpoint, especially since this standpoint formed part of a "classic enumeration" of "Platonist" views misrepresenting the soul's function of an informing form, much like the classic enumeration of "Platonist" theories denying the efficacy of material objects, which we met with in the previous section.[63] Thus, Voetius may have associated the question of the *ens per accidens* with Averroism. Nevertheless, I tend to think that Voetius and his colleagues were affronted by Regius' thesis for a far more general reason.

---

[60] Voetius, *ibidem*.

[61] For instance, that the "Subjectum adequatum & totale virtutis & peccati, & consequenter præmii & poenæ non esset compositum & suppositum humanum secundum animam & corpus, nec corpus secundum omnes corporeas facultates & partes: sed tantum Mens seu anima hominis [...]. Vide 1. Corinth. 5. 5. & 7. 34 & 6. 19. 20. 1. Thess. 5. 23. Rom. 6. 12. 13." And: "Non tam commode refutari possent Sociniani; qui negant corpus esse partem essentialem hominis, aut cum animâ constituere substantiam &suppositum humanum; consequenter idem numero corpus non esse resuscitandum." Voetius, *Narratio*, p. 48 / *Select. Dispp.* I, p. 879 / *Responsio*, p. 33 / *Querelle*, pp. 112-113.

[62] Theo Verbeek, "'Ens per Accidens': Le origini della *querelle* di Utrecht", in *Giornale Critico della Filosofia Italiana* 1992 (VI-12) herafter to be referred to as '*Ens per Accidens*', pp. 276-288.

[63] Verbeek, *Ens per Accidens*, p. 284, note 39 especially.

6.2.2.2 Conclusion: The Binding Element of Substantial Unity

As Verbeek himself puts it in his article on the question of the *ens per accidens*, a Universe in which the human individual is an entity by accident

> is a universe without precise boundaries, without hierarchy and without order. By and large, the objections against the thesis of man as an accidental being would therefore be the same as those against the abolition of the substantial forms.[64]

In fact, the objections against Regius' thesis are nothing but subclasses of the arguments against abolishing the substantial forms. In the paragraph in which Voetius deals with the *ens per accidens*-thesis, he emphasizes that the question is linked to "very many doctrines of metaphysics", and, amongst others, mentions those

> Of being, essence and existence, nature and supposition. [...] Of the one, unity, the composite, the whole, down to "by itself" and "by accident". [...] Of principles and causes, [...]; of the principle, internal and external; of the informing and assisting form. [...] Of substantial and accidental distinction; of the distinction between a natural and an artificial thing; of the distinction between a monster (which, as such, is an entity by accident) and human nature as properly arranged according to the laws of nature,[65]

in other words, of almost every aspect of the philosophy of forms which we have met with in the preceding chapters. The binding element in all is that on every level of Voetius' argumentation, we meet with the general underlying thesis that with the acceptation of the idea of innate principles informing the matter of material objects, all problems reaching from the inactivity of matter down to the question of the unity of body and soul, may be solved.

---

[64] Verbeek, *Ens per Accidens*, p. 283.

[65] Voetius, *Narratio*, p. 49 / *Select. Dispp.* I, p. 879 / *Responsio*, pp. 33-34 / *Querelle*, p. 113. The list resembles the inventory of metaphysical concepts of which Voetius, in his sermon given on the Sunday before the inauguration of the University of Utrecht, had said that they are of great use for theology: "als wesen ende wesentheyt / eenheyt / verscheydenheyt / goetheyt / waerheyt / perfectie / oorsaecke ende werksaemheyt / geduericheyt 'tsij tijt of eeuwicheyt / eenvoudicheyt ende compositie of t'saemenstellinge / nootsaeckelijckheyt / gebeurlijckheyt / mogelijckheyt / substantie of selfbestaende wesen / accident of toevallicheyt / qualiteyt of hoedanicheyt &c." Voetius, *Sermoen van de Nuttigheydt der Academien ende Scholen*, p. 40.

There is a well-defined textual basis for this position. The idea that natural bodies form a composite unity of matter and substantial form, is an idea which in the Renaissance commentaries on Aristotle's *Physics* was formulated in a variety of ways, all relating to the objections raised by Voetius against the New Philosophy. Thus, we find the refutation of Platonist, occasionalist and atomistic accounts of causation in the commentaries on *Physics* II side by side with the refutation of Averroism. In the commentary of the Conimbricenses for instance, the Averroists are involved when, with regard to *Physics* II, the question is put forward "whether the soul is included in the definition of nature". Against the Averroists, the soul, including its intellectual part, is said to be "the [true] form of a physical compound, namely man".[66] It is, in other words, the form of the human being. There is no reference to the Averroistic theory of a tripartite animating faculty where, in their commentary with respect to *Physics* I, the Conimbricenses raise the question "whether the substantial form and matter form a unity".[67] Yet the context is the same: the emphasis is on the composite character of substances: *Ex materia & forma substantiali fit vnum per se*. Here, the answer is weighed against the phantasies (*insomnia*) of the Platonists who hold that the natures of body and soul differ too much to form a single whole. From the commentators' point of view, the Platonic account misses the point in the same way as it did in the case of the efficacy of corporeal natures: it presents the body in the same way as it presents matter, that is to say, as wholly passive. If only we view both the body and matter in general not as complete natures by themselves, which have to be activated from without, but as part of composite natures made up of soul and body, or in possesion of both matter and form, paradoxes such as the inefficacy of natural objects and the accidental nature of man may easily be resolved.

The Aristotelian theory of causality, with its informing and moving forms has long resolved such paradoxes. No wonder then, that Voetius' students are annoyed at having to discuss them, when Regius brought back into the university such childish theses as that man would be a being only by accident, and that quantity might have some physical efficacy. Such paradoxes, theologically unsound as they are,

---

[66] Conimbricenses, *In Physicam*, II, qu. 4: "Vtrum anima in naturæ definitione contineatur?", pp. 212-213.

[67] *Idem*, I, qu. 11: "Vtrum, ex forma substantiali & materia vnum fiat?", pp. 182-184.

are, as far as physics is concerned, two sides of the same Platonic coin.

6.3 Moving Forces

Both the "Platonic" and the "Aristotelian" forms and intelligences—and in fact all other principles governing natural bodies either from within or from without—were supposed to give an answer to the question why it is that something changes, moves, or works. In modern physics, this question was first restricted as a result of the consideration that not movement, but only a change in the quantity of movement needed explanation. Later, mathematical equations were, at least in some scientific disciplines, entirely to replace the causal terminology of spirits, forms and forces.

However, in the seventeenth century, the quest for "causes" was still the main issue in physical investigation. It led to some curious interpretations of Cartesianism from the part of authors like Voetius and Schoock. It is with two of these, the charge of magic and of animism, that I shall round off this chapter.

6.3.1 Mathesis and Magic

One of Voetius' objections to the New Philosophy and one of the central questions of Martin Schoock's *Admiranda Methodus*, concerns Descartes' replacement of the accepted scientific method by a single "mathematical" one. Did not the author of the "French Method" pretend to have obtained in only a few months' time a very secure and certain knowledge of many things with the aid of geometrical and mathematical analyses?[68] Of course, says Schoock, referring to the fact that a complete exposition of the Cartesian system still had to appear,[69] "our new Archimedes" has yet to reveal how this is to be done. Schoock's mistrust of Cartesian success is obvious. Both the mathematical method and the objects of mathematics seem hardly to be of use in other fields at all.

---

[68] Schoock, *Admiranda*, p. 125 / Querelle, p. 242 and Descartes, *Discours*, AT VI, pp. 20-21: "Comme, en effect, i'ose dire que l'exacte obseruation de ce peu de preceptes que i'avois choisis, me donna telle facilité a demesler toutes les questions ausquelles ces deux sciences s'estendent, qu'en deux ou trois mois [..] ie pouvoit determiner, en [ces questions] mesme que i'ignorois, par quels moyens, & iusques où, il estoit possible de les resoudre."

[69] By 1643, the *Principia* had yet to be published. The *Discours* on the other hand, had expressly been left rather sketchy.

In itself, Schoock argues, a mathematical demonstration is, of course, of a greater certainty than the demonstrations of other disciplines. But then again, it is, amongst other things, the type of certainty by which the various disciplines are distinguished. The certainty of theological dogmas, which is founded in the authority of God as He speaks to us through the Scriptures, differs from the kind of certainty of what experience and reason teach us in the field of medicine. As a consequence, one makes an inexcusable error when demanding of every demonstration of physics the accuracy proper to the mathematical realm.

To present physical or metaphysical doctrines *geometrico modo* and thus draw up a compendium of science or of metaphysics in an orderly, "geometrical" way, would not in itself be harmful, provided one acknowledged the differences of certainty just mentioned. If one does not do so, the mathematical model is in fact quite dangerous. The Euclidian outer form does not guarantee any truth in itself. Inventing one's own axioms, one even loses all "remedy for error", since the relative consistency of propositions will never bring the unhappy deduction back on the path of truth.[70] Thus, Schoock warns the Cartesians that an axiomatic presentation will not lead to the indubitability of every scholium and proposition, but is, on the contrary, deceitful.

Schoock's warnings against misusing the Euclidian model is only one aspect of his remarks on Descartes' use of mathematics. Apart from applying the mathematical method to other fields of science, there seems to be a second way in which the Cartesians plan to extend the certainty of mathematical theorems to non-mathematical demonstrations: by not transferring the model but the *objects* of mathematics. When Schoock says that he does not see how geometrical lines and algebraic numbers could help us find the truths of Nature,[71] he is making a point about accidental properties not yielding genuine causes. The problem with mechanicism was, according to Voetius and Schoock, that by making a piece of

---

[70] *Admiranda*, pp. 123-126 / *Querelle*, p. 242. Note that whenever Schoock mentions the possibility of a presentation of metaphysics *methodo Geometriæ ac Algebræ*, he has in mind Descartes' "Rationes", or "Abrégé des principales raisons [etc.]", which, on Mersenne's request, had been added to the second set of "Replies" following the *Meditations*. Cf. AT VII, pp. 160-170 / AT IX, pp. 123-132. Schoock criticizes Axiom X of the Abrégé, see *Admiranda*, p. 124 / *Querelle*, p. 242.

[71] Schoock, *Admiranda*, p. 125 / *Querelle*, p. 242.

clockwork the norm of physiology, one erroneously takes "accidents" for intrinsic principles of motion.[72] The same may be said for the mathematical accidents of numbers and figures. Mathematical properties are only idealisations of the physical dimensions of natural, material objects. The dimensions of the objects themselves are, of course, real, but they are accidental in two ways. First, they are accidental in the sense that an object would not substantially be a different object if it was smaller or larger. Specific difference transcends difference of size. Of course, it is again the inherent form which defines the specific difference. Secondly, dimensions are accidental in a causal sense: the size or shape of a substance is wholly irrelevant for explaining the *cause* of its action. Though size and shape of valves and tubes may explain the way in which the heart works, it does not explain *why*. Again, some spirit, form or force must do the job. Or, to quote Voetius once more: "some [...] motor remains to be sought which turns the potency into act." The Aristotelian would opt for inherent forms, that is, for a philosophy in which active and passive principles are united in one whole. But what about the Cartesians,

> What will they have to offer? The Platonic-Vergilian world-soul, or intellgences, or God, atoms, or heavenly globules? Something ought at least to be named here.[73]

Voetius presents the Cartesians with a list of possible alternatives. However, instead of these, the Cartesians talk a lot of quantity and shape. Are these then expected to be the principles "animating" their mechanical world of matter? In the essay, Voetius warns:

> What is to be avoided is that by accepting [the view] that efficacy and motion can be assigned to quantity and shape, students will, as a result, one day unwittingly accept the axiom of magic, rejected until now in all Christian theology and philosophy, that "quantity and shape have a certain efficacy which, either by itself, or in combination with other [principles] concurs as an active principle of change."

---

[72] Voetius uses "motor internus" at *Narratio*, p. 45 / *Select. Dispp.* I, p. 876 / *Responsio*, p. 24 / *Querelle*, p. 110; and "motor intrinsecum" at *Narratio*, p. 40 / *Select. Dispp.* I, p. 873 / *Responsio*, p. 12 / *Querelle*, p. 106; Schoock has "principium intrinsecum" e.g. at *Admiranda*, p. 137 / *Querelle*, p. 248.

[73] Voetius, *Narratio*, p. 40 / *Select. Dispp.* I, p. 873 / *Responsio*, p. 12 / *Querelle*, p. 106.

Schoock draws the same conclusion. Being first and foremost a mathematician, Descartes will probably just stick to his "lines and figures" as do all "sons of magicians" and "associates of magic", who hold numbers to be the key to "the mysteries and secrets of Nature".[74]

Schoock's accusation must be read against its historic background, in which the project of universalising scientific methods had become a central topic of occult traditions.[75] The association of Descartes with magic however, is primarily based upon the consideration that mathematical formulae and even exact measurement of movements and actions do not yield physical causes. The charge of magic is based on the idea that mathematical properties might be regarded as physical causes. If so, shapes and numbers would be assigned the efficacy which is normally assigned to spirits, forms, forces and other causal agents—which is exactly what magicians do.

Schoock hence advises Descartes to abandon philosophy and confine his studies to the field of mathematics, in order not to be continually accused of Cabbalism and magic for his randomly playing with "lines and figures" as if thereby explaining Nature.

6.3.2 The Charge of Animism

The quest for moving forces of bodily action, whether these be forms, spirits, or the magic force of numbers, set the stage for a discussion of animism also. According to Descartes, lifeless things have no inclinations.[76] Descartes' criticism of the Scholastic theory of causes

---

[74] Schoock, *Admiranda*, p. 128 / *Querelle*, p. 243.

[75] Raimond Lull's "Ars Magna" for example, "[nam] ihrerseits kabbalistische und alchemistische Traditionen auf und [fand] zunächst eine eher okkulte als den rationalen Kern der Intentionen Lulls herausarbeitende Fortsetzung." J. Mittelstraß, "Die Idee einer Mathesis Universalis bei Descartes", in *Perspektiven der Philosophie* 4 (1978), p. 177.

[76] Except, that is, for such fundamental "inclinations" as the tendency of objects to pursue their movement in a straight line, to be moved to the Earth's centre etc. Descartes uses the phrase *inclination naturelle* for instance in *Le Monde* in relation to the phenomenon of the liquid in a glass, rising against its "*inclination naturelle*". Descartes, *Le Monde ou Traité de la Lumière*, AT XI, p. 18. Such "inclination" are, however, experimentally observed and/or commented upon by the use of analogies. Both in *Le Monde* and, later, in the *Principles*, Descartes for instance illustrates what—in the *Principles*—was to become his second law of Nature ("that all motion is, from itself, rectilinear") using the example of a gravel thrown from a sling. The gravel follows its path always in the direction of the tangent to the point at which it left the sling. This is because it is, in every single instance, "determined to continue its motion in a certain direction along a straight line". Descartes, *Principia*

is characterized in his letter to Mersenne of 26 April 1643 in the following way:

> I do not suppose that there are in Nature any *real qualities*, which are attached to substances, like so many little souls to their bodies,[77]

a clear indication that Descartes interprets the Scholastic theory as being animistic. Time and again, Descartes emphasizes that one must sharply distinguish between physical and mental notions. Souls are the privilege of spiritual beings only. It is presumably for this reason that all teleological argumentation must be rejected in physics. It is in any case for this reason that Descartes saw the Scholastic theory of forms as an anthropomorphic projection.[78] Voetius however, argued that neither the philosophy of forms, nor its finalistic aspects necessarily involve any form of animism. The idea that things operate to a certain end on account of some moving force does not force one to accept the view that every object has an inner soul.[79] With respect to the question

---

Philosophiæ II § 39, AT VIII-I, p. 64. Rectilinear motion is therefore taken to be a natural tendency of particles, for instance in *Principia Philosophiæ* IV § 19, AT VIII-I, p. 211, where it is said that celestial globules for instance will always continue their movent "along straight lines insofar as they can" (*semperque quantum possunt secundùm lineas rectas*). In this case, the obvious anthropomorphic way of expression should presumably be taken as rhetorical finery, validated by the empirical example of the sling and the metaphysical argumentation concerning the invariability of God's conservation of motion.

[77] Descartes to Mersenne, AT III, p. 648. The translation is from CSM III, p. 216.

[78] Daniel Garber, *Descartes' Metaphysical Physics*, Chicago (The University of Chicago Press) 1992, p. 99: "Indeed, Descartes thinks, the hylomorphic body of the Scholastic philosophers, form and quality joined to matter, is just the image of the Cartesian human being, immaterial soul united to extended body, projected out onto the material world." Garber gives further examples of instances in which, in Descartes' writings, this theme "comes up a number of times in the 1640s."

[79] Closely linked to Voetius' argument there is, however, another argument against the animistic interpretation of Aristotelian causal theory. As Murray Miles argues: "a very different estimate of the charge of concealed animistic tendencies would result if it were considered in the light of the characteristic late medieval employment of the substantial forms as the principal causes of motions." Murray Miles, "Descartes' Mechanicism and the Medieval Doctrine of Causes, Qualities and Forms", in *The Modern Schoolman* LXV, January 1988, p. 116. Following Fr. James A. Weisheipl, Miles argues that it is only in later scholasticism that the idea of substantial forms as "causes" or internal "motors" of action was accepted. The idea is not found as such in Aristotle or Aquinas, for whom the substantial form is merely the metaphysical structure underlying the physically active "principles",

of the movement of the heart, Regius had denied that the human soul had anything to do with it. If it were the soul that operates the heart,

> it would be conscious of its operations and would make use of the power of reason.[80]

According to Voetius however, this does not have to be concluded at all. The human soul in this case acts as an "informing form". In entirely the same way, other natural objects act according to their informing forms without the use of reason. Albeit then, that lifeless things have no conscious inclinations, they do operate in a regular manner according to the forms and ends with which they are endowed:

> A swallow for instance cures the eyes of its young without the use of reason by applying *chelidonia*, and our stomach, liver etc. digest without being subjected to the direction of reason.[81]

The mechanical account of bodily functioning may offer fine descriptions of the necessary processes involved, but, as Aristotle showed in *Physics* II 9, there may well be necessity in Nature without this necessity making it superfluous to take "ends" into account.

To achieve some end in art, certain materials are "necessary". The same is true in the case of natural processes. However, no product of art will necessarily come about from the materials themselves. And so it is with bodily mechanisms: though being necessary for some purpose, the purpose itself is what initiates the process and the materials being used for its fulfillment. Hence, Nature is governed by ends, to which inherent forms are directed. The materials are necessary also, but only in a hypothetical way. They are *necessary conditions* for the end to be achieved.

The source from which the action comes is not to be sought in the material conditions. Hence, it is as if all of Nature must be "animated". Yet according to Voetius and Schoock, Descartes' emphasis on the fact that lifeless things have no mental inclinations and Regius' objection that we are not conscious of the workings of our hearts, are very much beside the point. First, no Aristotelian has ever maintained that substantial forms are conscious faculties. Secondly, the internal principles of natural action should not be put on a par with

---

which are not themselves causes at all, the only real causes being generative causes. See above, section 2.4.4.

[80] Regius, *Responsio*, p. 27.

[81] Voetius, *Narratio*, p. 45 / *Select. Dispp.* I, p. 877 / *Responsio*, p. 24 / *Querelle*, p. 110.

external motors activating dead matter from without. In fact, the whole idea of internal principles is meant to solve the problem of inert matter in a non-animistic way. Accordingly, Voetius claims that the charge of animism should rather be directed against the Cartesians themselves. Since they do not accept the idea of composite substantial unities in which an inherent faculty informs the matter which, left to itself, would be inert, they will, in order to explain natural action, ultimately need to have recourse either to magic, or to animistic principles such as the world-soul, or to yet other intelligences and spirits.

This, Voetius warns the reader of his essay, is what the rejection of "sane and sober"[82] Aristotelian philosophy would lead to. Whatever the value of Descartes' charge of animism may be, in the Utrecht Crisis, the roles were actually reversed: Gisbertus Voetius associated *Cartesianism* with animism and argued that animism could only be avoided by saving teleology.

6.4 Conclusions

We may now assess why, as was said in our introduction, the New Philosophy could not be seen as anything other than a reappearance of age-old alternatives. Voetius' objections to the New Philosophy depended heavily on the Medieval and Renaissance traditions of commentaries on Aristotle's *Physics*. Hence they are formulated in terms of the problems which the *Physics* had set, including the demand that physics was to search for the ultimate causes of natural motion. From this point of view, any philosophy rejecting the idea of internal principles of motion would have to introduce some external mover. Alternative solutions were not only known to all students of philosophy, they had also, it was widely thought, been sufficiently dealt with. Thus, Aristotle's *Physics* had defined the project of natural philosophy in terms of a search for the αἰτίαι of natural change. The Cartesians on the other hand, were suspected of atomism, animism and magic and were pressed to come up with a solid answer to these objections. What *did* the mechanical philosophy have to offer?

---

[82] Voetius, *Narratio*, p. 44 / *Select. Dispp.* I, p. 875 / *Responsio*, p. 23 / *Querelle*, p. 109.

CHAPTER SEVEN

ULTIMATE CAUSES IN MECHANICAL PHYSICS

According to their Scholastic critics, the mechanical philosophy was incapable of giving conclusive explanations of isolated phenomena. Confronted with a mechanical account of natural, biological, or physiological phenomena, the question could always be put why it is that the clockwork causes operate at all. What source is there for mechanical motion? Voetius and Schoock expected that a rejection of the Aristotelian answer to this question would ultimately lead to the acceptance of some animated and external moving principle. Mechanistic philosophers themselves, however, offered various answers of their own. I shall first discuss Regius' idea of a principle of motion, followed by Descartes'.

7.1 Regius and the Forms

Henricus Regius did not at all share Descartes' caution with respect to discussing "the substantial forms of the philosophers". Where Descartes kept silence, Regius openly rejected the idea of substantial forms. Nor did he need the philosophical support of Descartes in this matter. Indeed, the Aristotelian philosophy had already met with an onslaught on the forms by various other writers. On 8 April 1642, the Leiden professor Adriaan Heereboord expressed his sympathy with Regius, who had been ordered to silence, while Voetius was still publishing under the name of a student in Leiden. This, Heereboord writes to his Dordrecht correspondent, would have been rather unfair,

> if it was not for the fact that the brothers Bootius have already demolished those substantial forms in such a way that I really do not see what more could be said about them.[1]

Criticizing the forms of the philosophers was a favourite pursuit of many, and few were so mindful as Descartes to keep such criticism to

---

[1] Heereboord to Andreas Colvius, 8 April 1642, AT VIII-II, pp. 196-197. Gerard and Arnold Boot, both Utrecht doctors of medicine serving the English Crown, had, in 1641, published a *Philosophia Naturalis Reformata. Id est, Philosophiæ Aristotelicæ Examinatio ac Solida Confutatio et Novæ ac Verioris Introductio*, Dublinij in Hibernia, Ex Officinâ Typographicâ Societatis Bibliopolarum.

themselves. Voetius himself mentions David van Goirle, or Gorlæus (1591-1612), who had once been a fellow-student of his in Leiden, and, despite his early death, had become a leading champion of atomism.[2] According to Voetius, Gorlæus and Basso had argued that there are no "substantive" forms, and that

> the ancients could easily show in what way, by what [means] forms arise and from what. For they said that the soul and the form are the instruments of whatever thing and that they consist of a certain arrangement and proportion of the parts of an object.[3]

The intellectual climate was favorable to opposition to the forms, and opponents of the Aristotelian philosophy could draw upon ancient sources in order to establish a history of ideas anticipating their own.

---

[2] Van Goirle was the author of *Excercitationes Philosophicæ*, of 1620 and of the posthumusly published *Idea Physicæ* (1651), the latter of which has recently been published with Dutch and Frisian translations as Davidis Gorlæi Ultrajectini, *Idea Physicæ* (1651) met vertalingen in het Nederlands en Fries, Tilburg (F. A. H. Peeters) 1986. On Gorlæus' atomism, see especially: Tullio Gregory, "David van Goirle e Daniel Sennert", in *Giornale Critico della Filosofia Italiana* 20 (1966), pp. 44-63. Further: Theo Verbeek, "Ens per accidens", p. 280, note 17, where other references may also be found.

[3] Voetius, *Narratio*, p. 44 / *Select. Dispp.* I, pp. 875-876 / *Responsio*, p. 23 / *Querelle*, p. 109. Voetius quotes Sebastianus Basso, an author about whom very little is known, apart from the fact that he was a man of medicine and an important link in the development of the idea of the molecule. *Jöcher* I, col. 847. In the Intentiones 3 and 4 of the first book of his *Philosophia Natvralis adversus Aristotelem*, Amsterodami (Apud Ludovicum Elzevirium) 1649, he is, as Voetius remarks, concerned with rehabilitating Aristotle's ancient critics and praising his modern opponents. However, he is only concerned with the nature of mixtures at this point. See Basso, *idem*, p. 27: "Qvod igitur experientia, atque inde ratio manifesta docuit Scaligerum, id ipsum Veteres cognoverunt, innumeri post ipsos etiam ex recentioribus, Philosophi atque medici clarissimi eadem experientia, rationeque duce sequi voluerunt. Quod scilicet de vini & aquæ mistorum partibus compertum est, eas etsi minutissimas, & alias cum aliis continuas, in sua quamque natura persistere, idem de omnibus mistis esse dicendum; discrimenque solum essse, quod alia aliis facilius dissolvantur", after which an enumeration of *Veteres* follows, apparently known to Basso through the intermediate of commentaries by the Conimbricenses and Franciscus Toletus. The passages in which the forms are rejected and which Voetius quotes from, occurs only much further, pp. 143: "Ergo nullæ sunt formæ substantivæ", and 145: "Prisci verò, qui veluti quoddam instrumentum formam cognoverunt, eam in certa rei partium compositione ac proportione consistere crediderunt. Hinc facile est Veteribus, quomodo forma fiat, à quo & ex quo fiat, ostendere: Ex certis partibus scilicet certa ratione coëuntibus paulatim componi: donec qui eam construxit Artifex jam aptam moveat."

Given these circumstances, Regius was cautious in another way. Despite his unequivocal rejection of substantial forms, Regius took pains, in the *Responsio*, to solve all problems voiced by Voetius. His strategy for doing so is to introduce alternative, mechanistic, interpretations for Scholastic terms as "nature", "inherent form", "cause". In fact, Regius even translates the idea of composite substantial unities of matter and form in mechanistic terms. All in all, this makes a clear identification of his position troublesome, to say the least. Yet, as I hope to show, Regius does, in the end, accept the full consequences of the mechanical philosophy. For all his talk of individual forms, his ultimate view is that of a clockwork Universe in motion.

7.1.1 The Universal Chain

In the sketch which Descartes wrote to Regius by way of an answer to Voetius' essay, there is no allusion to individual forms.[4] The idea must therefore be attributed to Regius. Even before the appearance of the *Responsio*, Regius had put forward his ideas on the principles of corporeal activity. At the start of the second disputation *On Some Famous Questions of Physiology* we find the proposition that

> *Nature* is, properly speaking, the internal principle of action, passion and cessation.[5]

This principle, or "nature", is surely not the substantial form of the Scholastics. All the same, Regius' explanation is framed in an Aristotelian idiom. As he explains:

> *Nature* is of two kinds: the *matter* of things natural, and the *form*.[6]

---

[4] Descartes mentions the *Anima Mundi Vniversale*, but only in the context of the question the human soul. Descartes to Regius, AT III, p. 503 / *Rodis-Lewis*, p. 86 / CSM III, p. 207.

[5] Regius, *De Illustribus aliquot Quæstionibus Physiologicis*, Vltrajecti, Ex Officinâ *Ægidii Roman*, Academiæ Typographi, 1641, Disputatio II, Thesis 2. The curious idea of regarding "cessation" as a principle may have its origin in Basso, or in a common, probably Italian, source. Sebastian Basso in any case considers the following in his *Philosophia naturalis adversus Aristotelem*, p. 3: "Rursus omnium ejusmodi motuum cessationem observamus, non minus ab eadem principio intimo, quam motum procedentem. Cessat enim animal augeri, cessat agere, cessat incedere, cessat incalescere: item & herba, & planta cessat crescere, cessat frondescere, florere, virescere; vi quidem ac facultate interna."

[6] *De Illustribus Quæstionibus* II, 5.

The same idea occurs in the general introduction to the *Responsio*, which was also written independently of Descartes. According to Regius, there are two principles to every natural object: matter and form. As in the disputation, matter, or corporeal substance, is characterized in good Cartesian fashion as three-dimensional extension.[7] Form, on the other hand, is of two sorts: either "general" or "special". The only special form is the human mind, or soul.[8] General forms are more abundant:

> In another, general, sense, I say that "form" is common to all natural things. In this sense [the form] is commonly called the "material form", which consists of a combination (*comprehensio*) of motion or rest, as well as of the position, shape and size of the parts of matter which are united in natural objects.

Regius emphasizes that such a form must be distinguished from the matter itself, since,

> although it may be that this [combination] is accidental to matter, it is essential to natural objects, since it constitutes them and separates them from others and since there can be no material form other than this one.[9]

In other words, although matter itself is not characterized by a specific combination of movement, rest, position, shape and size of its parts, individual objects *are* differentiated by them. Therefore, such combinations are said to be "essential" to individual substances.

Hence, the substantial form is substituted for an "essential", material, form by which specificity can equally be explained. This is not all, however. According to Regius, the combination of movement, position and the like, is also endowed with causal efficacy. In fact,

> our true essential forms have no less efficacy than those fictitious ones are said to have, which are commonly called substantial.[10]

The "essential forms" are thus represented as true causal agents. But how is this to be interpreted? In the disputation *De Illustribus aliquot Quæstionibus Physiologicis*, Regius' answer is clear enough. Movement is there defined as

---

[7] Regius, *Responsio*, p. 9 and *De Illustribus Quæstionibus* II, 6.
[8] Regius, *De Illustribus Quæstionibus* III, 8.
[9] Regius, *Responsio*, p. 10 / *De Illustribus Quæstionibus*, II, 16.
[10] Regius, *Responsio*, pp. 16-17.

> the course of a natural body from one place to another by impressed impetus (*per impressum impetum*).[11]

Referring to Descartes' *Dioptrique*, Regius adds that such a movement is imposed (*inditus*) upon the parts of matter by God at the moment of Creation and has remained constant ever since.[12] Accordingly, it is conluded that

> all movement firstly derives from an external motor and no new movement is ever produced by an external principle, although it may sometimes be produced by the internal principle of a certain body, for example when a man moves himself from one place to another.[13]

Hence, there is only local motion, propagated from one moving object to the next. As a consequence, there is no need for any principle of motion other than the total number of moving objects in the Universe transmitting their respective movements. The only independent and individual cause left in this mechanical Universe is man, who moves on account of an internal principle.

There are difficulties here. Regius' endorsement of external movers alone makes the idea of internal principles, other than the special principle of the human soul, superfluous. There is in fact no place for individual essential forms of a more general type possessing "no less efficacy" than those "which are commonly called substantial". The acceptance of mechanical principles leads to a *diffusion* of causality—which was exactly what critics like Voetius and Schoock objected to. Regius, for his part, tries to save some idea of a concentration of causality within individual substances, but in fact, the results are nominal.

7.1.2 Essential Natures: Internal or External?

The disputation *On Some Famous Questions of Physiology* has the following corollary:

> Is all that is moved, moved by its own force? It is.[14]

The corollary was probably drawn up by a student. Nevertheless, Regius must have approved of its defence in the public university

---

[11] Regius, *De Illustribus Quæstionibus* II, 18.
[12] Regius, *idem*, II, 22-24.
[13] Regius, *idem*, II, 25.
[14] Regius, *De Illustribus Quæstionibus*, Corollarium I: "An quicquid movetur, moveatur sua vi? Aff."

disputation for which it was written. The question is, what it would mean for an object to be "moved by its own force". Did Regius have an interpretation of individual forces of objects which was consistent with the fundamentals of mechanicism? The question is especially pressing since, in the second thesis of the disputation *On Some Famous Questions of Physiology*, mention is made of a "nature" which is supposed to be "the internal principle of action, passion and cessation." It would therefore seem that Regius seeks to uphold at least some kind of internal action which is independent of the chain of movement brought about by impulse or collision.

The difficulty reappears in the text of the *Responsio*, where Regius restates his position. Essential forms are to replace the substantial forms of Aristotelianism. Here, however, it is added that

> the same motion which is in a movable [object] and, also, the disposition of the movable to move—which originates from quantity, shape and position—is itself the activity of an efficient cause.[15]

This somewhat puzzling remark is meant as a reply to Voetius' objection that the mechanical principles are conditions for, rather than causes of, action. Regius seems to argue that motion, in as far as it is "in the movable [object]", exceeds the status of a mere condition.

Regius, however, does not leave it at that. Confronted with Voetius' objection that the mechanical philosophy does not make room for individual centres of causation, he tries to offer a way out by interpreting his essential form as a mechanical alternate to such intrinsic motors. One should, Regius argues, distinguish between two types of movement. On the one hand, there is the movement which one finds in the movable object. On the other hand, there is a movement that functions as external cause. It is this latter type which Regius regards as the "less principal cause", or *causa minus principalis*.[16] The movement within the movable is hence regarded as being the more important one with respect to causality. It is in this manner that Regius tries to save the idea of a type of causality which may still be localised within individual substances.

This position is clearly at odds with the viewpoint that there is only a universal chain of movement transmitted from one object to the other. If, as Regius himself put it,

---

[15] Regius, *Responsio*, p. 17.

[16] Regius, *ibidem*.

> the whole Universe of things is created by God in such a way that all its parts are in need of reciprocal assistance and cannot at all operate without it,[17]

it does not make any sense to save the notion of "internal principle",[18] or of an "efficacy" which is "proper" to secondary causes,[19] or, finally, of an "activity of the efficient cause"[20] which is to be distinguished from the *causa minus principalis* of universal movement. All these are Aristotelian fossils, in the mechanical terminology.

7.1.3 Educational Matters

In his letter to Regius of January 1642, Descartes offers the Utrecht doctor the following advice:

> I should like it very much if you would not put forward any new opinions, but, keeping to all the old ones in name, only bring forward new arguments, of which nobody can disapprove.[21]

This is exactly what Regius does. Prompted by Voetius' criticisms, he makes it his task to resolve the objections made by a more or less complete incorporation of the Aristotelian terminology. Though at all times denying and criticizing the Aristotelian forms, Regius at the same time tries to show that the mechanical philosophy could fulfill every requirement the Aristotelian questions imposed. The strategy is something of a pitfall, however. Trying to incorporate the Aristotelian terms, Regius implicitly gives full credit to the Aristotelian argumentation.

From an educational point of view, there was no way around this difficulty. In university teaching, the technical terminology was quite fixed—as it still is. As a consequence, discussions might be limited to very specific questions. In Dutch academic disputations throughout the seventeenth-century, we find for instance a profusion of theses concerning what was called "the principles of physics". Rather than discussing developments which, at the time, presented themselves outside the academic enclosure, such discussions were limited to a

---

[17] Regius, *ibidem.*
[18] Regius, *De Illustribus Quæstionibus* II, 2.
[19] Regius, *Responsio*, p. 17.
[20] Regius, *ibidem.*
[21] Descartes to Regius, January 1642, AT III, p. 491 / *Rodis-Lewis*, p. 72 / CSM III, p. 205.

formulation of arguments either for or against the acceptance of "matter", "form", "privation". Instead of confining his explanations to his discoveries in natural philosophy, we likewise find Regius discussing matter and form also, and trying to prove to Voetius that the New Philosophy accepts the ideas of an internal principle, an individual form, and an efficacy of secondary causes.

This makes Regius rather vulnerable in the dialectic context of academic dispute. As we have seen, Schoock and Voetius could only ridicule his remark that

> the hanging weight and the wound-up spring are internal causes of motion, since they are integral parts of the clock.[22]

Both Regius' formulations and his strategy for convincing the Aristotelians were rather inept. It may well have been for exactly this reason that Descartes not only decided for himself, but actually advised his Utrecht friend also, not to enter in a discussion concerning substantial forms and real qualities.[23] Regius could not help himself.

On the other hand, for all his talk of inherent principles, his position is in the end a purely mechanistic one.

7.1.4 Conclusion: Ultimate Causes of Mechanical Philosophy

Regius took to Gorlæus' view. The latter had been quoted by Voetius as saying that

> entities should not be multiplied without necessity, since the effects of natural objects can be sufficiently explained by, and reduced to, other principles [than the substantial form].[24]

Voetius notices with some satisfaction that anti-Aristotelians of ancient and modern times cannot come to an agreement about the nature of such alternative principles. Yet, whether or not they can, this makes the Occamist argument no less pressing. Regius, for his part,

---

[22] For Voetius' and Regius' discussion of the matter, see *Responsio*, pp. 24-27. See also: Schoock, *Admiranda Methodus*, p. 137 / *Querelle*, p. 248.

[23] Descartes to Regius, January 1642, AT III, p. 492 and p. 501 / *Rodis-Lewis*, pp. 72, 74 and 82 / CSM III, p. 205.

[24] Voetius, *Narratio*, p. 44 / *Select. Dispp.* I, p. 876 / *Responsio*, p. 23 / *Querelle*, p. 109. David Gorlæus, *Excercitationes Philosophicæ*, p. 251: "In prima acie collocamus hoc usitatum argumentum; entium numerum augeri non debere sine manifestâ necessitate. Gorlæus' atomism was based on the idea of only "one simple substance", so that there are "tot formæ numero distinctæ in terra, [quot] indivisibiles partes in ea." Gorlæus, *idem*, p. 251 and 276.

uses it with respect to the example of the heart. It will be worthwile to quote the passage in full:

> [It] should not simply be assumed that substantial forms act in all [things natural], but it should be proved in a most evident way. This, however, does not seem to be done here, [and cannot,] because it is impossible. As we have said, in things natural nothing can be intelligently observed or demonstrated apart from these observable principles of ours [i.e., motion, quantity, shape etc.] which are very clear, always obvious and unique. If someone ignores them, all explanation will merely be question-begging and, covering up his ignorance and feigning science, [such a peron] will in fact seek an unsafe sanctuary in substantial forms, occult qualities, sympathies and antipathies and other non-intelligible causes, which are far more obscure than the initial problems, [that is to say, than] the effects of natural objects themselves.[25]

To introduce substantial forms is merely question-begging. It is feigning science.[26] Gorlæus' argument that entities should not be multiplied beyond necessity, is, moreover, illustrated by Regius once more with the help of the example from which the machine metaphor derived so much of its attraction: that of the mechanical clock.

That all "forms" in Nature except for the mind consist of the configuration and movement of the parts, is shown by the fact that when the material form is in bad condition, the object operates equally badly, just as it does in the case of a broken clockwork. Therefore,

> it is not necessary to add yet another cause, which is commonly called the substantial form: for entities should not be multiplied beyond necessity.[27]

All operation is dependent on construction only. Adding a moving principle is unnecessary. It would seem that Regius thus takes the maxim of economical explanation to its extreme: the *description* of a well-disposed machine, of a machine-like physiological organ, or, indeed, of a complete organism, suffices in order to explain its action.

---

[25] Regius, *Responsio*, p. 27.

[26] Regius, *Responsio*, p. 26: "Nam si dicant [i.e., the champions of the forms] actionem aliquam procedere à formâ substantiali, idem est ac si dicerent illam proficisci à re à se non intellectâ, quod revera nihil aliud quam ignorantiam ipsorum explicat." The idea is taken over from Descartes: Descartes to Regius, January 1642, AT III, p. 506 / *Rodis-Lewis*, p. 92 / CSM III, pp. 208-209.

[27] Regius, *Responsio*, p. 26.

All reference to *motors* initiating such operation, is an unnecessary addition.

Yet Regius does not quite go this far. Although he does relegate substantial forms to the realm of super-abundant causal metaphors, he does not escape introducing new causal metaphors himself. It may be that internal motors form a multiplication beyond necessity. This is not, however, the case with *external motors*. Regius argues that it is no use saying that the example of the clock is inept because the clockwork is in need of weights and springs. At this point, he presents his much discussed idea that these are integral parts of a clock, since the clock would not be complete without them. However, even if the clock is moved by an external motor, Regius argues,

> this would nevertheless be similar to all natural objects—man excepted—inasmuch as these act in the same manner, with the assistence of an external motor. A most evident indication of this is shown in the case of plants and animals, none of which would be able to live or accomplish their other actions without an external motor, such as the Sun, the air, food, and other such things.[28]

The whole Universe of natural objects thus moves and works because things move each other. The world is one mechanical clockwork in which man alone walks about as and acts as an independent cause.

It is important to see that this way of representing Nature exceeds the methodology of economical reasoning. If a clock could be aptly explained by its construction, it would not stand in need of any motors, be they internal or external. Yet in the mechanical philosophy of Regius, exactly the same causal requirement is made as in Aristotelian thought: physics should not only describe Nature: it should explain the *why* of physical change. This can only be done by interpreting the mechanical metaphor, that is, the clockwork conception of Nature, in a wholly literal way. This is already evident in Regius, who, in the *Responsio*, makes the remark that clocks are indeed directly moved by the "ethereal matter", that is, by the airy particles surrounding them.[29] The literal interpretation of Nature as a giant clockwork is presented even more clearly in Descartes, to whom I shall now turn.

---

[28] Regius, *Responsio*, pp. 26-27.

[29] Regius, *Responsio*, p. 26: "Et deinde etiamsi horologium ab externo motore moveretur (*uti revera movetur ab æthereâ materiâ*, quod alias demonstrabitur) illud tamen omnibus rebus naturalibus, excepto homine, esset simile"; my italics.

7.2 Descartes' Theory of Causation

If Descartes had taken his maxim of reductive reasoning seriously, he might have disregarded the objections by Voetius and Schoock concerning the insufficiency of mechanical principles. In contrast, his philosophy meets the requirement of offering ultimate causes for natural change. Since this takes us to the heart of Cartesian thought, we shall have to proceed with caution. I shall here analyse three different aspects which are all involved in this matter: the questions of certainty, of empiricism and, finally, of causality.

7.2.1 Evidence and Certainty

Accepting only "mathematical" principles, Descartes claimed to offer explanations of natural processes which had the certainty of mathematical demonstrations. But did they? Opponents were in any case not convinced.

7.2.1.1 Schoock's Rejection of Hypotheses

In his *Admiranda Methodus* of 1643, Martin Schoock devotes many pages to the question of the evidence and certainty which the Cartesians claim for their philosophy. Schoock doubts that the new theories of natural philosophy have any evidence of truth. Indeed, who ever knew of the existence of "ethereal globules", which both Descartes and Regius hold to be the necessary medium for the propagation of the pressure that is supposed to produce the sensible phenomenon of light?[30] Or, for that matter, the existence of insensible particles, the movement of which should, in the eyes of the Cartesians, produce the sensible qualities of heat and cold?[31] No Cartesian ever

---

[30] Schoock, *Admiranda*, p. 67 / *Querelle*, p. 214 and Descartes, *La Dioptrique*, Discours premier, AT VI, p. 87: "il est necessaire que ces pores (i.e., the pores of natural bodies) soyent remplis de quelque matiere fort fluide, qui s'estend sans interruption depuis les Astres iusques a nous." Cf. CSM I, p. 154. Spinning round their axes in various ways, these material globules also produce, according to Descartes, the different colours of bodies. Descartes, *idem*, AT VI, pp. 91-92. See also: Regius, *Physiologia sive cognitio sanitatis* III, I, p. 37: "Color (propriè loquendo) est globulorum æthereorum perceptibilis in lumine processus & circumvolutionis proportio." Note that it is the action, or *processus*, of these globules that causes light; the particles themselves do not in any way form the constituent parts of the phenomenon.

[31] Schoock, *ibidem* and Descartes, *Les Météores*, AT VI, p. 235-236: "pour le froid & le chaud, il n'est point besoin de conceuoir autre chose, sinon que les petites parties des corps que nous touchons, estant agitées plus ou moins fort que de

succeeded in proving the existence of such particles, let alone in showing that either their existence or their effects are naturally known and that the particle theory is therefore less revolutionary than it seems to be. If the Cartesian dogmas are really "bestowed (*inditus*) by Nature upon the human mind", then why is it that no one ever knew of them before?[32] As far as Schoock is concerned, these Cartesian ideas are neither naturally known nor evident by experience or reason. They are imaginary and purely hypothetical.

In chapter 4, above, I mentioned the fact that Descartes apologized to the readers of the *Essays* which accompany the *Discours*, that he had to make use of what he calls *suppositions*. It is exactly for his introduction of a particle theory that the apology is made. In the *Météores*, Descartes defends his use of particle hypotheses on account of their simplicity. As for Schoock, however, the only thing simplicity guarantees is what he calls "evidence of demonstration", which should be sharply distinguished from "evidence of proof".[33] Thus, he is not very impressed by the Cartesian hypotheses. When evidence is restricted to evidence of presentation, the resulting philosophical theory can at best convince "old ladies", "inexperienced youngsters", or "the ignorant mob".

However, far from simplicity itself being the only guarantee for truth in Cartesian explanations, a certain way of connecting causes and effects is rather what characterizes them. No circularity is involved if, from the fact that the effects can be explained by their supposed causes, it is deduced that the latter are also proved to have been rightly hypothesized:

> For as experience makes most of these effects quite certain, the causes from which I deduce them serve not so much to prove them as to explain them; indeed, quite on the contrary, it is the causes which are proved by the effect.[34]

---

coustume, soit par les petites parties de cete matiere subtile, soit par telle autre cause que ce puisse estre, agitent aussy plus ou moins les petits filets de ceux de nos nerfs qui sont les organes de l'attouchement." And: Regius, *Physiologia sive cognitio sanitatis*, I, I, p. 6: "Calor actualis est varia agitatio insensibilium particularum: frigus autem est earum quies."

[32] Schoock, *Admiranda*, p. 66 / *Querelle*, p. 214.

[33] Schoock, *Admiranda*, p. 70 / *Querelle*, p. 216.

[34] Descartes, *Discours de la Méthode*, AT VI, p. 76, quoted from CSM I, p. 150. As Ferdinand Alquié puts it in a note to his edition of this text, being the effects of the laws of Nature, the facts themselves do not have to be proved "puisque précisément ils sont des faits, et que leur existence est constatée

Saving all relevant phenomena, the "suppositions" are "proved" by the facts. The evidence of the Cartesian particle theory is thus based upon the fact that it accounts for the phenomena actually experienced by the senses.[35]

A critic like Schoock would still not be impressed. The fact that the phenomena may be saved does not itself warrant a belief in the existence of small particles. Once again, Schoock distinguishes between two types of evidence, one being "solid", the other only persuasive.[36] Concerning the explanation of the tides, Schoock offers an interesting discussion of the Cartesian theory, at once arguing for its insufficiency and confessing his own ignorance on the point. Indeed, confronted with this miracle of Nature, even the greatest geniuses have had to admit their *erudita ignorantia*.[37] The Cartesian on the other hand, "not troubled by erudition", concocts a theory of his own. "Just listen to Doctor Regius,"[38] who holds that, as a result of its own movement and of the position of the Moon, "our particular part

---

expérimentalement." See René Descartes, *Œuvres philosophiques*, Textes établis, pésentés et annotés par Ferdinand Alquié, Paris (Garnier) 1963, Tôme I, p. 647, note 1.

[35] This scientific method is well-described as "hypothetico-deductive", although the term may perhaps better be reserved for later scientific theories. For a representation of Descartes as a precursor of the hypothetico-deductive method, see Ernan McMullin, "Concepts of Science in the Scientific Revolution", in David C. Lindberg, Robert S. Westman (edd.), *Reappraisals of the Scientific Revolution*, Cambridge (C. U. P.) 1990, pp. 27-92.

[36] Schoock, *Admiranda*, pp. 70-79 / *Querelle*, pp. 216-219. As far as Schoock is concerned, an elegant presentation of fictive causes may well do for the unlearned, or for the disciples of Cartesianism who accept the Pythagoric adage Αὐτὸς ἔφα— "he, Descartes, said it himself". But why should anyone take an *evidentia narrationis* for truth? The *Talmud* relates the story of Leviathan, a monstrous fish that by moving its tail causes earthquakes, and by taking up water with its gills, influences the tides; a fable which, according to Schoock, even more evidently accounts for the facts than any of the "Cartesian dreams". But could that ever be reason enough for believing in the existence of such a fish?

[37] Schoock, *Admiranda*, p. 234 / *Querelle*, p. 299. Accepting one's *learned ignorance*, instead of striving after a *pansofia*, was also what Voetius prescribed concerning phenomena (such as the tides) which could only be explained by "occult qualities". See *Narratio*, p. 46 / *Select. Dispp.* I, p. / *Responsio*, p. 28 / *Querelle*, p. 111. Voetius wrote a complete disputation "De docta ignorantia, included in the third part of his *Disputationes Theologicæ Selectæ* of 1648. See also: Theo Verbeek, "From 'Learned Ignorance' to Scepticism; Descartes and Calvinist Orthodoxy", in Richard H. Popkin / Arjo Vanderjagt (edd.), *Scepticism and Irreligion*, pp. 31-45.

[38] "Unus modo Medicus audiatur," *Admiranda*, p. 234 / *Querelle*, p. 299.

of the sky" will remove the Earth from its place and thereby be compressed in two diametrically opposed places: between the Earth and the Moon, and between the Earth and that part of the sky in the direction from which the Earth was pushed away. Since there is a constant celestial flux of insensible particles around the Earth, the flux will also be compressed at two opposite sides of the Earth, thus forcing the waters in these parts to rise. Next, the diurnal rotation of the Earth causes a decrease of pressure in these areas, as the waters are slowly turned away from the compressed parts.[39]

Schoock compares this explanation of the tides with the arguments put forward by astronomers in support of the Copernican system. Both Cartesians and Copernicans make use of hypotheses which enable them to successfully save the phenomena. However, who would grant that their reconstructions have anything in common with the true nature of things?

> Hypotheses which are contrived concerning things natural are only inventions of the human mind and by no means themselves things of Nature.[40]

Schoock anticipates Newton in objecting to the Cartesian anticipation of Nature. His point, as far as the differences between the Ptolemaic and Copernican systems are concerned, is to argue that a mathematical model never has the force of a physical theory concerning the true state of things. Eccentric orbits and epicycles can be endlessly introduced in order to save celestial phenomena, but

> who would therefore conclude that there are real eccentric orbits, or epicycles in heaven?[41]

---

[39] Regius' theory of the tides may be found in the *Fundamenta Physices*, pp. 90-91. It is completely identical to the one Descartes had given in the unpublished work of *Le Monde*, AT XI, pp. 80-83, and which was later to reappear in *Principia* IV 49-56, AT VIII-I, pp. 232-238. Schoock on the other hand, quotes the now lost *Dictata Physica* by Regius, a work Regius is said to have composed after having read Descartes' *Dioptrique* and *Météores*. The explanation of the tides however, occurs in neither of these works. Nor do we find any such discussion in either Regius' *Physiologia* or his *Responsio*, although in the *Responsio* Regius mentions the *æstus maris* as an example of phenomena that can be mechanically explained instead of through the use of occult qualities; see *op. cit.*, p. 29. It is also in the *Responsio* that Regius first mentions *Le Monde* (idem, p. 20), which, according to Theo Verbeek, he had not yet read when writing the first series of disputations collected in the *Physiologia*. Theo Verbeek, "Regius's *Fundamenta Physices*" in *Journal of the History of Ideas* 55 (1994), pp. 542-545 esp.

[40] Schoock, *Admiranda*, pp. 238-239 / *Querelle*, p. 301.

The mathematical model is only refined to save the phenomena. Since both mathematical hypotheses saved the relevant phenomena, the decision which of the two, Ptolemaic or Copernican, corresponded to the true nature of things, would have to be made on physical grounds.[42]

According to Schoock, the same argument may be brought up against the Cartesian hypothesis. Introducing ethereal globles or insensible particles is as irrelevant to physics as introducing new mathematical devices to save celestial phenomena. In the case of the Cartesian particle theory, a "physical" way of seeing things, would be to show that these "Democritean atoms", as Schoock unjustly calls them,[43] actually

---

[41] Schoock, *Admiranda*, p. 238 / p. *Querelle*, p. 301.

[42] Note that the Ptolemaic model was no less accurate in predicting the orbits of heavenly bodies. Motives for accepting the Copernican system may largely have been of an aesthetic nature. Although E. A. Burtt's classic book on *The Metaphysical Foundations of Modern Physical Science*, has, in recent years, met criticism on account of its simplification and overestimation of "the metaphysical viewpoint", its exposition of both the mathematical and the physical arguments for and against the acceptance of a heliocentric model still stands firm. Cf. Burtt, *op. cit.*, London (Routledge and Kegan Paul) 1932², chapter II, §§ A and B, pp. 23-39. It should, however, be noted that in Schoock's days, important empirical evidence against the Ptolemaic system had already been found. Although Galileo was unable to disprove the Tychonian alternative on empirical grounds, his discovery of the precise character of the phases of Venus, published in the *Siderius nuncius* of 1610, was a downright falsification of the Ptolemaic system. As for the impact of metaphysical ideas accompanying the scientific revolution, see the recently published collection of articles on *Reappraisals of the Scientific Revolution*. David C. Lindberg and Gary Hatfield for instance take their starting-point in the works of authors such as Ernst Cassirer, Alfred North Whitehead, Edwin A. Burtt, Herbert Butterfield and Alexandre Koyré. Hatfield criticises Burtt in particular for having relied too heavily on the powers of the "nonscientific" and the "motivational", the danger being "of dissolving text and author into a set of background influence." Gary Hatfield, "Metaphysics and the new science", in David C. Lindberg and Robert S. Westman (edd.), *Reappraisals of the Scientific Revolution*, Cambridge (C. U. P.) 1990, pp. 93-166. In the same volume, Ernan McMullin goes into the conceptions of science in Descartes, Galileo and others. See above, note 35.

[43] Schoock, *Admiranda*, p. 127 / *Querelle*, p. 243. Descartes' conception of matter does not allow the existence of atoms, as his identification of matter and three-dimensional space makes matter infinitely divisible. Descartes, *Principia* II 20, AT VIII-I, p. 51-52 / CSM I, pp. 231-232.

> flutter about in the whole world and by their varying confluence, number, and consequent shapes, produce the differences of all things.[44]

Again, Schoock's criticism may be read as a prelude to later Newtonian critiques.[45] However, not only does he condemn the ease with which the Cartesians invent corpuscular hypotheses for saving natural phenomena: his disapproval of the Cartesian hypotheses is, at the same time, a demand for genuine demonstration.

Perhaps Schoock was willing to endorse at least some hypotheses in natural philosophy. In the Introduction to the *Admiranda Methodus*, in any case, he reproaches the new philosophers for having misused "all the nice, and, to a certain extent, plausible" parts of Democritean atomism.[46] He does not elaborate the point, but seems not to object to corpuscular theories as such. As we have seen, Schoock was, moreover, generally quite open-minded towards empirical investigations. The problem with Cartesianism however, was one of method. Scientific hypotheses might well save the phenomena, but as soon as corpuscular models are presented as physical facts, empirical corroboration is needed. Physics is a demonstrative science. Without demonstrative proof, therefore, hypotheses may not be presented as truths.[47]

---

[44] Schoock, *Admiranda*, p. 127 / *Querelle*, p. 243.

[45] As Roger Cotes would write against the Cartesians in the 1729 Preface to the *Principia*: "But when they take the liberty of imagining at pleasure unkown figures and magnitudes, and uncertain situations and motions of the parts, and moreover of supposing occult fluids, freely pervading the pores of bodies [...] they run into dreams and chimeras, and neglect the true constitution of things." See Alexandre Koyré, *From the Closed World to the Infinite Universe*, Baltimore (John Hopkins U. P.) 1968, p. 230, from which the translation is taken. It should be noted however, that Cotes' words conceal a censure of the mathematical inaccuracy of the Cartesian models which is absent in Schoock.

[46] Scoock, *Admiranda*, p. 4 / *Querelle*, p. 182-183.

[47] It may therefore have been the Cartesians' emphasis on the physical *truth* of their scientific conjectures that conflicted most with Schoock's empiricist and common-sensical outlook. A similar point has been raised with respect to Voetius, who might have been more inclined to accept the mitigated scepticism of someone like Pierre Gassendi. Theo Verbeek has shown such considerations to be useful for explaining the Gassendist and even Cartesian tendencies in later Scholastic authors. See his "Voetius en Descartes", in J. van Oort, C. Graafland, A. de Groot, O.J. de Jong (edd.), *De onbekende Voetius: Voordrachten wetenschappelijk symposium Utrecht 3 maart 1989*, Kampen (J. H. Kok) 1989, p. 218. Further: Theo Verbeek, *Descartes and the Dutch*, pp. 9, 63 and 88. Also of interest in this respect is Edward G. Ruestow's *Physics at 17th and 18th Century Leiden*, The Hague

Regius indeed presented his corpuscular hypotheses as physical facts. He did not, however, make a point of demonstrating their actual truth. In fact, Regius' position is in the end a rather sceptical one, which, for different reasons, was also something Schoock and Voetius were afraid of.[48] It was Descartes however, who claimed certainty for his scientific hypotheses.

7.2.1.2 Descartes' Metaphysical Argument

In chapter 4, we saw how Descartes sought to invest natural philosophy with a mathematical type of certainty. Explanations of physics and physiology should not only be restricted to as few and as simple explanatory elements as possible: they should also make use of only such "mathematical" notions as size, shape, movement and disposition of parts. This would indeed guarantee certainty, but only in a subjective way: the force of the demonstrations would acquire the

---

(Matinus Nijhoff) 1973. Ruestow offers various examples of later Aristotelians who proved to be more open-minded in their scientific investigations than some of the Leiden Cartesians, who were less inclined to do experimental work. Nevertheless, Voetius probably did not grasp any of this. He did try to incite Marin Mersenne to write against Descartes. Referring to this fact, Bernard Rochot has said that "Voet n'a pas tout à fait tort d'opposer Mersenne à Descartes, et entièrement raison de rapprocher Gassend du premier. [...] Mersenne, Gassend, Galilée aussi, ne veulent que tirer de l'observation et de l'expérience [...] des lois applicables aux phénomènes, sans toucher aux essences." See *Correspondance du P. Marin Mersenne* XI, pp. 376-377. At the same time, it seems fair to conclude, with Rochot, that the difference between Descartes and his more empirically minded fellow-*novateurs* must in the end have escaped Voetius entirely. *Correspondance du P. Marin Mersenne*, p. 377: "Mais la nature profonde d'un désaccord Mersenne-Descartes devait échapper à un traditionaliste outré." We do not in fact know what Voetius wrote to Mersenne concerning Gassendist philosophy, since most of what Voetius wrote to Mersenne is now lost. Descartes kept no less than five letters, to which he refers in the *Lettre apologétique aux Magistrat d'Utrecht*. See *Querelle*, pp. 409-410 and notes. See *Correspondance du P. Marin Mersenne* X, p. 166 esp. and Vol. XI (1970), pp. 363-377. Presumably, Voetius saw Mersenne and Gassendi simply as outright opponents of Descartes because of their *Objections* to his *Meditations*.

48 Regius holds Terence's motto that there are as many opinions as there are human beings to be applicable to scientific reasoning as well. In fact, he bases this view on a physio-psychological theory concerning the variety of bodily temperaments. Henricus Regius, *Fundamenta Physices*, Amstelodami, Apud Ludovicum Elzevirium, 1646, p. 306. For a discussion of issues involving scepticism in relation to the Utrecht Crisis, see Verbeek, Theo, "From 'Learned Ignorance' to Scepticism; Descartes and Calvinist Orthodoxy", in Richard H. Popkin / Arjo Vanderjagt (edd.), *Scepticism and Irreligion*, pp. 31-45.

force of those with which we are acquainted in the mathematical sciences. It is this type of certainty which, in Schoock's terminology, guarantees "evidence of presentation": the hypotheses precisely explain the phenomena. This certainty is "subjective" in the sense that it remains a matter of conviction that the hypotheses actually apply in the world itself. We may be convinced by mechanical hypotheses in the case of the heart, where we can dissect and experiment and thus test the hypotheses empirically. However, in the case of corpuscular hypotheses, such as ethereal globules and insensible particles, another type of certainty is required, namely an "objective" certainty that these phenomena actually obtain in the microsopic world.

When Descartes claims that the mechanical explanation of hidden processes by means of corpuscular hypotheses may be taken *pro Mathematicâ demonstratione*,[49] he claims objective certainty as well. Right at the end of the *Principia* of 1644, Descartes discusses the question of certainty as regards explanations hypothesizing things we cannot see. He states the problem by comparing the author of Nature, God, with a skilled clockmaker who might make two clocks of identical appearance, but with completely different mechanisms inside. Likewise, God could have made Nature in an infinite number of ways even though its outward shape would be identical. How then do we know that our explanations are correct? Descartes first agrees that there is no self-evident reason for believing that we do know the exact causes of things natural. However, he puts forward a pragmatic argument, indicating that there is actually no need to always explain things as they are. As for "insensible processes", it suffices, as Aristotle had said in the *Meteorology*, to render explanations that "agree" (*respondent*) to natural phenomena.[50] Desmond Clarke has interpreted this passage as one in which Descartes professes a "crude instrumentalism". Since it is not Descartes' own position, he would merely be

> using references to [the saving-the-phenomena] tradition to excuse his apparent failure to achieve the kind of certainty he promised at the outset of his career.[51]

---

[49] Descartes, *Principia* II 64, AT VIII-I, p. 79 / CSM I, p. 247.

[50] Descartes, *Principia* IV 204, AT VIII-I, p. 327 / CSM I, p. 289 and Aristotle, *Meteorologia* I 7, 344^{a}5-7: "ἐπεὶ δὲ περὶ τῶν ἀφανῶν τῇ αἰσθήσει νομίζομεν ἱκανῶς ἀποδεδεῖχθαι κατὰ τὸν λόγον, ἐὰν εἰς τὸ δυνατὸν ἀναγάγωμεν."

[51] Desmond Clarke, *Descartes' Philosophy of Science*, Manchester (M. U. P.) 1982, p. 116.

In fact Descartes does not make an epistemological point here. He is merely putting forward the pragmatist view that as long as one's theories of medicine, mechanics and the like are useful in practice, epistemological considerations are irrelevant. It does not matter to the sick whether they are cured by imaginary rather than true causes. Likewise, it does not matter much that God might have made things appear to us in the way they do by making use of an infinite number of different causes.

However, although there is no self-evident reason for believing in the truth of our hypostheses concerning the microscopic world, there are other reasons. The first concerns probability. Our models may be at once so simple and so universally applicable, that it would be unreasonable to doubt them. This is what Descartes calls *moral certitude*, which, he says, certainly applies to the mechanical philosophy.[52] Besides the sufficiency of moral certitude there is however yet another argument for believing that the mechanical explanations are correct. This "metaphysical argument" actually secures *absolute certitude*. It is the point Descartes had introduced in his *Meditations*, namely that we have a firm foundation for our subjective convictions in God's goodness:

> namely that God is supremely good and in no way a deceiver, and hence that the faculty that He gave us for distinguishing truth from falsehood cannot lead us into error, so long as we are using it properly and are thereby perceiving something distinctly. Mathematical demonstrations have this kind of certainty, as does the knowledge that material things exist; and the same goes for all evident reasoning about material things.[53]

It seems that with the appearance of the *Principia* in 1644, Schoock was proved right in his general claim that Cartesian method, epistemology and metaphysics are only meant to convince the reader of the truth of Descartes' hypotheses in physics. This does not mean, to be sure, that for Descartes, every mechanical visualisation of hidden processes, such as those concerning the causes of the tides, must of

---

[52] Descartes, *Principia* IV 205, AT VIII-I, pp. 327-328 / CSM I, pp. 289-290. Descartes compares the case with that of a cipher. If we can produce a complete and sensible interpretation of a text written in code with the help of a simple formula, we cannot reasonably doubt that this formula was indeed used for writing it.

[53] Descartes, *Principia* IV 206, AT VIII-I, p. 328. Translation from CSM I, p. 290.

necessity be true. We could have been mistaken in our observations and not know which phenomena to save. Or we could have overlooked some irregularities as a result of which our explanation turns out to be false. Yet what, according to Descartes, we can be certain of in physics, is that as long as we do not invoke any other explanations than purely mechanical ones, and stick to propositions that are reducible to mechanical notions of movement, shape, position etc., our demonstration will have the evidence of a mathematical demonstration.[54] This evidence being too convincing to us to possibly believe in its falsity, we have a metaphysical foundation of its truth, even in the case of particular physical hypotheses.

Consequently, mechanical hypotheses do more than just save the phenomena. But if the evidence of the mechanical hypotheses indeed guarantees their truth, then why is it that, at the end of the *Discours de la Méthode*, Descartes excuses himself for using such hypotheses? The answer must be, as Étienne Gilson rightly says in his *Commentaire Historique*, that for Descartes himself, the hypothetical character of his particle theory was due to his not yet having deduced them from *a priori* principles, rather than to the validity of particular theorems depending on the use of such hypotheses.[55] We have, however, yet to see what, in the Cartesian sense, such *a priorism* consists of. For as regards the question of *a priorism*, Descartes' philosophy has been represented in some extremely unfortunate ways.

7.2.2 Cartesian Empiricism

Historians of philosophy have traditionally distinguished between the two modern schools of rationalism and empiricism. The distinction is made in order to contrast opposing philosophical and scientific methods, the one emphasizing the use of the faculty of reason, the other that of sense. Though interesting from an analytical point of view, the distinction has proved very misleading when it is interpreted historically as a distinction between different schools of thought. A famous example of a writer interpreting the distinction in this way—and one who may have been an important source for the

---

[54] See also the letter to Mersenne of 11 March 1640, where Descartes says: "Pour la Physique, ie croyois n'y rien sçauoir, si ie ne sçauois que dire comment les choses peuvent estre, sans demonstrer qu'elles ne peuuent estre autrement; car l'ayant reduite aux lois des Mathematiques, s'est chose possible, & ie croy le pouuoir en tout ce peu que ie croy sçauoir [...]." AT III, p. 29 / CSM III, p. 145.

[55] See R. Descartes, *Discours de la Méthode*, texte et commentaire par Étienne Gilson de l'Académie Française, Paris (Vrin) 1976[5], p. 471.

misapprehension—is Kant, who thought of his own "critical way" not only in philosophical terms as the third way transcending the faculties of reason and sense, but also as crowning past philosophic history, which he supposed to have been divided according to two persuasions: of *bloß Sensual-* and *bloß Intellektual-philosophen.*[56] Intellectual strategies being thus represented with the aid of just one pair of possibilities, there is no doubt that Kant's reconstruction has inhibited a serious interpretation of historic figures and their thought.[57] Likewise, the distinction between rationalist and empiricists has misrepresented early modern philosophy, especially when, as in the case of Descartes, the distinction is used in relation to physics.

7.2.2.1 Cartesian Expériences

Indeed, what could a rationalist philosophy of Nature be like? If rationalism is taken in the strict sense as a repudiation of empirical fact, then a rationalist physics would be a physics without observation. Of course, no seventeenth-century physicists ever held such a thing possible, but despite the work of Alan Gewirtz and others who have pointed to the role of experience in the Cartesian method,[58]

---

[56] Immanuel Kant, *Kritik der reinen Vernunft*, B 881. The idea of the *kritische Weg* occurs at B 884. Note that, in strictly Kantian terms, only the intellectual faculty is a *facultas* in the active sense of a *Vermögen*, the sensible *Empfänglichkeit* being only passive. Cf. Immanuel Kant, *Antropologie in pragmatischer Hinsicht* § 7, in *Kant's Gesammelte Schriften*, Berlin (Georg Reimer) 1917, Band VII, p. 140.

[57] As John Cottingham rightly observes, "it is probably fair to say that the six great philosophers [i.e., Locke, Berkeley, and Hume on the one hand and Descartes, Spinoza, and Leibniz on the other] would all have seen themselves as engaged on a recognizably similar kind of enterprise." However, as Cottingham also observes, anachronistic projection is not the only problem with the empiricist / rationalist distinction. John Cottingham, *The Rationalists*, in *A History of Western Philosophy*, Volume 4, Oxford (O. U. P.) 1988, pp. 2-4. For all the caveats however, the distinction between "rationalists" and "empiricist" remains to be used as a denotation, if not for schools of thought, than at least for groups of philosophers supposedly exhibiting similar ideas. Likewise, Rainer Specht on the one hand observes that there are "bemerkenswerte Argumente gegen die Brauchbarkeit dieser Termini [i.e., *Rationalist* and *Empirist*]", only to argue on the other hand, that "ein stärkeres Argument ist aber offenbar die Feststellung, daß [diese Termini] akzeptiert sind." *Rationalismus*, in *Geschichte der Philosophie in Text und Darstellung*, Band 5, Stuttgart (Philipp Reclam jun.) 1986^2, p. 17.

[58] Alan Gewirtz's 1941 article on "Experience and the Non-Mathematical in the Cartesian Method" offers a fine review of the parts "played by experience, both as sense ofr imagination, and as experiment, in the sequence of operations which the

there is a persistent tendency to attribute the idea of a "rationalist" physics to Descartes.

A reason for doing so may have been that, for Descartes, the domain of truth in general is identical to that of the evident type of truth which we find in the mathematical sciences. Accordingly, Descartes' examples are often mathematical ones.[59] If physical truth is in essence no different from mathematical truth, it may be equally independent of observation.

However, we have already seen how physical truths may be of a "mathematical" type, namely, when they are formulated in mechanistic terms. Also, such truths may acquire the certainty of mathematical ones. Neither of these considerations, however, is of any consequence with regard to the manner in which these truths are found. Nor do we find grounds in any of Descartes' works for the interpretation of Cartesian physics as solely consisting of the application of some formal technique upon a limited set of axioms.

There is, on the other hand, ample evidence for Descartes' empirical attitude. A lot is said for instance in the *Regulæ* about the importance of intuition and deduction,[60] yet both the intuition of particular truths and the consequent composition of a body of scientific knowledge are expressly said to be governed by experience.[61]

---

method prescribes for scientific discovery." Gewirtz, op cit., in *Journal of the History of Ideas*, 2^2 (1941), pp. 183-210, and esp. pp. 189-201.

[59] In the *Meditations* for instance, the argument for relying on the certainty of what seems evident, that is, the argument that God does not deceive us, is formulated in such a way that it seems as if all evident truths might be deducible from mathematical essences such as the fact that three angles of a triangle equal two right angles. Descartes, *Meditations*, AT VII, p. 71 / AT IX, p. 56 / CSM II, p. 49, where the issue of certainty and truth is said to concern *all sciences*. At the start of the sixth *Meditation* moreover, Descartes summarizes that he now knows everything that there is to know about material things except that they exist, for he knows them clearly and distinctively in as far as they are objects of "pure Mathesis". Cf. AT VII, p. 71 / AT VII, p. 57 / CSM II, p. 50.

[60] See for instance the concise formulation of *Rule* III: "quid clarè & evidenter possimus intuiri, vel certò deducere, quærendum est." Also: *Rule* V: "ex omnium simplicissimarum intuitu ad aliarum omnium cognitionum"; *Rule* VI: "directè deduximus"; VII: "continuo & nullibi interrupto cogitationis motu"; VIII: "[quod intellectus noster nequeat] satis bene intueri"; IX: "veritatem distinctè & perspicuè intueri"; XI: "distinctè concipere: ita enim & cognitio nostra longè certior fit, & maximè augetur ingenij capacitas." And so on, *Regulæ*, AT X, *passim*.

[61] Descartes, *Regula* XII, AT X, p. 410: "Denique omnibus vtendum est intellectûs, imaginationis, sensûs, & memoriæ auxilijs, tum ad proposiones simplices distinctè intuendas; tum ad quæsita cum cognitis ritè componenda, ut

Hence, Cartesian physics cannot be reduced to the sheer intellectual construction of the material world by means of *a priori* reasoning from evident notions alone. Indeed, to do so would surely provoke the biting criticism of Descartes himself, who, in the Regulæ, ridiculed philosophers

> who take no account of experience (*neglectis experimentis*) and think that truth will spring from their brains like Minerva from the head of Jupiter.[62]

Nor did Descartes neglect experiments himself. In fact, he made observations for the sake of increasing his knowledge of Nature throughout his philosophically active life.[63]

---

agnoscantur; tum ad illa invenienda quæ ita inter se debant conferri, ut nulla pars industriæ humanæ omittatur."

[62] *Regulæ*, AT X, p. 380. Translation from CSM I, p. 21. The *Discours* is no less clear about the role played by observation and experiment in the natural sciences: "Or, ayant dessein d'employer toute ma vie a la recherche d'vne science si necessaire, & ayant rencontré vn chemin qui me semble tel qu'on doit infailliblement la trouuer, en le suiuant, si ce n'est qu'on en soit empesché, ou par la brieueté de la vie, ou par le defaut des experiences, ie iugeois qu'il n'y auoit point de meilleur remede contre ces deux empeschements, que de communiquer fidellement au public tout le peu que i'aurois trouué, & de conuier les bons esprits a tascher de passer plus outre, en contribuant, chascun selon inclination & son pouuoir, aux experiences qu'il faudroit faire, & communiquant aussy au public toutes les choses qu'ils apprendroient, affin que les derniers commençant où les precedens auroient acheué, & ainsi ioignant les vies & les travaux de plusieurs, nous allassion tous ensemble beaucoup plus loin, que chascun en particulier ne sçauroit faire." Descartes, *Discours de la Méthode*, AT VI, pp. 62-63 / CSM I, p. 143.

[63] See for instance the introduction to the *Traité de Méchanique*, in Descartes to Huygens, 5 October 1637, where Descartes tells his Dutch friend that he is less in the mood for writing than ever, since the publication of the *Discours* gave him so many grey hairs that all that he now does, is slowing down their growth. Apart from that, Descartes says, "ie tasche a supleer par industrie le defaut des experiences qui me manquent, a quoi i'ay tant de besoin de tout mon tems que i'ay pris resolution de l'y employer tout." AT I, pp. 434-435 / CSM III, p. 66. The correspondence offers a great many other references to Descartes' experimentations and anatomical investigations. See e.g. the letter to Mersenne dated 13 November 1639, in which Descartes tells us about his days in the Kalverstraat in Amsterdam: "i'ay esté vn hyuer à Amsterdan, que i'allois quasi tous les iours en la maison d'vn boucher, pour luy voir tuer des bestes, & faisois apporter de là en mon logis les parties que ie voulois anatomiser plus à loisir; ce que i'ay encore fait plusieurs fois en tous les lieux où j'ay esté, & ie ne croy pas qu'aucun homme d'esprit m'en puisse blâmer." AT II, p. 621. See also: Descartes to Mersenne, 15 April 1630, AT I, p. 137 / CSM III, p. 21, where Descartes claims that he daily finds things he could never have

Still, there may be a way of saving the rationalist interpretation by redefining it. Thus, one might argue with Rodis-Lewis, that in Cartesian physics, observation is at least subordinate to deduction.[64] Rodis-Lewis claims that the unity that the rationalistic method demands can only be achieved by ignoring particular facts that do not fit the system. Consequently, precise laws would, in Cartesian physics, be "rare and often mistaken".[65] However, this being granted, it remains to be shown that the rarity and falsity of precise laws in Cartesian physics is due to the priority given to deduction over observation. Rodis-Lewis offers the example of the void, the existence of which was denied by Descartes entirely on the basis of clear and distinct ideas. However, since the question is conceptual rather than empirical, it is not clear what empirical stance Descartes could have

---

found in books. It is not without reason that Adrien Baillet, in his biography of the French philosopher, compares these passages with the texts of the *Discours*, AT VI, p. 63 / CSM I, p. 143, where Descartes urges the establishment of a scientific community communicating their findings to each other. Adrien Baillet, *La Vie de Monsieur Des-Cartes*, A Paris (Chez Daniel Horthamels) 1691, pp. 196-197. Neither is Baillet's defence of the philosopher's passion for dissection inapproriate, seeing that Descartes himself has to convince Mersenne of the fact "[que] ce n'est pas vn crime d'estre curieux de l'Anatomie". AT II, p. 621. As for anatomical researches, see also the letter to Mersenne of 31 March 1638, concerning Descartes' successful dissection of a cow's eye, AT II, pp. 86-87. Descartes not only attended, but even took part in an autopsy on a woman by professor Adriaan Valkenburg at Leyden University. Cf. AT III, pp. 48-49 / CSM III, p. 146. The autopsy however, was not a success. Descartes could not find the pineal gland, an organ which he had so often observed in animals and which was of great importance to both his physiology and psychology and his metaphysical theory of the real distinction of mind and body.

[64] G. Rodis-Lewis, *Descartes et le Rationalisme*, Paris (P. U. F.) 1970, p. 57: "la déduction [...] commande l'observation."

[65] Rodis-Lewis, *idem*, p. 57. Note that the main goal of *Descartes et le Rationalisme* is to present Descartes as the founder of the rationalistic tradition. In order to achieve this goal, Rodis-Lewis first analyses the method of various Cartesian disciplines, arguing that Cartesian physics tries to deduce all *vérités de fait* from *a priori* evident notions. I do not find this in Descartes, nor in the other rationalist authors mentioned. Leibniz for instance actually denied the possibility of such rationalism in science on account of the fact that "certitudo et perfecta ratio veritatum contingentium soli DEO nota est". Leibniz, *De rerum originatione radicali*, in *Die philosophische Schriften von G. W. Leibniz* (ed. Gebhardt), Bd. VII, p. 309. For Leibniz's solution to the problem of the raison suffisante of particular truths, see G. H. R. Parkinson, *Logic and Reality in Leibniz's Metaphysics*, Oxford (Clarendon Press) 1965, esp. chapter IV, "Logic and Theism".

taken in order to accept the possibility of empty space. The problem is not in any simple way one of reason versus observation.[66]

There are certainly examples in Cartesian physics which better correspond to the picture of deduction having priorty over observation. A famous one is Descartes' theory of collision. In the *Principia* Descartes formulated seven laws of collision of which only the first was not immediately rejected afterwards for not being in accordance with the facts. Descartes himself may have been aware of this, since he immediately cautions that

> In fact it often happens that experience may appear to conflict with the rules I have just explained.[67]

Descartes attributes the incongruity between laws and experience to the fact that the laws only apply to a pair of bodies which are perfectly "hard" and free to move independently of all other bodies. The situation described is, in other words, an ideal situation which is never found in the real world of natural objects, since this is in fact entirely filled with an infinite number of infinitely divisible bodies. Thus, Descartes does away with observation in favour of a rational reconstruction of the way in which bodies would behave were they to collide in such a way that no other bodies would influence their course. Clearly, observation is subjected to deduction in this case and in such a manner at that, that it results in a set of laws which are entirely at odds with observation. A finer example of rationalist physics according to Rodis-Lewis' definition is hardly conceivable.

---

[66] In spite of profound differences, there is a perfect resemblance between Descartes and Newton on the point of regarding the question of the void as a conceptual one. Empiricist though he may have been, Newton regarded the question of matter and its relation to space no less a question of axioms. See, e.g., Ernst Cassirer, *Das Erkenntnisproblem der Philosophie und Wissenschaft der neueren Zeit*, Band II, Berlin (Cassirer) 1911, p. 463: "So zurückhaltend Newton in der Aussprache der allgemeinen Folgerungen aus seiner Lehre von Raum und Zeit ist, so entschieden fixiert er den Gehalt dieser Begriffe selbst, wo es sich darum handelt, lediglich ihren empirisch-wissenschaftlichen Sinn und Gebrauch zu bestimmen." However, determining the content of these notions, Newton also made use of *a priori* arguments concerning the absurdity of an infinite matter occupying space. According to Newton and Clarke, such matter being infinite, it would have to exist of necessity and thus be uncreated—a consequence quite unacceptable to these philosophers. For a survey of these points, see Alexandre Koyré, *Newtonian Studies*, London (Chapman & Hall) 1965, pp. 164-169.

[67] Descartes, *Principia* II 53. The line only occurs in the French edition, AT VIII-I, p. 93. The translation is from CSM I, p. 245.

And yet it may again be asked what, in this case, the terms "deduction" and "observation" are meant to signify. As Descartes himself explains, the laws which are supposed to be "evident" to reason are not meant to apply to observable collisions. This however, does not mean that observation is simply neglected by Descartes. In fact, in the case of the laws of collision, we are dealing with one of the examples in which a hypothesized reconstruction of microscopic phenomena—in this case, the impact of small particles—is supposed to explain observable macroscopic events. The collisions are not meant to be observable at all, but they are supposed to explain observable effects caused by them.[68]

Hence, although in one sense, deduction may seem to overrule observation, in another sense, it is supposed to explain it. However, in lieu of creating further confusion with respect to the Cartesian method of science, let us see what Descartes himself has said with regard to the notions of deduction and observation.

7.2.2.2 The A Priori *in Descartes*

In *Le Monde*, Descartes self-assuredly announces that he will offer

> *a priori* demonstrations of everything that can be produced in this new world.[69]

---

[68] Martin Kalmar, *Some Collision Theories of the Seventeenth Century: Mathematicism vs. Mathematical Physics*, PhD. dissertation John Hopkins University, Ann Arbor (University Microfilms International) 1981, p. 136: "Descartes never intended his rules to apply to such situations [of sensible objects involved in actual collisions]. If Descartes failed to find the correct solution to the problem of impact, it was because that was not his purpose. [...] Descartes' rules of impact were never devised to describe the same phenomena which are discussed in the works of Harriot or Huygens. Descartes' theory is meant to apply only to theoretical microscopic particles in order to justify a mechanistic explanation of macroscopic phenomena." Besides his recognition of the true function of Descartes' laws, Kalmar's analyses are also of great importance for understanding the rationale behind the laws themselves, in which the "principle of greater force" plays an important part: "If change always arises from a "stronger force", then there will be many instances in Descartes' rules in which one body in the collision fails to affect the situation of the other body." Kalmar, *idem*, p. 103. For Kalmar's presentation of the ways in which other scholars have tried to come to terms with the Cartesian rules of impact, see *idem*, pp. 92-105.

[69] that is, in the "imaginary world" described by Descartes. See *Le Monde ou Traité de la Lumière*, AT XI, p. 47. The translation is from CSM I, 97.

It is such statements that must have led post-Kantian readers of his texts to the assumption that Descartes planned to devise a physics without the use of empirical data, especially since the project of Cartesian metaphysics was also designed to secure a degree of truth that sense perception could not offer. In defence of his experiment of universal doubt, Descartes argued that its benefit lies

> in freeing us from all our preconceived opinions, and providing the easiest route by which the mind may be led away from the senses.[70]

As sense-perceptions deceive us on some occasions, we should never rely upon them completely, since

> it is prudent never to trust completely those who have deceived us even once.[71]

In the *Regulæ* moreover, Descartes contrasts the "deductive way" to the "way of experience", only to block the latter, since experiences of things are "often false".[72] Finally, referring to the text of *Le Monde*, Descartes in the *Discours* tells us how he deduced a physics of the heavens, the stars, the Earth and a host of terrestial phenomena from "first causes" and that in doing so, he "considered" two things only: 1) the idea of a God-Creator and 2) some "seeds of truth which are naturally in our souls".[73]

---

[70] Descartes, *Meditations*, AT VII, p. 12 / AT IX, p. 9. The translation is taken from John Cottingham's edition. See René Descartes, *Meditations on First Philosophy. With selections from the Objections and Replies*, Translated by John Cottingham with an introduction by Bernard Williams, Cambridge (C. U. P.) 1986, hereafter to be referred to as "COT", followed by page numbers. The same translation is taken up in CSM II. For the reference, see COT / CSM II, p. 9.

[71] Descartes, *Meditations*, AT VII, p. 18 / AT IX, p. 14. The translation is from: COT / CSM II, p. 12.

[72] Descartes, *Regulæ*, AT X, pp. 364-365: "notandum est, nos duplici viâ ad cognitionem rerum devenire, per experientiam scilicet, vel deductionem. Notandum insuper experientias rerum sæpe fallaces, deductionem verò, sive illationem puram vnius ab altero, posse quidem omitti, si non videatur, sed nunquam malè fieri ab intellectu vel minimùm rationali." Cf. CSM I, p. 12.

[73] Descartes, *Discours de la Méthode*, AT VI, p. 64 / CSM pp. 143-144. As for the "semences de Veritez", these are none other than the "primitive notions" which Descartes regards as "the patterns on the basis of which we form all our other conceptions". Besides the common laws of logic, that is, besides what is known by using the "bon sens" every human being is equally gifted with and the "most general conceptions" of "being, number, duration etc.", the primitive notions of physical science are the "mathematical notions" of extension, shape and motion etc. which are needed for mechanical explanation. See above, section 5.4.3, especially

Such methodological maxims have all given rise to the interpretation of Cartesian physics as a matter of "*a priori* deduction". Apparently however, neither in advocating the use of *a priori* deductions in science, nor in stressing the fallaciousness of the senses, did Descartes intend to ban observation and experiment from physics. Indeed, how could Descartes have been committed to deduction instead of observation when in practice he refused to discuss what he had not yet observed? Thinking he could reconstruct the formation of the embryo for instance, Descartes nevertheless postponed his explication of the phenomenon since he had not yet been able to make enough observations with which he could verfy his thoughts on the matter.[74]

The obvious role experiment plays in Descartes' scientific treatises has led many commentators to renounce the rationalist interpretation. Some have emphasized Descartes' use of analogy in scientific explanation.[75] Others have boldly affirmed Descartes' empiricism.[76] Finally, Desmond Clarke has tried to assess the experiential content of both Descartes' methodological theory and his scientific practice, clearing the philosopher of the traditional, "rationalist" charges.[77] Yet

---

the letter to Princess Elisabeth of Bohemia—referred to in capter 5, notes 81 and 82—of 21 May 1643, AT III, p. 665 / CSM III, p. 218.

[74] Descartes, *La Description du Corps Humain*, AT XI, pp. 252-253: "Et bien que ie n'aye pas voulu iusques icy entreprendre d'écrire mon sentiment touchant cette matiere, à cause que ie n'ay pû encore faire assez d'experiences, pour verifier par leur moyen toutes les pensées que i'en ay eu: ie ne puis neantmoins refuser d'en mettre icy en passant quelque chose de ce qui est le plus general, & dont i'espere que ie seray le moins en hazard cy-aprés de me dédire, lors que de nouuelles experiences me donneront dauantage de lumiere." Cf. CSM I, p. 321.

[75] In a book which bears the significant title *L'Expérience de la pensée dans la philosophie de Descartes*, Paris (Vrin) 1978, Nicolas Grimaldi for instance, says that Cartesian scientific explanation proceeds by way of an "'assimilation' de toute réalité naturelle aux choses 'connues par les sens'." Grimaldi paraphrases a remark from the *Cogitata Privata*, AT X, p. 219. Grimaldi, *op. cit.*, p. 45.

[76] See for instance: Jean Laporte, *Le Rationalisme de Descartes*, Paris (P. U. F.) 1945 / 1950², p. 477: "si nous voulons à toute force caractériser la philosophie de Descartes par un nom, le nom qui lui sérait le mieux serait, tout paradoxe à part, celui d'empirisme—empirisme radical et intégral." Peter Schouls associates Descartes' ideas in the *Regulæ* with those of the father of empiricism, John Locke, in the *Essay*. Schouls observes that the fact that "'empiricist' elements must be considered at certain points in a discussion of Descartes' methodology is clear, for example, from the role 'sense' plays in the construction of an 'applied' science." P. A. Schouls, *The Imposition of Method*, Oxford (Clarendon Press) 1980, p. 25.

[77] Desmond Clarke, *Descartes' Philosophy of Science*, Manchester (M. U. P.) 1982, p. 17, where it is argued that although "Descartes has consistently

what this discussion shows is not so much that Descartes is an empiricist, but rather that the rationalist / empiricist distinction, in spite of being useful for describing psychological theories, is without any use with regard to the analysis of methodological procedures in seventeenth-century natural philosophy.[78]

Accordingly, neither should Descartes' allegiance to "*a priori* deduction" be interpreted along traditional lines as being a method which excludes empirical observation. When Descartes claims that many truths concerning the heavens right down to terrestrial things can be deduced from the rational premisses of God and some seeds of inborn truth alone, he is not advocating a deduction independent of observation. To show this, we may refer to a passage in one of Descartes' letters to Father Mersenne. Descartes informs his correspondent about his preparations for the work which was later to be entitled *Le Monde*.[79] He asks the Parisian divine:

> If you know of any author who has particularly collected the various observations which have been made of comets, you would oblige me if you could inform me about them.

---

communicated the impression to his readers that he undervalues the significance of empirical evidence in science and that he is anxious to substitute 'rational arguments' in place of empirical research", Descartes is "unquestionably innocent at least of traditional charges that he ignores or distrusts experiential evidence as a basis for knowledge claims about physical nature."

[78] An historical basis of the methodological distinction may however, be of eighteenth-century origin. Claiming for Newton's achievements a solid foundation in experience and condemning the Cartesian system for being no more than an hypothesis, newtonian propaganda aimed at establishing the view that the epistemology advocating the existence of "innate ideas" and the epistemology of a mental *tabula rasa*, is paralleled by an *a priori* and an experimental method of science. See for instance Roger Cotes' words in note 45, above and Voltaire's characterization of Cartesianism in his *Lettres philosophiques sur les Anglais*, ed. Gustave Lanson, Paris (Didier) 1964, tôme II, p. 6: "alors [la philosophie de Descartes] ne fut plus qu'un roman ingénieux, et tout au plus vraisemblable pour les ignorants." Despite the fact that he made it his aim to point out the experiential content of many of Descartes' rules of method, a preoccupation with Newtonian standards must also have been the main reason for Alan Gewirtz to conclude that "the rationalist character of the method is evident in the nature of the reductive process, in the use made of hypotheses, and in the prescription of necessity which the deduction must fulfill." Alan Gewirtz, "Experience and the Non-Mathematical in the Cartesian Method", in *Journal of the History of Ideas*, 2^2 (1941), p. 209.

[79] Descartes had actually just started working on the treatise: "depuis deux ou trois mois, ie me suis engagé fort auant dans le Ciel." Descartes to Mersenne, 10 May 1632, AT I, p. 250 / CSM III, p. 37.

As Descartes explains, the observational data with respect to comets might yield a certain "order" by which the positions of the fixed stars could be explained. Knowing this "order" is said to be the key to a science with which

> one could know *a priori* all the divers forms and essences of terrestrial bodies, whereas without it, we must content ourselves with *a posteriori* conjecture.[80]

This indicates that the distinction between *a priori* and *a posteriori* routes is not one of intuition versus observation. Indeed, Descartes asks Mersenne for *observations* in order that he may know *a priori* what must otherwise be conjectured in an *a posteriori* manner. And observations are necessary even where it involves seeking some of the most general causes in the order of the Universe.

The letter also shows that the Cartesian type of *a priori* deduction is to be seen as a linking of causal chains in a "natural order". Descartes' *Le Monde* illustrates what type of deduction was intended. With respect to comets, Descartes had to explain their trajectories. Since he supposed movement within the solar system to be generally circular, the linear movement of comets appeared to him to be exceptional. Having all relevant data at hand, Descartes formulated his theory of celestial streams of "air" (the so-called *second element*), rotating quite independently from one another around particles of the first, "fiery", element. Left over after all of the intervals between the particles of a greater size had been filled, these fiery particles had gathered at the centre of each flux.[81]

The theory thus accounts both for the fixed position of the stars, which are formed by the clusters of fiery particles, and for the trajectories of bodies of the third, "earthy" element, such as planets and comets. The anomaly of comets passing through various fluid "solar systems", in a more or less rectilinear way, is accounted for by referring to the analogous movements of bodies floating in a river and approaching the point where another stream is met.[82] With the use of this model, the mechanical interpretation of the trajectories of comets and planets passing through and rotating within the various solar systems, are defended by an appeal to analogy[83]: as a result of their

---

[80] Descartes to Mersenne, 10 May 1632, AT I, pp. 250-251 / CSM I, p. 38.

[81] Descartes, *Le Monde*, AT XI, p. 52.

[82] Descartes, *Le Monde*, AT XI, pp. 58-59.

[83] Since experience is consulted by Descartes in order to be able to describe both sensible and hidden processes in a purely mechanical way, using analogies to

respective sizes, the former are drawn away into the next stream, whilst the latter are held close to the centre of one and the same celestial fluid.

Imagination and conjecture are at the heart of Descartes' explanations. What matters, however, is that observation is to produce the facts which are to be explained. Provided that only mechanical priciples of action are taken into account, that is, only such "mathematical truths"

> in accordance with which God Himself has taught us that He has ordered everything by measure and number and weight,[84]

and some *Règles* or axioms concerning the preservation of the total quantity of movement in the Universe, a deduction of observed phenomena is presented from hypothesised causes by means of observable variables and analogies. Instead of particle fluids, Descartes could for instance have chosen for an atomistic model. The important thing is, that the choice of mechanical model is not in itself a choice for or against the use of empirical data. Without invoking anything but the idea of God and the mechanical notions and logical laws which are naturally in our soul, Descartes reconstructs the most general physical processes neither rationalistically nor in an empiricist way, but by observations and conjectures concerning the causes of the observed phenomena.

In the *Discours*, Descartes observes that

> when I decended to more particular things, I encountered such a variety that I did not think the human mind could possibly distinguish the forms or species of bodies that are on the Earth from an infinity of others that might be there if it had been God's will to put them there. Consequently I thought the only way of making these bodies useful to

---

explain the latter, such analogies are numerous in all of Descartes' scientifical writings. Visualising mechanical processes behind sensible phenomena, they are never meant to prove any physical law in the modern sense of the word, but only the aptness of some "hypothetical" mechanical model. Likewise, Descartes defends his third axiom of movement, which holds movement to be the combined effect of particular inclinations to move in a rectilinear way, by referring to the empirical fact that stones rectilinearly leave the sling. Descartes, *Le Monde*, AT XI, pp. 45-46, and *Principlia* II 39, AT VIII-I, p. 63-65 / CSM I, pp. 241-242.

[84] Descartes, *Le Monde*, AT XI, p. 47 / CSM I, p. 97. The reference is to the *Wisdom of Solomon*. See above, chapter 5, note 79.

> us was to progress to the causes by way of the effects and to make use of many special observations.[85]

Reading this passage, even the most severe critics of the rationalistic exposition of Descartes' writings have thought it worthwhile to climb down from their empiricist interpretation.[86] The standard reading of this passage is that, since the deduction cannot be completed, Descartes acknowledges that he has to have recourse to experiment. However, as we have seen, even in the case of the most general phenomena, such as the heavens and the stars, observation was consulted. It is only that, in the case of the heavens, stars and planets, the reconstruction of the way in which particles will behave is, according to Descartes, more or less obvious. Confronted with the richness of the more particular objects with which Nature is endowed, it cannot easily be reconstructed how these could have come to be from the currents of earthy, airy and fiery particles. Therefore, says Descartes, one must have recourse to "many special observations".[87] Descartes does not introduce experiment at this point because the deduction has failed. He merely acknowledges that precise and special observations of the effects will have to provide starting-points for explaining them mechanically. Thus, the *a priori* route has (temporarily) to be abandoned. The difference between *a priori* and *a posteriori,* however, does not lie in the exclusion or acceptance of observational data as such. Nor has Descartes' distinction between *a priori* and *a posterioiri* routes anything in common with the Kantian distinction between rational and empirical faculties. For Descartes, the difference between the two routes is one of order and of sequence only. *A priori* deduction starts with ultimate causes, *a posteriori* theories go the other way, starting from the effects.

---

[85] Descartes, *Discours de la Méthode*, AT VI, p. 64, CSM I, p. 144.

[86] Jean Laporte for instance, notwithstanding his calling Descartes an empiricist, acknowledges some truths of Cartesian physics to be directly deducible from metaphysics. "De ces principes se peuvent former—et par conséquent déduire—des cieux, des astres, de l'air, des minéraux, et beaucoup 'd'autres telles choses qui sont les plus communes et les plus simples'." J. Laporte, *Le Rationalisme de Descartes*, Paris (P. U. F.) 1945, p. 206.

[87] Descartes, *Discours*, AT VI, p. 64: "plusieurs experiences particulieres": it is to the credit of the translators of the CSM edition, that "experiences particulieres" is here translated as "*special* observations". CSM I, p. 144.

7.2.3 *Ultimate Causes*

Descartes' project of mechanistically interpreting the material Universe, from the stars, comets and planets downwards to particular terrestrial objects, is not meant as a rationalistic conjecture only. In fact it is a causal account of the Mechanism of Nature.

In relation to the movement of the heart, Descartes had supposed that there is a certain heat inside the heart and that the blood is of a nature that it expands upon heating. With the help of these two postulates alone, Descartes had managed to replace the Scholastic principles of action by mechanical ones. With the help of the two postulates of God and of some innate "seeds of truth", he likewise explains the whole clockwork of Nature at once, without introducing any other effective principle apart from God causing and maintaining movement in a determinate quantity of matter.[88] Synthetically reasoning from this Prime Cause, that is, reasoning in an *a priori* way in the sense in which Descartes uses this term, a complete "deduction" of all natural mechanisms could be made. The deduction includes not only the greater material objects, but also terrestrial objects such as minerals, and could, in principle, be further developed to include every particular natural object, living creatures as well:

> From the description of inanimate bodies and plants I went on to describe animals, and in particular men.[89]

The entire description is in fact a generic one, streams of material particles moving and combining in such a way that they will ultimately form—and explain—the functioning of any material body.[90]

It is in this generic way that the Aristotelian demand for a genuine principle of action is met. Like Regius, Descartes argues that all mechanical processes can be deduced from the only *principium effectivum* which apart from human volition may act as an external motor. There is only a single efficient cause for all natural mechanisms in Cartesian philosophy: God. For the rest, nothing

---

[88] See e.g. *Principia* II 36, AT VIII-I, pp. 61-62: "putemus ex hoc solo, quòd Deus diversimodè moverit partes materiæ, cùm primùm illas creavit, jamque totam istam materiam conservet eodem planè modo eâdemque ratione quâ priùs creavit, eum etiam tantundem motûs in ipsâ semper conservare." See also section 9.4, below. Cf. CSM I, p. 240.

[89] Descartes, *Discours de la Méthode*, AT VI, p. 45 / CSM I, p. 134.

[90] G. Lewis, *L'individualité selon Descartes*, p. 64: "Ainsi un enchaînement purement méchanique d'actions et de réactions explique à la fois la formation des organes et leur fonctionnement."

produces motion, since the entire clockwork already is in motion. This however, at the same time means that, as Regius had put it:

> the whole Universe of things is created by God in such a way that all its parts are in need of reciprocal assistance and cannot at all operate without it.[91]

While all motion stems from God, it is transmitted from one part of the Universal Machine to the other. Regius indicated this when, in order to rebuke Voetius, he claimed that clocks are moved by an external motor. In fact, it is "ethereal matter" that causes the movement of their counterweights.[92] In any case, what matters is that even if a clockwork is dependent on an external motor for its action, this is really no argument against mechanical explanation. Except for human beings, the same is true for all things natural: everything works on everything else as external motors of action. Only man is an independent cause and only God is the ultimate source of all action.

And as far as the human body alone is concerned, even man is taken up in the chain of mechanical causality. In the *Discours de la Méthode*, Descartes formulates the strategy he followed in the *Traité de l'Homme*:

> I supposed, too, that in the beginning God formed the body of a man [...] [without] any rational soul or any other thing to serve as a vegatative or sensitive soul, but rather that he kindled in its heart one of those fires without light which I had already explained, and whose nature I understood to be no different from that of the fire which heats hay when it has been stored before it is dry, or which causes new wine to seeth when it is left to ferment from the crushed grapes.[93]

Without supposing anything but the presence of a certain heat, the mechanical principles suffice to explain the workings of the human body and supplant the Aristotelian principles of the vegatative and sensitive soul. Descartes draws an analogy with other instances of "fires without light". The idea is, that this peculiar type of heat, which ultimately explains the workings of the human body, is itself a natural, i.e. *mechanical*, phenomenon.

---

[91] Regius, *Responsio*, p. 26.

[92] Regius, *ibidem*. See above, note 29. The point is not elaborated. It may be that Regius has some particle theory of gravity in mind. For his idea of gravity, see *Fundamenta Physices*, Amstelodami, Apud Ludovicum Elzevirium, 1646, pp. 55 ff.

[93] Descartes, *Discours de la Méthode*, AT VI, p. 46 / CSM I, p. 134.

Descartes however, is unable to provide an *a priori* explanation for the fact that such a particular kind of heating is actually in the heart in the way that, in *Le Monde*, he had been able to provide an explanation for the actual existence of the stars, the comets etc., since, as he informs us in the *Discours*,

> I did not yet have sufficient knowledge to speak of them in the same manner as I did of the other things.[94]

In the case of the movement of the heart, Descartes therefore works in an *a posteriori* way, examining the process in detail and explaining it with the help of a few suppositions only. But his aim in physiology was to attain a complete generic account of the human body and its workings. It was only in 1648 that Descartes offered a mechanical description of the way in which the male and female semens "ferment and warm each other" in such a way that

> some of their particles, acquiring the same agitation as fire, dilate, push aside (*presser*) others and arrange them bit by bit in the manner required for the formation of the limbs.[95]

The observational basis was admitted to be a meagre one, which is why Descartes says that he can only present some very general ideas concerning the process. Nothing very precise could be concluded with respect to the "shape and arrangement" of the microscopic parts of the semens themselves.[96] However, observation of these was not in the first place what was thought necessary. What Descartes did observe, were the first phases of embryonic formation, that is, of the *punctum saliens* developing into the heart, brains and other limbs of the embryo. Since these are thought to develop from the fermentation of the semens, a mechanical process of dilation of small particles is hypothesized to account for the formation of the body.

Just as in the case of the comets, macroscopic observation is taken as the starting-point for explanation. Mechanical hypotheses concerning microscopic processes are next formed in analogy to macroscopic phenomena, in this case, the phenomenon of fermentation. Having then explained how the embryo is formed by the

---

[94] Descartes, *Discours de la Méthode*, AT VI, p. 45 / CSM I, p. 134.

[95] Descartes, *Description du corps humain*, AT XI, p. 253 / CSM I, p. 322.

[96] Descartes, *Description du corps humain*, AT XI, pp. 252-253 / CSM I, p. 321. That the type of "knowledge" that was wanting was empirical knowledge, is shown in the *Description du corps humain*, where Descartes writes: "ie n'ay pû encore faire assez d'experiences." AT XI, p. 252 / CSM I, p. 321.

"two liquids" of semen "fermenting" in such a way that they form a human heart, Descartes might have resumed work on the project of explaining animal life as a whole as originating from purely natural, that is, mechanical, principles.[97] An important point in the account of embryonic formation is that it not only explains the *genesis* of the human body, but also its *operations*. In particular, the fermentation of the semens is the source of a dilation which is preserved in the form of a natural warmth. It is this warmth which could earlier only be hypothesized as the external motor of pulsation.[98] Now that it is explained as deriving from semens readily formed, the *a priori* route could be given in more detail.

It may thus become clear to what degree Descartes took the mechanical metaphor to the extreme. Nature is, according to Descartes, quite literally a giant item of clockwork. Physical phenomena are only then fully explained when they are explained by the reciprocal influence of bodies upon each other which ultimately derives from God's causing motion in the mechanical machine. To explain bodily operations is ultimately to explain the *genesis* of the same bodies within the mechanical framework. To fully explain Nature would be to finish the *a priori* route of material evolution from the first formation of the stars onwards to every particular object. The causally complete account of physical phenomena is their reduction to the Cartesian Big Bang.

7.2.4 Cartesian Suppositions

In the *Discours*, Descartes apologizes for the use of certain "suppositions", which, although they explain the facts, are not themselves explained:

---

[97] Note that, for all his emphasis on teleological argumentation, Aristotle in fact accepted the description of the male semen passing its motion onto the embryonic matter in the female as an *efficient* cause. See above, chapter 2, note 77.

[98] Descartes, *Description du corps humain*, AT XI, p. 254: "Et c'est en cette dilatation, qui se fait ainsi à diuerses reprises, que consiste le battement du coeur, ou le poulx." Cf. CSM I, p. 322. See also the *Passions of the Soul*, where Descartes claims that the heart is formed and that the blood circulates the body on account of a natural warmth which is preserved by the blood itself. AT XI, p. 333 / CSM I, p. 331.

> I have called them "suppositions" simply to make it known that I think I can deduce them from the primary truths I have expounded above.[99]

What are these primary truths? They are not general laws of impact from which mechanical suppositions regarding particular effects of particle movement are still to be deduced. Even in the *Principia*, Descartes was not at all concerned with the mathematical undertaking of deducing effects of particle collisions from more general laws of motion. This type of mathematical enquiry is simply not what Cartesian science is about.

The primary truths may be of a metaphysical nature. Descartes for instance apologizes for the fact that particular corpuscular explanations are yet to be deduced from the general hypothesis of corpuscularism, that is, from the hypothesis that visible effects may be explained by particle pressure and collision. Referring to a letter to Père Vatier of 22 February 1638, Étienne Gilson has argued that Descartes' "primary truths" are indeed to be interpreted metaphysically.[100] Descartes writes the following to Father Vatier:

> assuming that I could deduce [my ideas] in an orderly manner from the first principles of my Metaphysics, I have decided to neglect all other type of proof.[101]

It is metaphysical arguments by which the suppositions ultimately have to be proved. The metaphysical account of the real existence of a material universe with primary, that is, "mathematical", qualities alone, and of the real existence of what is hypothesized in corpuscular explanations, are both needed in order to validate Descartes' suppositions. We have seen that Descartes did later offer such a metaphysical proof for the objective certainty of what is subjectively evident. But this was not all.

The apology Descartes makes for his use of suppositions and *a posteriori* demonstrations is elsewhere formulated in terms of an apology for not yet having presented his "principles".[102] Descartes had

---

[99] Descartes, *Discours de la Méthode*, AT VI, p. 76. The translation is from CSM I, p. 150.

[100] See R. Descartes, *Discours de la Méthode*, texte et commentaire par Étienne Gilson de l'Académie Française, Paris (Librairie philosopique J. Vrin) 1976^5, pp. 470-471.

[101] Descartes to Père Vatier, 22 February 1638, AT I, p. 563 / CSM III, p. 87.

[102] See for instance the letter to Mersenne of 11 March 1640, AT III, p. 39: "Pour la Physique, ie croyois n'y rien sçavoir, si ie ne sçavois que dire comment les

doubts as to whether he would ever finish the project of presenting them,[103] but finally, in 1644, he did so, by publishing the *Principia Philosophiæ*. It is in this book that all metaphysical and physical questions relating to the corpuscular philosophy are dealt with. The first part of the *Principia* raises the general question of Cartesian metaphysics. The second part is significantly entitled "Of the Principles of Material Things". Here, Descartes elaborates ideas which he had earlier broached in the first seven chapters of *Le Monde*, concerning the material essence of the physical world and the behaviour of material objects and particles.[104]

These then, are presupposed in the specific questions which Descartes deals with in, for instance, *Les Météores*. But there is more. At the end of the *Discours*, Descartes says concerning the hypotheses which occur at the start of *Les Météores*:

> I have called them "suppositions" simply to make known that I think I can deduce them from the primary truths I have expounded above.[105]

However, both the metaphysical considerations regarding the validity of particle explanations and Descartes' general ideas about the behaviour of particle matter, are wholly lacking in the *Discours*. What Descartes did "expound above", were the first *effects* of particle movement. It thus seems that something more is presupposed in the Essays accompanying the *Discours*: the formation of the material Universe from particle chaos.

7.2.5 *The* A Priori *Route*

The purpose of the "principles" of Descartes' natural philosophy in *Le Monde* chapters I-VII and *Principia* Part II, is to introduce the fundamentals of corpuscular philosophy. In other words, Descartes introduces some general ideas concerning the behaviour of bodies and

---

choses peuuent estre, sans demonstrer qu'elles ne peuuent estre autrement; car l'ayant reduit aux lois des Mathematiques, c'est chose possible, & ie croy le pouvoir en tout ce peu que ie croy sçavoir, bien que ie n'aye pas voulu y donner mes Principes & ie ne voy encore rien qui me conuie à les donner à l'auenir." Cf. CSM III, p. 145. Descartes also mentions his "principles" at AT VI, p. 76 / CSM I, p. 150.

[103] See the end of the quotation from the letter to Mersenne in the former footnote.

[104] Descartes, *Principia*, AT VIII-I, pp. 40-79 / CSM I, pp. 223-247 and *Le Monde*, AT XI, pp. 3-48 / CSM I, pp. 81-98.

[105] Discartes, *Discours*, At VI, p. 76. The translation is from CSM I, p. 150.

particles in collision or mutual contact. These ideas concern the existence and the primary qualities of matter; rarefaction and the void; the identity of space and matter; the relativity of motion; the indefinite divisibility of matter and the rejection of atomism; the explanation of all formal diversity by local motion; inertia; the laws of movement and the laws of collision; and finally, the essence of solidity and fluidity and the behaviour of particles in a fluid. All these prepare the hypothetical deductive explanations of mechanical philosophy. The principles, in other words, lay the foundations for the idea that microscopic models of particle collision may serve as causal explanations for physical events.

Whether or not they qualify as such is, however, only proved by the success with which the principles may be applied to the world of sense perception. Descartes' way of proving their explanatory strength is in the *Principia* of 1644 the same as it was in *Le Monde*:

> Thus we may be able to think up certain very simple and easily known principles which can serve, as it were, as the seeds from which we can demonstrate that the stars, the Earth and indeed everything we observe in this visible world could have sprung.[106]

Descartes next recapitulates the ideas concerning the movement of material fluids of *Principia* II, adding that we cannot rationally intuit the sizes and speeds of any of the material particles from which the world was formed. Nevertheless, Descartes continues, we may assume anything

> with the sole proviso that all the consequences of our assumption must agree with our experience.[107]

The effects, in other words, need not to be proven. It is the hypothesized causes which are proved by the effects. Thus we find exactly the same argument as appears in the *Discours* concerning the "suppositions" Descartes made in *Les Météores*. The important point is, that what in fact proves the hypothesized causes of mechanicism, are not the principles themselves, but their application in an evolutionary reconstruction of the *genesis* of the physical Universe:

---

[106] Descartes, *Principia* III 45, AT VIII-I, p. 100. The translation is from CSM I, p. 256.

[107] Descartes, *Principia* III 46, AT VIII-I, p. 101. The translation is from CSM I, pp. 256-257.

> So, if we may, we will suppose that the matter of which the visible world is composed was originally divided by God into particles which were approximately equal, and of a size which was moderate, or intermediate when compared with those that now make up the heavens and stars.[108]

What follows is the imaginary reconstruction of the stars, planets, comets and vortices, all of which are "deduced" from the "chaos" of the initial particle fluid. What proves the particle hypothesis is, in other words, the success of the *a priori* deduction which Descartes had originally set out in *Le Monde*. What would happen, Descartes had asked, if God imparted motion to a certain material chaos in an imaginary Universe? According to Descartes, this imaginary Universe would evolve into an exact duplicate of ours.[109] The heavens and stars and planets would be formed by the sole impact of variously shaped parts of matter.[110] The conclusion must be, that the reason for Descartes' following the *a priori* route was not so much that he sought certainty for the possibility of explaining physical events from principles concerning the behaviour of material bodies. Such metaphysical certainty would, in any case, apply to the particular explanations of *Les Météores* just as much as to the general principles of particle motion. The reason for taking the *a priori* route is not that the so-called "suppositions" need to be *proved* or *demonstrated*: what matters is that they need to be *deduced*—deduced, that is, from the *a priori* route starting with the first flow of generic particle movement when God activates the initial particle chaos.

7.2.6 The Reduction of Physical Phenomena to an Initial Big Bang

In the *Principia*, the explanation of mechanical principles is followed by a deduction of the first hypothesized effects of particle motion from chaos. Both in *Le Monde* and in the *Principia*, the two elements of (1) general ideas concerning the behaviour of matter and (2) the first steps in the development of an imagined universe together form the basis for deducing all natural phenomena from first causes *a priori*. Descartes

---

[108] Descartes, *Pincipia* III 46, AT VIII-I, p. 101 / CSM I, p. 257.

[109] Descartes, *Discours*, AT VI, p. 45 / CSM I, pp. 133-134.

[110] Note however that, in the *Principia*, Descartes assumes the original particles to be of the same magnitude and to have the same amount of motion, since a complete chaos like the one that he had formerly presented in *Le Monde* and in the *Discours* seems to agree less "cum summâ Dei rerum creatoris perfectione". Descartes, *Principia* III 47, AT VIII-I, pp. 101-103 / CSM I, pp. 257-258.

in fact does not always seem to distinguish very sharply between these two aspects of his *a priori* deductions. In the Cartesian reconstruction of the material Universe, evolution and explanation somehow coincide. The first flow of particles is so easy to understand that anyone can comply with the corpuscular hypotheses in the case of the formation of the stars. Yet simplicity is not what is missing in particular corpuscular explanations. What is missing is a causal deduction.

In section 4.1.1 we have seen how, in Descartes' essay on *Meteorology*, the boundary between general principles of Nature and applied theory is rather diffuse. Indeed, there is no discernible borderline between general suppositions of corpuscular philosophy and mechanical interpretations of specific macroscopic phenomena. We may now see why. Descartes never finished the kind of *a priori* deduction of the world system he started with in *Le Monde*, and which was to explain even living Nature from the sole idea of God imparting motion on a mass of inert matter. Every time that, as in the *Météores*, Descartes intends to discuss individual physical processes, he is confronted with the problem that the explanation will be incomplete. To a certain extent, therefore, Descartes shared Voetius' and Schoock's criticism that the mechanical account of natural phenomena would only yield secondary instead of ultimate *principia effectiva*. Whenever he examines isolated processes, he accordingly points out that his account must remain incomplete. In particular explanations therefore, he makes use of what he calls "suppositions". It is in order that he may escape the charge of incomplete causal demonstration that Descartes "supposes" for the time being what misses in an as yet unfinished description of the world in which Nature is represented in the very literal sense as a gigantic machine. Cartesian physics is thus "reductionist" in two ways. It is firstly reductionist in its manner of explaining observable phenomena by referring to the behaviour of elementary particles. There is, however, also the "generative reductionism" of deducing all observable phenomena from first events in the physical Universe. This rounds off our search for ultimate causes in Cartesian physics.

7.3 Conclusions

The Cartesian account of the movement of the heart presupposes only that there is a certain heat inside the heart and that the blood is of a nature that it ferments, or "boils" in a certain way. These facts are (1) not further analysed mechanically and (2) held to be dependent on empirical observation. The account provoked the ridicule and scorn of

Voetius and Schoock, both of whom criticized the notion of such a piece of clockwork without a weight or spring. Whereas Regius tried to solve things by reformulating the mechanical explanation in the Scholastic terms of his critics, Descartes' solution of the problem of insufficient causal explanation was to offer an *a priori* route, in which motion was explained by the motion of surrounding elements, leading ultimately to the primal motion imparted by God upon the clockwork of Nature as a whole.

CHAPTER EIGHT

THE CARTESIAN WORLD

Misera anima, quæ socium in toto naturæ theatro non habet.

Martin Schoock, *Admiranda Methodus Philosophiæ Renati des Cartes*[1]

Since the concept of individual forces was abandoned, the mechanical philosophy had to accept the idea of a "dispersion" of causality, thereby making conclusive explanations of isolated phenomena impossible. It also had to give up the idea of individual substances. As we have seen, Gisbertus Voetius defended the idea of individual substantiality against Descartes and Regius primarily on Biblical grounds, referring to the occurrence of the concepts of *species* and *genus* in the Book of *Genesis*. Voetius also foresaw that the mechanical philosophy would lead to a world in which

> all created substances would merely be accidental beings, collections, aggregates, and not essences or unique natures by themselves.[2]

---

[1] Schoock, *Admiranda* p. 205 / *Querelle*, p. 284.

[2] Voetius, *Narratio*, p. 41 / *Select. Dispp.* I, p. 873 / *Responsio*, p. 13 / *Querelle*, p. 107. Recall also Voetius' demand to explain and defend "the distinction between the entity of substance and accident. For in [a philosophy without forms], there cannot be any substantial difference between a wolf, a sheep, a whale, an elephant, a snake a stone, a tree, a turnip, an aconite, wheat, the Sun, the Moon, the Earth..." See above, chapter 2, note 67. In her excellent study *L'Individualité selon Descartes*, Geneviève Rodis-Lewis has traced the concept of individuality in Descartes' writings. Analysing a great number of relevant texts, she has shown Descartes' allegiance to the ideas of substantiality and individuality in the mental realm. As for the physical realm however, Lewis concludes: "Ainsi l'individualité apparantedes objets matériels n'est finalement fondée sur aucune identité substantielle." And: "L'individualité des corps matériels—plus ou moins organisés—repose donc toujours sur la détermination fragile de parties liées par un ensemble de mouvements unifiés." Geneviève Lewis, *L'Individualité selon Descartes*, Paris (Librairie phiosophique J. Vrin) 1950, pp. 60 and 65. Voetius' insight in these matters and his analysis of the consequences of the New Philosophy is remarkable. It was only two years later, in the beginning of 1645, that Père Mesland was to discuss similar topics with Descartes in connection with the

Though wholly formulated in Aristotelian terms, the characterization nicely fits the Cartesian world-view, since, in the clockwork of Nature, individuality is indeed lost: diversity is merely the effect of movement and inertia.[3] In fact, the only individual beings left in the mechanical Universe are human beings, which form an exceptional species indeed—and God, whose creativity and goodness are known only very indirectly.

8.1 Matter in Motion

Defending the substantial forms of Scholastic philosophy, Voetius argues that there must at least be some principle for the action and "perfection" of natural objects. If the principle would be matter, no distinction could be made between the various operations of different things, since matter is common to all. Nor can operations be the result of accidents, since these "cannot form an ultimate principle".[4] It must therefore be the form which is responsible for the action. The same is true with regard to essence:

> There is some first root and first conception of every entity, e.g. of humanity, horseness etc., which constitutes a thing in its proper being and distinguishes it essentially from others. This, however, is not matter, since this is common [to every object alike]; nor accidents, because these cannot compose or constitute a substance, and give it its being. It is therefore what we call form, idea, essential nature, actuality of the perfect, or nature *par excellence*, since it actuates and informs matter and constitutes a compound being with it.[5]

In both examples, the alternative possibility that accidents may be responsible for the operation or essential nature of a thing, is renounced purely on account of the definition of substance and accidents. The reason why the difference between substance and accident was so important, comes out more clearly in Voetius' third

physical explanation of transsubstantiation. See in general, Geneviève Lewis, *L'Individualité selon Descartes*, pp. 56 ff.

[3] Geneviève Lewis, *L'Individualité selon Descartes*, p. 55: "[le] seul jeu des lois du mouvement et de l'inertie assure [...] la diversité des corps."

[4] Voetius, *Narratio*, p. 43 / *Select. Dispp.* I, p. 875 / *Responsio*, p. / *Querelle*, p. 108. Voetius advises students to carefully read some recent disputations of physics or metaphysics, by authors such as Suárez, Pererius, Ruvius, Toletus, the Conimbricenses, the Complutenses, Ariaga, or Mendoza. Of these, Suárez is said to "[defend] his cause in the most eminent and most subtle way."

[5] Voetius, *ibidem*.

argument in favour of the forms: the argument from corruption. If a horse or a dog ceases to be, it cannot be on account of its matter. For the matter is incorruptible and is not destroyed. Therefore,

> it must be the form, by which it is brought about that the *compositum*, that is to say this horse, this dog etc., dissolves and becomes a *non-being*.[6]

Again, the alternative might be given, that accidental factors are the cause of the corruption. In that case, the end of a substance would happen by destruction (*per interitum*). However, in that case, a dead man would differ only accidentally and not substantially from a living one, that is, no more than an ill man differs from a healthy man, or someone sitting down from someone standing up. This is what would happen when the forms are done away with and "all created substances" are regarded as "beings *by accident*, collections, aggregates" and not as "single essences or natures *by themselves*."[7]

The rejection of the forms is at once a rejection of individuality. Regius indeed made an effort to redefine individuality on the basis of his concept of "essential form":

> there is no "corruption of the substantial form," but only of the essential form, which consists of the aforementioned suitable combination of accidents.[8]

If the essential form is defined as the specific combination of accidents—of measure, shape and motion and the like—that is particular to a substance, a recombination of these accidents comes down to a corruption of the specific essence. In that way, accidents *can* "constitute or change the essence of things". Therefore, dead and living objects do

---

[6] Voetius, *ibidem*.

[7] Voetius, *Narratio*, p. 41 / *Select. Dispp.* I, p. 873 / *Responsio*, p. / *Querelle*, p. 107. Voetius uses quotation marks and may be paraphrasing David Gorlæus, *Excercitationes Philosophicæ quibus Vniversa fere discutitur Philosophia Theoretica. Et Plurima ac præcipua Peripateticorum dogmata evertuntur*. In Bibliopolio Commeliniano (Sumptibus viduæ Ioannis Comelini) 1620, pp. 265-266: "Neque in hisce rebus [the context concerns the human mind as the informing form of the human body] ita compositus datur ullum ens per se unum, quod indigeat aliquâ substantiâ, à quâ in unitate contineatur; sed entia sunt per aggregationem. Ita in rebus animatis datur corpus & anima, ex quibus componuntur, in rebus mixtis elementa."

[8] Regius, *Responsio*, p. 22.

> differ essentially even if no substantial difference stands between them.[9]

Thus, there may yet be more difference between a dead man and a living one than between the same man standing up and sitting down, since neither of these are essential for a human being.

To the standards of common sense however, the argument is still debatable. Voetius could argue that in both cases, the difference is one of "position" only—an accidental difference therefore. Descartes had no mind to go into the question of individuality. In fact, in his letter to Regius of January 1642, he has extremely little to say on the matter. He merely advises Regius to return the argument and say that

> [each of Voetius'] arguments in favour of the forms may be applied to the form of a clock, of which, however, no one will say that it is substantial.[10]

Clocks, in other words, operate, have individual natures and can be destroyed. Yet no one will say that they possess substantial forms.

The discussion however, could be prolonged infinitely. Voetius would agree that clocks do not have substantial forms and say that this is exactly because they do not operate without external motors. The clock only operates because it is forced to do so by outer conditions. That is why individual centres of causation are lacking when all of Nature would be compared to a clockwork. Accordingly, individual substances could neither be distinguished from the accidental parts of the universal mixture of mechanical Nature.

Whether it is the stars arising out of the juxtaposition of fast and fiery particles, or a human heart begotten in the fermentation of the male and female "semens", it was all part of one evolving mechanical Universe. Again we may point to the fact that Descartes took the machine metaphor quite literally. On the face of it, the Cartesian world was more a swirling mass of insensible particles than a neat clockwork of springs and wheels—more like a soup than a machine. What matters is, that it had the unity of a single material substance. If

[9] Regius, *ibidem*.

[10] Descartes to Regius, January 1642, AT III, p. 505 / *Rodis-Lewis*, p. 90 / CSM III, p. 208. Regius gives this argument in the *Responsio* as well: "Omnes autem hujusmodi rationes, quæ ad formas substantiales probandas adduci solent, exactissimè formis horologiorum aliorumque αὐτομάτων applicari possunt, quas tamen nemo substantiales esse dixerit." Regius, *Responsio*, p. 22.

it were not for the exception of individual human souls, Spinoza's fully determined holism was never far away.

However, the human mind did make a difference—which brings us back to an aspect of Cartesian method which we have so far ignored.

8.2 A World of Reason and a World of Sense

That the Cartesian method of natural philosophy does not exclude the use of observation, has been sufficiently demonstrated above. Nevertheless, we are still confronted with the fact that Descartes is time and again playing down the use of the senses, especially in the metaphysical writings. The question is: why are the senses said to be unreliable, when observation is the starting-point of physical inquiry?

8.2.1 Primary and Secondary Qualities: A World Undressed

It has been argued that Descartes' criticism of the use of the senses is related to a scepticism concerning the use of experiments. According to Desmond Clarke,

> the Cartesian distrust of *expériences* is, in part, a suspicion about the complexity and uncertainty of experiments [...] [and its] preference for reason rather than experience is, at least in physical science, a preference for one kind of experience rather than another.[11]

Thus, Descartes would prefer ordinary experience to experiments under artificial conditions.[12] Yet Descartes does not exactly say so. What he does say, is that the complexity of experiments is a reason for distrusting the outcome of experiments *done by others*.[13] This may be

---

[11] Desmond Clarke, *Descartes' Philosophy of Science*, p. 209.

[12] See also *idem*, p. 39, where Clarke argues that the "relative complexity [of experiments] makes them more liable to errors and misinterpretations than the common experience of unscientific observers of Nature."

[13] See e.g. Descartes to Mersenne, 11 March 1640, AT II, p. 38. Descartes there suggests that a bullet shot from a distance of 1 foot against a wall may hit the wall with less speed than a bullet shot from a distance of 15 or 20 yards. The idea is, that a bullet cannot easily displace the surrounding air if it is shot too close to the wall. However, Descartes adds: "Toutefois, c'est à l'experience à determiner si cette difference est sensible, & ie doute fort de toutes celles que je n'ay pas faites moy-mesme." See also: Descartes to Huygens, 18 or 19 February 1643, where Descartes mentions that he has ordered a twelve feet high glass tube with a narrow horizontal side-piece at the bottom, in order to measure the jet of water form the side at different water-levels in the tube. Cf. AT III, p. 617. Here also, Descartes

regarded as a rather unprofessional or at least not very generous standpoint. Nevertheless, the quotations referred to do not suggest that Descartes meant to nullify the importance of experiments as such. Indeed, in the *Discours*, Descartes even makes a rather "Baconian" proposal by proposing payment to unprejudiced artisans, who could faithfully and accurately experiment within well-defined conditions.[14]

Decartes also defended his ideas against the reproach that they were not confirmed by experience, at least in such a way that they could be accepted in the Schools, by arguing that he had in fact explained

> all observations (*expériences*) which may be done with respect to inanimate bodies, and that, on the contrary, none has ever been well explained by the principles of the vulgar philosophy.[15]

Without any hesitation, Descartes says that in contrast to those of the School philosophy, his own principles are affirmed by experience. Thus, it is against Scholastic doctrines that Descartes vindicates the senses. On the other hand, the issue of the *détachement des sens* is equally directed against the traditional philosophy.

---

says that he does not trust "experiences que ie n'ay point faites moy-mesme". Both examples are cited by Clarke, *op. cit.*, pp. 39 and 46.

[14] Descartes, *Discours de la Méthode*, AT VI, pp. 72-74 / CSM I, pp. 148-149. Descartes was, by the way, fully aware of the possibility of intervening factors hampering exact measurement in experimental situations. In the short *Treatise of Mechanics*, or *Explication des engins par l'aide desquels on peut avec une petite force lever un fardeau fort pesant*, he mentions all sorts of intervening forces causing certain quantifications—as, e.g. the determination of the force needed to pull a body up a plane inclined by 45° being half the force needed to lift it vertically—to be less exact when verified by experiment. Cf. Descartes to Huygens, 5 October 1637, AT I, pp. 435-447 / CSM III, pp. 66-73. In his article on "Descartes' Use of Empirical Data to Test Hypotheses", Spyros Sakellariadis has given a clear exposition on the problem of "interference forces" hampering the establishment by experience of ideal laws of physics. He concludes that "interference arguments themselves [i.e., arguments for neglecting those observations in which physical laws, though valid, cannot be verified because other, hypothetical, forces interfere] [...] do not necessarily preclude data from having a function in science in general and in Cartesian science in particular." S. Sakellariadis, "Descartes' Use of Empirical Data to Test Hypotheses", in *Isis* 73 (1982), pp. 68-76.

[15] See Descartes' letter to an unknown correspondent, dated June 1645, AT IV, p. 225 / CSM III, p. 252. The letter is thought to have been adressed to Constantijn Huygens, but left out by Leon Roth in his edition of the correspondence between Huygens and Descartes.

The paradox may be explained by referring to what we have earlier said concening the qualities of sound and heat. Given the fact that in the mechanical philosophy, all such sensible qualities may be reduced to the "mathematical properties" with which the material substance of Nature is endowed, sense experience is, to some extent, fallacious. On the other hand, experience itself teaches us what mechanical processes lie behind the sensible phenomena. Since everyday observations teach us for instance that friction causes heat, we may extrapolate this finding and say that all heat should be attributed to analogous causes.[16]

Invoking mechanical analogies to visualise processes that are hidden from the senses, Cartesian physics dispenses with attributing other qualities to material objects than "mathematical" ones. The Cartesian maxim to detach oneself from the senses should be read in this light: far from being a plea against the use of experience, it prepares the way to a new ontology of sensible qualities. The emphasis Descartes lays on the question of the *détachement des sens* is motivated by the contrasting way in which sensible qualities play a role in his philosophy and in the School philosophy from which he was emancipated.

8.2.2 Science and the Common Life

A new appraisal of scientific knowledge is the direct result of the Cartesian *détachement des sens*. For the common-sensical philosophy of Scholasticism, sense experience is defined as observations made by many individuals alike. As Martin Schoock illustrates: when I take a grain of pepper and eat it, I most clearly perceive it has a "heating force" or "power" (*vis*), and others perceive the same thing.[17] In the modem world of the mechanical philosopher, however, objectivity cannot so easily be obtained.

8.2.2.1 Schoock on Sense and Reason

Experience, as Schook pointed out, guides us in our daily lives. It informs us of the nutriciousness of edible substances for instance, and

---

[16] Descartes, *Le Monde*, AT XI, p. 10: "Mesme plusieurs experiences favorisent cette opinion: car, en se frottant seulement les mains, on les échauffe; & tout autre corps peut aussi estre échauffé sans estre mis auprés du feu [pourvue] seulement qu'il soit agité & ébranlé, en telle sorte que plusieurs de ses petites parties se remuent, & puissent remuer avec soy celles de nos mains." Cf. CSM I, p. 84.

[17] Schoock, *Admiranda*, p. 107 / *Querelle*, pp. 233-234.

not only us, but other animals as well, who learn from experience what is best for them to eat. On the other hand, since brutes do not have the capacity to reason from particular to universal knowledge, they just eat the right portion of food they think fit for themselves, without being able to correct this "knowledge" through an "experience of reason". Only human beings know what different amounts of hay should be distributed amongst different hay-eating animals. And doctors too, can write out prescriptions for others on the ground of their vast knowledge of particular experiences. What is more, only humans can correct sense experience whenever it is troubled by illnesses for example. Thus, although to a feverish person honey may taste bitter, his reason tells him that this is caused by the fact that bad humours have penetrated his organs of taste, and that it ought not to be a reason for concluding that bitterness is a property of honey.[18]

In universalising particular qualitative data, reason is, according to Schoock, guided by the number of cases in which a phenomenon is experienced. Of course, bitterness of honey is no less evidently experienced in any commonsensical way when one does have a fever, but the lack of universality which characterizes this experience will not make a reasonable person conclude that honey really is bitter. Experience may, on the other hand, lead one to agree to the proposition that, whenever a person is feverish, honey *seems* bitter to the person involved. Doctors of medecine may even deduce a regulative principle from this experience. The role of reason in such a deduction is yet entirely to generalize from a sufficient number of particular instances and the reasoning is entirely phenomenal.

8.2.2.2 Descartes on Sense Delusions

In the *Meditations* of 1640, Descartes uses exactly the same medical paradigm in order to evaluate sensible data. Yet although the examples are similar, notorious dissimilarities lay hidden in their theoretical use.

In Schoock's case, the man who erroneously judges that honey has the quality of tasting bitter could well be said not to be deceived by his sense of taste. After all, he is ill, and might well benefit from his lack of appetite. Hence, if we reformulate the example in terms of final causality, and say that the nutriciousness of hay for hay-eaters and of honey for humans is the final cause of their respective natural tastes, the same thing might be said with respect to the ill, who misjudge the universally accepted quality, but, being ill, misjudge it for their own

[18] Schoock, *Admiranda*, pp. 107-108 / *Querelle*, pp. 233-234.

good. In other words, if the sick man rejects honey because it seems bitter, he is wrong concerning the honey, but perhaps right concerning himself.

In Descartes' example, however, the deception is far more radical: a dropsical person, being genuinely thirsty, might beg for water, although his body, which is already supplied with an overdose, would, according to contemporary medical theory, be even more weakened by letting him drink.[19] The senses in this case deceive us in a way that cannot be justified by referring to the functionality of the experience. Arguing that persons suffering from dropsy make the wrong judgement just because "their nature is disordered",[20] does not alter the fact that relying on sense experience would in this case be quite harmful.

In Descartes' sixth *Meditation*, this inconsistency of Nature and other misjudgements caused by a delusion of the senses—such as the famous example of the man feeling pain in a missing limb[21]—are introduced primarily in the metaphysical context of proving the existence of a material world which is independent of our sensation of it—a question which need not concern us here. One that does concern us however, is the "real distinction" of mind and body—the other subject-matter of the sixth *Meditation*.

In the example of the dropsical person, the idea of a real distinction of mind and body is taken for granted, since Descartes had presented his proof for it in the earlier parts of the work. Descartes claims that this "real distinction" is the ultimate source for the misjudgements of the senses. The reason is, that in the grand clockwork of Nature, the introduction of the *res cogitans*, a being endowed with thought and perception, brought about difficulties that somehow never procured a perfect solution. Indeed, the mechanical world is a world full of particle matter and automata deprived of sense perception. In such a world, sensible misjudgements would have no place. The animal automat can never be thirsty, neither "subjectively", since automata have no feelings, nor "objectively", since the "sole disposition of its

---

[19] Descartes, *Meditation* VI, AT VII, p. 84 / AT IX, p. 67: "ex quo morbus augeatur" / "& ainsi d'augmenter son mal & se nuire à soy-mesme." Cf. COT / CSM II, p. 58.

[20] Descartes, *ibidem*: "corrupta" / "corrompuë".

[21] Descartes, *idem*, AT VII, p. 77 / AT IX, p. 61 / COT p. 53 / CSM II, p. 53.

organs"[22] causes it to do whatever is necessary for the maintenance of the machine—in this case, to search for water as a result of a dry throat. We could suppose that, for some reason or another, the feedback mechanism does not work and that the animal automat breaks down, but, says Descartes, continuing to obey the laws of Nature, the "nature" of the automat would in no way be deceived.[23] The same would be true of human beings insofar as their bodies alone are concerned.[24] But now suppose that the Deity were to link a "thinking substance" to these machines, thereby making them self-conscious. The cerebral information could then properly be called deceptive whenever it misleads the mind into thinking that, for instance, the dropsical body is in need of water. God thus permits His flock to be deceived.[25] Of course, God chose the most perfect and efficient way to connect the human machinery with its finite number of bodily functions to the human mind or soul. Nevertheless, the spino-cerebral system can only pass on information by way of a certain disposition or movement of the nerves which connect the throat, for example, with the brain. Despite the fact that in extraordinary cases, as when a person suffers from dropsy, this information is of no use, it is still, according to Descartes, far better to be deceived in some extraordinary cases than in all normal cases of thirst.

Like Schoock, Descartes thus offers a finalistic explanation of the senses.[26] Sensations of pain, thirst, and the like, which are said to be

[22] The quotation is from *L'Homme*, AT XI, p. 202 / CSM I, p. 108, a passage perfectly resembling the text added in the French translation of the *Meditations*, AT IX, p. 67. Cf. COT / CSM II, p. 58.

[23] Descartes, *Meditations*, AT VII, p. 84 / AT IX, p. 67 / COT, p. 58 / CSM II, p. 58. The automat would continue to obey the laws of Nature just as a clock "non minus accurate leges omnes naturæ observat, cùm male fabricatum est & horas non recte indicat, quàm cùm omni ex parte artificis voto satisfacit."

[24] See also Martial Gueroult, *Descartes selon l'ordre des raisons*, Paris (Aubier) 1953, vol. II, p. 172: "si on réduit 'la nature de l'homme' [...] à l'animal machine, il est impossible de parler d'un manquement."

[25] Descartes, *Meditations*, AT VII, p. 88 / AT IX, p. 70: "Ex quibus omnino manifestum est, non obstante immensâ Dei bonitate, naturam hominis ut ex mente & corpore compositi non posse non aliquando esse fallacem." Cf. CSM II, p. 61.

[26] Descartes, *ibidem*: "cùm potu indigemus, quædam inde oritur siccitas in guttore, nervos ejus movens & illorum ope cerebi interiora; hicque motus mentem afficit sensu sitis, quia nihil in toto hoc negotio nobis utilius est scire, quàm quòd potu ad conservationem valetudinis egeamus, & sic de cæteris."

"confused ways of thought",[27] are a means for survival. Both Descartes and Schoock would confirm that, accustomed to a lifelong practice, we may attribute, for instance, a heating power to pepper, which will help us avoid eating too much of it at once. Relying on the senses is thus useful in most cases. In fact, the very meaning of sensation is their practical use.

For Descartes, however, the kind of knowledge which the "confused thoughts" of sensation offer, has only a very limited range. God connected certain bodily dispositions to mental phenomena, not because the sensations resulting from them are theoretically right, but because they are practically useful. Sensation is thus only a practical substitute for genuine knowledge. Happily, God also gave us a capacity for correcting faulty judgements of the senses.[28] Hence we can adjust our immediate judgements by examining the bodily causes of a particular delusion of the senses by reasoning and experimenting on the spino-cerebral information system. The common-sense knowledge which is good for practical life should thus be ignored once the mechanical redescription of bodily processes is completed.

Though functional in normal life, sense perceptions are theoretically confused. If we start attributing to pepper the quality of a "heating force", then what are we to say when we are feverish?—That the quality has left the substance? That only people with a certain ratio of bodily humours are entitled to determine the real qualities of things? Accepting the possibility of being deceived by the senses, Descartes time and again emphasizes the limited role subjective sensations may have. His forewarnings concerning the delusions of the senses again have no connection to some sort of "rationalist method", nor are they meant to exclude observation or experiment from science. What is excluded is the philosophy of real qualities. Not identifying science with the every-day practice of attributing sensible qualities to natural objects in the way the Scholastic philosophy did, Descartes argued that scientific practice should not codify and catalogue perceptions, but explain them and possibly correct them.

---

[27] Descartes, *idem*, AT VII, p. 81 / AT IX, p. 64: "nihil aliud sunt quàm confusi quidam cogitandi modi" / "certaines façons confuses de penser." Cf. COT / CSM II, p. 56: "confused modes of thinking".

[28] Descartes, *idem*, AT VII, p. 80 / AT IX, pp. 63-64: "[...] hoc tamen ipsum quòd Deus non sit fallax, quòdque idcirco fieri non possit ut ulla falsitas in meis opinionibus reperiatur, nisi aliqua etiam sit in me facultas a Deo tributa ad illam emendandam, certam mihi spem ostendit veritatis etiam in iis assequendæ." Cf. COT / CSM II, pp. 55-56.

8.2.3 Conclusion: Cartesian Dualism

Descartes' method of interpreting observational qualities in a mechanistic way, unveiling the material processes that lay hidden behind the phenomenal world of every-day experience, nicely reflects his dualistic ontology. Besides—or rather, behind—the world we see, there is a world which can be known by reason. For Descartes, science is essentially the search for the real, mechanical world behind our subjective perception. The *détachement des sens* and even the famous Cartesian project of hyperbolic doubt cannot be read apart from this project of finding the right mechanical descriptions for natural processes. Although Descartes had no wish to enter a discussion concerning the real qualities of the philosophers and argued that he could simply do without them, he actually does much more than that and supplants the qualitative ontology with a view of the world in which—it is well worth repeating—the image of the machine was not merely a metaphor. The Cartesian world was literally a world of interconnected automata; a world moreover, in which individuality and quality would be lost if it were not for the presence of human beings—or, rather, human souls, the deplorable ontological status of which was lamented by Schoock in the *Admiranda Methodus*:

> miserable mind, having no mate in all of Nature's theatre.[29]

8.3 Final Causality and the Clockwork of Nature

If individuality is lost in a world in which everything but the human mind works according to the blind mechanism of clockwork receiving its motion from God, the question of final causality again raises its head. In the first place, a certain type of final causation is introduced where Descartes explains sensations as functional interactions between mind and body.[30] However, as Gary Hatfield has recently pointed out, final considerations not only occur in relation to this type of well-functioning composite. To illustate Hatfield's point, let us take

---

[29] Schoock, *Admiranda*, p. 205 / *Querelle*, p. 284.

[30] See Gary Hatfield, "Descartes' physiology and psychology", in John Cottingham (ed.), *The Cambridge Companion to Descartes*, Cambridge (C. U. P.) 1992, p. 360: "There are lessons that come unbidden from mind-body interaction, as when a dryness in the throat causes a jiggle in the brain which in turn changes the flow of spirits and, via the institution of nature, makes one feel thirsty and hence directs one towards drink. These "teachings of nature" are instituted by God for the preservation of the body."

the example of the animal automat to which we referred above. According to Descartes, such an automat may either function properly or, like a broken clock, stop doing what it was made for. In either case, says Descartes, it continues to obey the laws of Nature. However, the whole idea of well-functioning machines introduces the finalistic idea of their being made for a certain end. As Hatfield argues,

> "mechanism" means not only blind causation according to natural law—it also means machinelike. [...] But machines are artifacts; the structure of a machine is identified by virtue of a conception of what counts as its proper functioning and what counts as its being broken.[31]

Accordingly, especially in the case of animals described as well-functioning wholes, the mechanistic account steps over from the domain of blind, efficient causation to the level of finality.

Hatfield thus contrasts the "efficient causality of impact" with the finality of "God's handiwork". Descartes would very much agree that God is to be praised for having invested Nature with such natural laws that all of its perfections may come about from such a simple general cause as the first motions of elementary matter.[32] Seen in this way, we might say that the whole machinery of Nature is, to Cartesian eyes also, ultimately planned as a way for God to achieve His plans with the Universe. When, at a certain point of natural development, some particles of semen

> acquire as much agitation as fire has, and expand (*se dilatent*), and press on other particles, thereby putting them little by little into the state required for the formation of the parts of the body,[33]

it is hard not to interpret this formation in terms of the end towards which the motions of the particles were aimed. If the particle flux created by God leads to such a wonderful combination of well-

---

[31] Gary Hatfield, "Descartes' physiology and psychology", p. 361.

[32] Descartes, *Le Monde*, AT XI, pp. 34-35 / CSM I, p. 91, and Geneviève Lewis, *L'individualité selon Descartes*, p. 65, where this passage is quoted alongside others equally that conclude to a divine origine of natural laws.

[33] Descartes, *Description du corps humain*, AT XI, p. 253. The translation is from CSM I, p. 322. Note however that the particles are said to acquire not only as much, but actually the same (*la mesme*) agitation as fire; the fermentation of the semens, acting on each other as "a kind of yeast" (CSM I, p. 322), being exactly the of the kind of those "fires without light" mentioned elsewhere. See above, chapter 5, note 46.

functioning automata, it must have been God's intention in the first place to make them. Hence, according to Hatfield, the mechanical causes of Cartesianism may just as well be redefined in terms of final causality.

Yet there are good reasons for not doing so. First, Descartes does not himself use this terminology. He does not, in other words, describe the purposefulness of sensation and the well-functioning of (animal-) machines in terms of final causality. Secondly, what he does say about the function of causation belongs to metaphysics rather than to physics proper. Both of these arguments are important for historical reasons and should therefore not lightly be discarded. Indeed, even if there are analytical reasons for ascribing to Descartes some concept of final causality, we shall miss important historical issues if we do. Both Descartes and contemporary Scholastic writers made use of the ambiguous concept of final causality. We have already seen that Descartes banned the notion from physics entirely.[34] We shall have to take his word for it if we are to understand his position in its historic context. In this case, Descartes' rejection of final causality should be read against the background of its acceptance in the philosophy he emancipated from. Only then may we understand whether, and if so, in what way, the rejection of final causation made a difference.

Descartes rejected final considerations in order to pave the way for a type of explanation which was purely "naturalistic". His account of, for instance, the formation of the animal body might well be redescribed in finalistic terms. Thus, one may say that the perfection of the animal machine is the end, or final cause, to which the fermentation of the semens is directed. Descartes would not have made much of a problem if one reformulated his theory in this way and even considers it pious to ascribe to God a conscious motivation of investing Nature with such wonderful laws. However, he does not consider this "scientific". To put it in his own words:

> it is plainly ridiculous and inappropriate to consider this in Physics.[35]

---

[34] See section 5.2.2, above and *Principia Philosophiæ* III 1-3, AT VIII-I, pp. 80-81 / CSM I, pp. 248-249.

[35] Descartes, *Principia Philosophiæ* III 3, AT VIII-I, p. 81 / CSM I, p. 249. It should be noted that Descartes here criticizes the thesis that everything is made for our benifit alone. The prohibition against using final causality in Physics however, is also argued from the general standpoint of not supposing that we understand the aims God had in creating the world. Cf. *Meditationes* IV, AT VII, p. 55 / COT, p. 35 / CSM II, p. 35.

What *was* appropriate in physics, was to reconstruct the physical world as an *a priori* mechanistic evolution from a primal flux of matter. Thus, although finalistic accounts may be put forward and considered metaphysically, the aim of *physical* science is to reconstruct without the use of final causes, and to explain physical processes by the use of natural causes alone. As for our example of the formation of the limbs, Descartes might well—and in fact *does*[36]—agree that God has made and planned things beautifully. However, he reckons the mechanistic deduction of the formation of the limbs from the embryo, of the embryonic heart from the semens and of the semens from the parents etc., to be of more importance to "physics". The question is, thus, whether Descartes' use of the idea of God as the ultimate craftsman is of much importance in the light of the considerable naturalistic turn which his rejection of final causality brought about. The *a priori* deduction not only led to a world in which human individuals stood alone in a completely mechanical clockwork environment. It also excluded God from interfering with His very own work.

8.4 The Cartesian God

In the *Discours de la Méthode*, Descartes tells of the deductive project on which he had embarked in *Le Monde*. Preparing a book in which everything concerning the nature of material things would be included,[37] he started out not by describing the actual world of things, but by describing the possible evolution of a fantasy world evolving in an imaginary Universe:

> I therefore supposed that God now created, somewhere in imaginary spaces, enough matter to compose such a world; that He variously and randomly agitated the different parts of this matter so as to form a chaos as confused as any the poets could invent; and that He then did nothing but lend His regular concurrence to Nature, leaving it to act according to the laws He established.[38]

---

[36] See section 5.3.1, above, especially the quotation from the *Discours de la Méthode*, AT VI, pp. 55-56 / CSM I, p. 139. Further: G. Lewis, *L'individualité selon Descartes*, pp. 64-65.

[37] Descartes, *Discours de la Méthode*, AT VI, p. 41: "I'ay eu dessein d'y [i.e., in *Le Monde*] comprendre tout ce que ie pensois sçauoir, auant que de l'escrire, touchant la Nature des choses Materielles." Cf. CSM I, p. 132.

[38] Descartes, *Discours de la Méthode*, AT VI, p. 42 / CSM I, p. 132.

According to Descartes, this primordial chaos of matter in motion would develop into the exact duplicate of the world we live in. The world thus forms itself purely on account of the regular movement of matter according to the laws of Nature. Descartes realises that this picture might seem to be at odds with the religious representation of creation. However, he argues that his imaginary description does not rival the Biblical account:

> I do not wish to infer from all this that our world was created in the way I proposed, for it is much more likely that from the beginning God made it just as it had to be.

The evolutionary reconstruction on the other hand, is justified on methodological grounds, since

> [the] nature [of material things] is much easier to conceive if we see them develop gradually in this way than if we consider them only in their completed form.[39]

This is surely problematical in the face of Descartes' own conviction that it is the facts we may rely on, that is, the objective truth of what is evidently explained by mechanical hypotheses. According to the *Discours* itself, "the causes are proved by the effects",[40] that is to say, if they do indeed explain them. This is exactly what Descartes claims his evolutionary reconstruction does. Accordingly, if the world we perceive may be succesfully explained by an imaginary evolution from previous mechanical causes, these causes should, at least according to Cartesian standards, be regarded as being proven by observable fact.

However, it does not seem appropriate to discard Descartes' methodological motivation entirely. Although, as A. Vartanian has argued, Descartes' imaginary reconstruction was

> capable of serving, into the future, as motivation for the trend to scientific naturalism that made rapid head-way in eighteenth century France,[41]

the literal idea of an evolving Universe must have been as psychologically infeasible to Descartes as it was outlawed by religion.

---

[39] Descartes, *Discours de la Méthode*, AT VI, p. 45 / CSM I, pp. 133-134.

[40] Descartes, *Discours de la Méthode*, AT VI, p. 76 / CSM I, p. 150.

[41] A. Vartanian, *Diderot and Descartes; A Study of Scientific Naturalism in the Enlightenment*, Princeton (P. U. P.) 1953. Of special importance in this respect is chapter II, "From Descartes' *Monde* to the Worlds of Diderot and Materialist Science", pp. 47-131. The quotation is from p. 48.

In any case, it is not found in any of his writings, nor is it in those of his contemporaries.

And yet, the whole project of *Le Monde* was theologically threatening in other ways. Descartes' reconstruction differs from the account we find in the book of *Genesis*, firstly because of the fact that it does not consider the hexaemeric order in which God is said to have completed the Universe. Indeed, in his *Conversation with Burman*, Descartes actually maintains that the six-day Creation might well have to be interpreted "metaphorically".[42] What is more, Descartes is said to have abandoned the project of writing a commentary on the Book of *Genesis* himself, since he thought it better to leave the matter "to the theologians".[43] Such indifference to Holy Writ would surely have infuriated Voetius as much as it once shocked Anna Maria van Schurman.[44]

---

[42] Descartes, *Entretien avec Burman*, ed. Ch. Adam, pp. 90-93 / AT V, pp. 168-169 / CSM III, pp. 349-350.

[43] According to Frans Burman, Descartes remarked that he would regard as a "mighty Apollo" anyone who could explain the book of *Genesis* (or the *Song of Solomon* and the book of *Revelation* for that matter). See *Entretien avec Burman*, ed. Ch. Adam, pp. 92-93 / AT V, p. 169 / CSM III, p. 349. Descartes is also reported to have once started work on a commentary of the first chapter of *Genesis* himself, *ibidem*. The fact is also mentioned elsewhere, both by Descartes and others. See AT III, pp. 71 and 296, and AT IV, pp. 698 and 700-701.

[44] Descartes abandoned his work on *Genesis* since, if we are to believe the indirect evidence, he could not find anything in the book which might be "clearly and distinctly" perceived. This at least is what Descartes is reported to have said to Anna Maria van Schurman, Gisbertus Voetius' female student, when, during a visit in Utrecht, he found her reading the Bible. Having first shocked Van Schurman by asking her why she dedicated her time to "something of such small importance", Descartes explained himself by saying that since he did not understand what Moses had to say, he had abandoned studying the Bible altogether. Van Schurman interpreted the event as a divinely pre-ordained trial of her faith and decided never to see the French philosopher again (AT IV, pp. 700-701). Descartes nowhere mentions the incident, but there is no doubt that he did know Anna Maria van Schurman. In a letter to Mersenne of 11 November 1640 from Leiden, Descartes describes Van Schurman as someone who might have excelled in poetry, painting or similar *gentillesses*, had she not been "posessed" by Voetius and since lost contact with the world (AT III, p. 231). Van Schurman used to follow Voetius' lectures at Utrecht University from a box especially made for her in order not to disturb the male audience. Descartes once proposed to make use of this provision in order to be able to appear incognito at Regius' *Disputatio medico-physica* on the circulation of the blood of 10 June 1640. Cf. Adrien Baillet, *La Vie de Monsieur Des-Cartes* II, pp. 59-60; AT III, pp. 71-72 and *Thijssen-Schoute*, p. 239. Nothing came of it, however.

The most fundamental difference between the Biblical and the Cartesian accounts, however, is that, in Descartes' world, God does not complete His Creation Himself. If only He "[lends] His concurrence to Nature", He may *leave* it "to act according to the laws He established."[45] The Sun and the stars, the Moon and the Earth, the planets and comets, water, air, fire and minerals, mountains, rivers, plants, animals and even man himself are all formed by the natural flow of particle matter.[46] Descartes does not completely rule out divine interference in this process: indeed, he stresses that

> it is certain, and it is an opinion commonly accepted among theologians, that the act by which God now preserves [our world] is just the same as that by which he created it.[47]

Theologians agreed, just as they agreed with Descartes that God

> [lends] His concurrence to enable Nature to operate as it normally does.[48]

Such avowals could not conceal that the *Discours* actually brings divine interference in Nature to a minimum. Gisbertus Voetius for one, warned the reader of his 1641 essay that "there would" in fact "be no divine concurrence" in the New Philosophy.[49] In the next chapter, we shall see in what way Cartesianism threatened theological dogma. First, however, let us have another look at what we found the Cartesian world to look like.

8.5 Conclusion

Along with the literal interpretation of Nature as a giant clockwork mechanism, the acceptance of mechanical principles resulted in a dispersion of causal forces. Human individuals form a rare species in this mechanical world. They are at once (1) the unique receivers of sensible data which no longer have any ontological status apart from being perceived, and (2) the only remaining causal centres of action

---

[45] Descartes, *Discours de la Méthode*, AT VI, p. 42 / CSM I, p. 132: "[...] & que, par apres, Il ne fist autre chose que prester son concours ordinaire a la Nature, & la laisser agir suiuant les Lois qu'il a establies."

[46] Descartes, *Discours de la Méthode*, AT VI, pp. 42 and 64 / CSM I, pp. 132 and 144.

[47] Descartes, *Discours de la Méthode*, AT VI, p. 45 / CSM I, p. 133.

[48] Descartes, *ibidem*. See also AT VI, p. 42 / CSM I, p. 132.

[49] Voetius, *Narratio*, p. 40 / *Select. Dispp.* I, p. 873 / *Responsio*, p. 12 / *Querelle*, p. 106.

within the clockwork. As such, they are in fact rather better off than the Deity, to Whom no further influence is granted once He has left Nature "to act according to the laws he established." The result is that, along with substantial unities and real qualities, teleological arguments also become superfluous. The Cartesian explanation of natural phenomena simply does not give rise to the isolated explanations on which the teleological and finalistic accounts of Nature are based.

CHAPTER NINE

CAUSES, CONSERVATION AND CONCURRENCE

Wie niet en heeft te recht den Schepper
leeren kennen
Die sal sijn herte noyt tot rechte wijsheyd
wennen

Anna Maria van Schurman[1]

Voetius saw great dangers for theology. Having for some time played a waiting game, he first struck at the New Philosophy in December 1641.[2] Displaying an extraordinary capacity for foreseeing what was to be expected from the mechanical standpoint, Voetius listed a catalogue of possible consequences of the New Philosophy in his essay *On the Natures and Substantial Forms of Things*. Whether it was a case of Platonizing, occasionalist, Spinozist, or Leibnizian radicalisations of Cartesianism, all were more or less foreseen and rejected by Voetius in 1641.

At the same time, these interpretations of the New Philosophy were based on old antagonisms. We have already seen in what way objections to the occasionalist theory of the Islamic Mutakallimûm, which had for centuries formed part of the received commentary tradition, re-surfaced in Voetius' essay against Descartes and other anti-Aristotelians of his kind. Such objections might well be relevant to seventeenth-century discussions,[3] but they had previously been

[1] Two lines from a poem written by Anna Maria van Schurman at the occasion of Gisbertus Voetius' Opening Sermon of Utrecht University, on 13 March 1636: "He who has not rightly learned to know the Creator [of things], will never accommodate his heart to proper wisdom." The poem was printed at the end of the finer edition which was printed of Voetius' *Sermoen* by Ægidius and Petrus Roman, the University printers. See also: *Duker* II, p. 134. On Van Schurman, see chapter 8, note 44.

[2] See section 1.4, above.

[3] For instance, Voetius rightly observes that in Cartesian philosophy, there would be no *causes secundas*. His warning is formulated in terms of the objections made to the Mutakallimûm and a variety of Platonic and atomist thinkers. It nevertheless does anticipate the ideas Malebranche was later to draw from Cartesian philosophy. In the case of the Mutakalllimûm moreover, it is interesting

formulated with respect to other philosophies rivalling Aristotelianism. Cartesianism was accordingly seen as a revival of old errors.

9.1 Aristotelian Causes, Common Sense and Christian Theology

If the merit of new philosophical systems and schools could be evaluated in terms of a position for or against Aristotle, it was the singularity of the Aristotelian theory of causes that made this possible. The great question of Greek philosophy had been how to explain natural change. It was still alive in the discussions between Voetius and Descartes in the sense that if individual forms—which put potency into act—were rejected, natural change would have to be explained by alternative sources of action.

Such alternatives were not thought to seriously rival the "sane and sober philosophy" of Scholasticism. Indeed, what would be won by the re-introduction of Platonic Forms, of separate intelligences, of the God of the "occasionalist" philosophies, of fantastic atoms displaying a force of their own, or, finally of a world-soul moving the matter of its body? The anti-Aristotelians are of the opinion that one could do without the forms

> since the effects of natural objects can be sufficiently explained by, and reduced to, other principles.[4]

However, says Voetius, none of these philosophers agree as to what should replace them:

> For the ancients, whom Basso praises and Aristotle refutes in Book II of the [*Physics*][5], explain them in one way, modern philosophers in

---

that Aquinas argues that their theory is at odds with what experience teaches us. If God directly causes all natural action, we would be deceived by our judgement of sense. This argument strikes us because of its obvious Cartesian ring. In fact it hints at the opposition that exists between Aristotelian and non-Aristotelian philosophies as regards the truthfulness of common sense experience. As in the case of the Mutakallimûm, likewise in that of Descartes, perception of sensible qualities is thought to have become problematic since it is no longer immediate.

[4] Voetius, *Narratio*, p. 44 / *Select. Dispp.* I, p. 876 / *Responsio*, p. 23 / *Querelle*, p. 109.

[5] Voetius does not mention the *Physics* as such, but refers to "lib. 2. *Akroas.*" Now Aristotle's "akroamatic" or *esoteric* writings are those which were supposedly not written for a large audience, but solely for use in the Lyceum. The *Physics* form only one example of the numerous texts of this type which have survived. Accordingly, Theo Verbeek for instance also mentions the second book of the *Metaphysics* as a possible reference for what Voetius has in mind (*Querelle*, p. 473, note 80). However, not only does the context of finding the right principles of

> another—although most of these viewpoints have some common aspect, the denial and refutation of which also leads to the collapse of the [viewpoints] themselves. We shall not here unearth the loathsome and long discredited [opinions] of both old and modern Paracelsists and Hermeticists [...], but merely aim to describe the [theory] which has suddenly emerged and in which it is held that everything derives from quantity, shape, position or situation, motion [and] rest, and that all secrets of Nature can be perfectly explained and demonstrated by them—which we deny.[6]

Voetius goes on to refute Regius' mechanicism, but it is clear that he does not regard the new Cartesian viewpoints as really so much different from rival philosophies to Aristotelianism that went before. The reason is, that all, be they ancient Platonists or modern mechanicists or magicians, must come up with an alternative to the substantial forms, and this will most likely be a spiritual one.

Of course, Descartes had tried to do without any alternative, and this by keeping silence with respect to the question of the forms. In practice however, he did not confine himself to a scientific explanation of isolated phenomena on the basis of efficient mechanical explanations alone, but showed as much allegiance as his Scholastic adversaries to demonstrative explanations of change.[7] Even though Voetius cannot in 1641 have had much of an insight into the Cartesian world-system of interconnected particle fluids, nor of the unfinished *a priori* deduction of natural processes from an initial Cartesian Big Bang—what he did know was enough for understanding what such a

---

change—and indeed the subject-matter of the essay "On the Natures and Substantial Forms of Things" generally—point to the second part of Aristotle's *Physics*: the *Physics* were, in Greek editions, also known as the Φυσικὴ Ἀκρόασις or *Lectures on Nature*, which is no doubt what Voetius had in mind when referring to the "lib. *Akroas*."

[6] Voetius, *Narratio*, p. 44 / *Select. Dispp.* I, p. 876 / *Responsio*, p. 23 / *Querelle*, p. 109.

[7] Desmond Clarke has even characterized Descartes as an "innovative Aristotelian" on account of this fact. Desmond Clarke, *Descartes' Philosophy of Science*, Manchester (M. U. P.) 1982, pp. 197 ff. This however is overdoing it, for although the similarities between Aristotelian and Cartesian philosophies may be of use in the context of contrasting Cartesian and Aristotelian theories of method with the methodologies of modern science, dissimilarities between Aristotelian and Cartesian methodologies are equally unmistakable. Consequently, to depict Descartes as an "innovating Aristotelian" will imply that important features of Cartesianism, such as the distrust of sense-experience, are left unaccounted for.

world would look like: it would be a world devoid of individual substances. Or, in Voetius own words:

> It will have to be seen in what way—given this position—one can conveniently explain and defend the distinction between the substance and accident of some entity. [...] Maybe they would point out that [substances] differ essentially on account of those five accidents of motion, rest, position, shape [and] quantity. My answer would be, that, if this were so [...], difficulties would still not be removed. What remains is that substances are not distinguished from substances in any other way than are accidents from substances and accidents amongst themselves; for all [would be] distinguished by accidents.[8]

No logic based on essential definitions would be possible.[9] Indeed, the concepts of unity and individuality would no longer apply in a world devoid of forms, since it would also follow

> that all created substances, even man himself [...] are beings by accident, collections, aggregates, not, however, single essences or natures by themselves.[10]

The clockwork conception of causality implies an ontology devoid of individual forms. Voetius might have thought that this in itself is reason enough to see to it that his students would not

> drift away along these lines and support the subversion and weakening of all sane and sober philosophy.[11]

Philosophy however, was only part of what was in danger. According to Voetius, the Aristotelian account of causality was justified by its philosophical sobriety and its accordance with common sense opinion, all of which could not be said of the Platonic and mechanical alternatives. The fact that these philosophies were thought to be mistaken would, however, not be as important if it were not for the fact that theological teaching was explained in terms of the

---

[8] Voetius, *Narratio*, pp. 39-40 / Select. Dispp. I, p. 872 / *Responsio*, pp. 11-12 / *Querelle*, pp. 105-106.

[9] Voetius, *Narratio*, p. 41 / *Select. Dispp.* I, p. 873 / *Responsio*, p. 13 / *Querelle*, p. 107: "Ex hâc opinione sequitur, nullas dari definitiones substantiarum: cum non dentur formæ, aut propria prima & interna essentialia, unde differentia desumi possit."

[10] Voetius, *ibidem*.

[11] Voetius, *Narratio*, pp. 43-44 / *Select. Dispp.* I, p. 875 / *Responsio*, p. 23 / *Querelle*, p. 109.

Aristotelian perception of causality, which meant that rival theories were not only in danger of being mistaken, but of being heretical as well.

Voetius' reaction to Descartes and other anti-Aristotelian writers of his day was to an important degree inspired by the belief that Aristotelian philosophy agreed with finality, with God's influence in Nature and with theological issues such as predestination, better than theirs. This viewpoint had for centuries been the claim of every Christian commentary on the second book of Aristotle's *Physics*. We have already seen how Christian commentators interpreted Aristotle in a devoutly finalistic way. But there were other ways in which Christian theologians judged the philosophy of internal causal faculties and Christian theology to be two sides of the same coin.

9.2 Divine Conservation

One specifically Christian idea which had entered the commentary tradition was the notion of a divine conservation of the natural world. The idea that God actively conserves the world is not, as Saint Augustine had already noticed in other writers, a self-evident idea. Suppose God once made the world: this does not mean that He should make any effort concerning it afterwards. Works of art are equally fabricated and afterwards left to themselves. Moreover, there are certain beings, such as angels, which "subsist by themselves" and do not need anything in order to conserve themselves in their being. Conservation would for them be superfluous. But even if we consider ordinary things, a divine conservation would seem unnecessary. For are we to conclude that without such conservation, that is, by themselves, they would strive after their own annihilation? That would be contrary to Nature. Finally, what would this divine conservation consist of? If God were to conserve heat in heated water, He would, in the first instance, create the heat along with the fire on which it is put, but conserving it would be quite a different action once the fire was extinguished. Both the fire and the conservation are nevertheless aimed at heat as their common *terminus*, which would mean that the action of conservation and the action of the fire itself would be one and the same. That God would immediately conserve the existence of everything seems to be an akward superfluity, especially since He would conserve things such as the matter and form of objects and, of the matter, its parts, which are already mutually dependent on each other for their being.[12]

Nevertheless, following a host of authorities both Christian and pagan, the Conimbricenses emphasize that God must yet actively perform a constant deed of conservation ever since He first created the world. God's conservation is in fact the major function of His omnipresence. Far from being superfluous, the presupposition of God conserving the world is necessary for making the comparison of God as the ultimate artisan or architect. Just as the works of human art presuppose Nature itself as that by which these works are supported, so do the works of Nature depend on the "divine art" of creation:

> a house in fact gets its duration in time from the solidity of the stone, the firmness of the wood and the cohesion of the lime.[13]

It does not, in other words, endure because of the craftsmanship of the builder or the architect, but stands in need of something extra—in this case, Nature, which supplies the matter of the artifact. Likewise, Nature itself, being the work of God, still stands in need of the support of something else in order to preserve it. This something else is God's act of conservation. If there was no such thing, things might endure without God willing it. He would then have to make a special effort to do away with an object of creation, which is clearly at odds with the power and dignity of the Supreme Being. In fact, God has only to wash His hands, so to say, of something in order for it to exist not a moment longer. Even more than light depends on a luminous body do the objects of creation depend on God for the continuation of their being.[14] This being said, it must be concluded

> that God immediately conserves all objects of creation. For albeit that He has first brought things about in conjunction to what is required

[12] This whole passage is a paraphrase of the arguments against conservation as it is found in the commentary on Aristotle's *Physics* by the philosophers of Coimbra. See *Commentarii Colegii Conimbricensis S.J., in octo libros physicorum Aristotelis Stagiritæ*, In II, Cap. VII, Quæst. 10, Art. 1, Lyons (Svmptibvs Ioannis Baptistæ Bvysson) 1594 / reprint Hildesheim (Georg Olms Verlag) 1984, I, pp. 265-266. St. Augustine, who is quoted by the Conimbricenses, dealt with the question in the fifth book of his commentary on the book of *Genesis*, viz. *De Genesi ad Literam, Corpvs Scriptorvm Ecclesiasticorvm Latinorvm* XXVIII-I, Prague / Vienna / Leipzig (F. Tempsky / G. Freytag) 1894.

[13] Conimbricenses, *Commentarii in octo libros physicorum* I, p. 266.

[14] *Idem*, pp. 266-267.

for their preservation, He nevertheless conserves them in their being by Himself by way of an intimate influence (*illapsus*).[15]

The force and significance of natural causes notwithstanding, God immediately conserves His Creation from every moment to the next. This is especially true of things that have no natural causes, such as angels—which answers the objection that no conservation be required in their case.[16] It is, however, no less true for all other objects in Nature, which, although they do not indeed strive after their own annihilation, would nevertheless instantly fall back into nothingness if God were to withdraw his existential sustenance.[17]

9.3 Divine Concurrence

In the commentary on Aristotle's *Physics* by the philosophers of Coimbra, the critique of "Platonist" philosophies which deprive "secondary"—that is, ordinary, everyday, physical—causes, of their individual action, immediately follows upon the *Quæstio* dealing with God's conservation. The reason is obvious: if God conserves all being from every moment to the next, what place is their for individual causation? Is it not so that God might continuously create the world in succesive steps which, though seemingly of a regular pattern, would nevertheless have no connection at all? This is exactly the position of the Mutakallimûm who sought to augment God's omnipotence by completely depriving natural objects of their internal powers. We have already seen that Christian Aristotelians following St. Thomas Aquinas objected to such views.[18] Independent action by natural, secondary causes should be upheld despite the fact that God is immediately involved in the continuation of their being. However, this does not mean that whereas they are dependent on God for their being,

---

[15] *Idem*, p. 268.

[16] *Idem*, pp. 268 and 269, "Sol. 2".

[17] It is thus a *continuous action* by which God preserves everything, which is to be distinguished from the singular, natural action by which a fire for instance produces heat in water. Although both the action of God creating the heat and His action conserving it are *realiter* the same in as far as they produce numerically the same heat in the heated water, the *actio conservatrix* is nevertheless formally different from the physical action of heating in which God and the fire co-operate in order to produce the natural effect of heating. This, however, already brings up the question of divine concurrence, which I shall deal with presently. For the remaining solutions to the objections raised against the idea of a divine conservation, see Conimbricenses, *idem*, I, pp. 268-270.

[18] See section 6.2.1, above.

natural objects would be completely independent as regards their activities and operations. Even granting that natural objects display a genuine action of their own and that neither God, nor separate intelligences, nor, finally, the atoms of the "Democritians" are needed to explain activity in the material world,[19] it is yet claimed that God is also involved in their action:

> God actually concurs with secondary causes to elicit whatever of their actions.[20]

Natural objects do, in other words, have an inner source for operation. Still, God *concurs*, or operates along with them.

It might be asked what need there is to postulate such co-operation, and some exceptional Scholastic doctors did in fact pose this very question.[21] If God, the Prime Cause of the Universe, is rightly said to operate by means of secondary causes, it would not be fit that He immediately bring forth the effects of their action. There is no contradiction in saying that God might create things with independent powers of action, since the Sun causes gold,[22] lion begets lion and all things operate according to the faculties imposed by them.[23] Again,

---

[19] Conimbricenses, *Commentarii in octo libros physicorum*, In II, Cap. VII, Quæst. 11: "Utrvm cavssæ secvndæ reuera aliquid agant, an non?", I, pp. 270-273.

[20] Conimbricenses, *idem*, I, p. 275.

[21] Much like the famous Gabriel Biel became the exemplary authority for the uncommon, occasionalist, view that secondary causes do not of themselves exhibit any action (See above, chapter 6, notes 44, 47 and 48), the nominalist Durandus, or Durand of Saint-Pourçain (d. 1332) became the standard example of an exceptional Christian Scholastic who had denied divine concurrence. See e.g. Conimbricenses, *Commentarii in octo libros physicorum* I, p. 273; Ruvius, *Commentarii in octo libros Aristotelis de Physico*, pp. 322-323. On Durand, see *Dictionnaire de Théologie Catholique*, Tome XIV-II, cols. 1964-1966. However, the question had already been raised by previous authors against "alijque nonnulli [...] quos referunt Henricus Gand. quodl. 14. q. 1. M. Albertus in 1. d. 37. art. 1. Vdalricus in sua summa lib. 2." Conimbricenses, *ibidem*. And even St. Augustine had had to defend God's concurrence against opponents who held the "deist" position that, once created, the world had been "propagatum [...] atque [...] multiplicatum in posterum per solam virtutem secundarum causarum, ita vt Deus nihil deinceps actu fecerit." Ruvius, *idem*, p. 322 and Augustine, *De Genesim ad litteram*, V, 20: "Sunt qui arbitrentur tantummodo mundum ipsum factum à Deo, cætera iam fieri ab ipso mundo, sicut ille ordinavit, & iussit, Deum autem ipsum nihil operari."

[22] On the generation of gold and other metals from congealed Earthly evaporations, see Aristotle, *Metereology* III 6, 378^{a}15-378^{b}6.

[23] The Conimbricenses, playing the advocate of the devil in their enumeration of arguments against the idea of a divine concurrence, add that the text of *Isaiah* 26:12 ("omnia enim opera nostra operatus es nobis" / "for thou also hast wrought all our

what would this concurrence consist of? Is God's operation identical to that of the object? It would seem not, since two similar operations are only current in cases in which the one supplements the other, as when someone helps somebody else carry a weight. God's powers, however, are of a wholly different order from those of natural objects. Would God's concurrence result in one and the same effect?[24]

As in the case of conservation, divine concurrence also leads to questions based on the idea that, once the world is created and all objects are endowed with the proper powers to produce their effects, further divine interference seems excessive. Yet as we have already noticed, from the writings of the Fathers onwards, the Christian tradition had made it the prime objective of physics to illustrate divine government and administration.[25] The idea of a divine *concursus* with Nature was a cornerstone for this view. Had not Aristotle said that the prior cause "moves more strongly and more effectively" than the latter?[26] But God is the Supreme and First Cause. Just as in the case of conservation, it would not fit the Supreme Being to have to work against the objects of His own Creation. Thus, if He were to decide that a certain natural cause should not take effect, He would not have to make any extra effort, but simply withdraw His co-operation. This is what God did in the case of the Babylonic furnace which, as it is told in *Daniel* 3, did not burn Shadrach, Meshach and Abed-nego, since the fire had no power upon their bodies.[27] Finally, as was said

---

works in us") would, in this interpretation, bear only on the fact that the cause, in this case God, has given the effect its form and method (*ratio*) for operation. The operation should nonetheless be ascribed to the effect itself. Conimbricenses, *Commentarii in octo libros physicorum* I, p. 274.

[24] Conimbricenses, *Commentarii in octo libros physicorum* I, pp. 274-275. The Conimbricenses add a fourth argument against the concurrence of God with secondary causes. In this argument however, which deals with the status of secondary causes as the "instruments" of God, various theological dogmas are involved, which it would take too far to explain here. See however also below, section 9.5.2.

[25] See section 3.1.1, above.

[26] Conimbricenses, *idem*, I, p. 276. Aristotle however, had in fact held that a prior cause in a series of movers is more the cause *of its own movement*, than that which comes later. Aristotle, *Physics* VIII 5, 257^{b}16-17.

[27] Conimbricenses, *Commentarii in octo libros physicorum* I, p. 274 and *Daniel* 3:27 esp. The example was a classic one in discussions on divine concurrence. Aquinas explains the incident as one in which God holds back (*retinet*) natural operations. See the *Quæstiones Disputatæ De Potentia*, quæst. 3, art. 7; ed. Marietti 5-II, p. 159, where Aquinas spells out the various ways in which God can act *præter naturam*: "vel quantum ad esse, in quantum aliquam novam formam

with regard to God's conservation, the effects of secondary causes could not exist for one moment without an immediate "influx" (*influxus*) from God guarding (*tuentis*) them, neither could they produce anything without the same immediate influx. For certainly, it does not require less effort to produce a thing than to keep and conserve it.[28]

Many a Biblical verse could be quoted in support of the idea, not to mention the famous words of St. Paul preached on the Areopagus in Athens, that in God

> we live, and move, and have our being.[29]

God in fact directs and conserves not only us, but all parts of His Creation. By an "occult potency", He moves the Universe,

> and while it is kept moving by this motion, while the angels accomplish their tasks, the stars revolve, the winds turn, while the abyss of the waters is engaged in a steady motion, while green places sprout and their seeds grow, while animals are begotten and the unjust are permitted to harass the righteous, [while all this is happening,] the ages unfold [in the way] He authorized when things were first established, although they would not develop their course if He Who established them would hold off administering them by a provident motion.[30]

---

inducit rebus naturalibus quam natura inducere non potest, sicut formam gloriæ; aut huic materiæ, sicut visum in cæco; vel quantum ad operationem: in quantum retinet operationes rerum naturalium ne agant quod natæ sunt agere, sicut quod ignis non comburat, ut patet Daniel. III, 24, vel quod aqua non fluat, ut patet de aqua Iordanis [Iosue, III, 16]." The example of the Babylonic furnace was also used by Voetius in relation to the question of divine concurrence. Voetius, *Select. Dispp.* III, p. 299.

[28] Conimbricenses, *Commentarii in octo libros physicorum* I, pp. 275-276.

[29] *Acts* 17:28. Other verses mentioned by the Conimbricenses apart form *Acts* 17 and *Daniel* 3, are: Job 10:8 "Thine hands have made me and fashioned me together round about"; *Psalms* 147:8 "[Sing unto the LORD,] Who covereth the Earth with clouds, who prepareth rain for the Earth, who maketh grass to grow upon the mountains"; *John* 1:3 "All things were made by Him; and without Him was not any thing made that was made." The Coimbra philosophers might well have added *Isaiah* 26:12 "for thou also hast wrought all our works in us", which they quoted in their list of counter-arguments. See above, note 23.

[30] Conimbricenses, *Commentarii in octo libros physicorum* I, p. 275 and St. Augustine, *De Genesi ad Literam* V, cap. 20, in *Corpvs Scriptorvm Ecclesiasticorvm Latinorvm* XXVIII-I, Prague / Vienna / Leipzig (F. Tempsky / G. Freytag) 1894, pp. 164-165.

The operations which, according to Aristotle, natural objects display by themselves, thus require a second, "concurrent", motion not only for the reason that, as finite causes, their operations would be wholly ineffective without it: the *concursus divinus* is also the instrument of Providence, by which the Prime Cause administers and directs the secondary causes. As Bishop Robert Mossom put it in 1657:

> all secondary causes are links together in one chain of divine providence, which the *Heathens* feigned to be fastened at *Jupiter's* chair, and we Christians believe to be held in God's hand.[31]

But the Christian God did not only, so to speak, pull at one end of the chain of causes. Every link in fact was helped along individually by Him.

The fact that theories of conservation and concurrence were coupled to questions concerning the causal efficacy of natural objects as these were discussed in commentaries on Aristotle's *Physics*, meant that new theories of causal mechanisms in Nature would modify the idea of divine interference also. Descartes' mechanistic reconstruction of the material Universe did just that.

9.4 The Exclusion of Concurrence in Mechanical philosophy

In the previous chapter, we argued that by deducing all physical phenomena "*a priori*" from a prior particle chaos, Descartes reduces divine interference to a minimum. He does, however, maintain the notion of a *concours ordinaire de Dieu*. In fact, Descartes expressly states that God

> [lends] His concurrence to enable Nature to operate as it normally does.[32]

Still, the fact of Descartes' commitment to the Scholastic concept cannot conceal the ambiguous way in which he uses it.

9.4.1 Descartes' Conception of a Concours Divin

Referring to a divine *concours* with Nature, Descartes does not completely rule out divine interference in things natural. But what, exactly, does he mean by a divine concurrence? A first thing to notice,

---

[31] *The Preachers Tripartite* II 67, quoted in C. A. Patrides, *Premises and Motifs in Renaissence Thought and Literature*, Princeton, New Jersey (P. U. P.) 1982, p.34.

[32] Descartes, *Discours de la Méthode*, AT VI, p. 42 / CSM I, p. 132.

is that on one of the two occasions that Descartes uses the term in the *Discours*, his reference to God's concurrence is meant to diminish the apparent inconsistency of the Biblical view of Creation in the book of *Genesis* and his own "evolutionist" reconstruction of how the world came into being. What Descartes says is that since God's act of keeping the world as it is, is identical to the act by which he created it—which, according to Descartes, is "an opinion commonly accepted among theologians"[33]—, it does not make a difference whether one describes Nature as something established at once, or something developing gradually. Thus, Descartes appeals to God's continued creation and concurrence in order to free himself from the accusation that his presentation of the material world as having evolved from a primordial chaos would challenge the account given in the book of *Genesis*. In other words, the fact that he mentions the theory at all is motivated by considerations of circumstance. It is not clearly spelled out—nor does Descartes aim at spelling out—how to interpret the notions of *concours* and *création*.

The thesis to which all theologians are supposed to agree, namely that the action by which God now preserves the Universe is exactly similar to God's initial action of creation, seems to develop the Thomistic idea of a *creatio continua*, on which theories of divine conservation were based. Accordingly, this is how scholars have often interpreted the text of the *Discours*.[34] However, the need for a divine preservation of the world from every moment to the next, saving it thereby from annihilation, has also been interpreted as both Descartes' and the Scholastics' idea of a divine concurrence.[35] This does not do

[33] Descartes, *Discours de la Méthode*, AT VI, p. 45 / CSM I, p. 133: "Mais il est certain, & c'est vne opinion communement receuë entre les Theologiens, que l'action, par laquelle maintenent il (i.e., God) le conserue (i.e., our world), est toute la mesme que celle par laquelle il l'a creé."

[34] See, e.g.: Émile Bréhier: *La philosophie et son passé*, Paris (Alcan / P.U.F.) 1940, pp. 127-137, and: René Descartes / Étienne Gilson: *Discours de la méthode*, Paris (Vrin) 1976[5], pp. 340-342. Consequently, the *Enciclopedia Filosofica* for instance, does not name Descartes where it has a lengthy discussion of the *concorso divino*, but does in fact group not only Descartes, but also Malebranche and Spinoza under the heading of *creazione continuata*, identifying the Cartesian theory however, since it is based upon an idea of discrete moments in time, as "*creazione iterata*" rather than "*continuata*".

[35] See for instance, André Lalande, *Vocabulaire technique et critique de la philosophie*, Paris (P.U.F.) 1972[11], p. 163, where a comparison is drawn between Scholastic and Cartesian philosophy on this very point under the heading of *CONCOURS ordinaire de Dieu*: "On appelle ainsi, dans la scolastique et dans l'école cartésienne, l'opération par laquelle Dieu conserve le monde dans l'existence;

justice to the Scholastic view, but as regards Descartes, the situation is more complex. The question is whether Descartes' *concours ordinaire* is to be seen as a reformulation of the Thomistic idea of a *creatio continua*, or, rather, as a modern version of the Scholastic *concursus*. In the *Discours*, Descartes apparently does not use the term *concours* to refer to God's continuous action of conservation. What is said there is that, "provided that God has established the laws of Nature", Nature is guided by a divine concurrence, "to operate as it normally does". Are conservation and concurrence therefore to be distinguished? Although many passages in the Cartesian oeuvre do seem to point to such a distinction,[36] two passages in Descartes' correspondence clearly point to an identification of conservation and concurrence. In a letter to Mersenne, Descartes affirms his allegiance to the idea of a divine concurrence, referring however to the *Primæ Responsiones*, in which he speaks of divine conservation only.[37] Moreover, in the letter *ad Hyperaspistem*, he uses the term *concursus* five times in the context of explaining his theory of divine conservation, using the term *conservari* also.[38] Hence, it seems safe to conclude that Descartes does not distinguish between a divine *concursus* or *concours* and God's continuous conservation.[39]

---

l'indépendance admise par ces écoles entre les moments du temps, ayant cette conséquence que l'universe cesserait d'exister aussitôt que Dieu cesserait de vouloir actuellement en maintenir la réalité."

[36] In other passages in which Descartes mentions the idea of discrete moments of time in connection with continuous creation (for instance in the third *Meditation*, AT VII, pp. 48-49 / AT IX-I, p. 39 / CSM II, p. 33; in *Principia* I 21, AT VIII-I, p. 13 / CSM I, p. 200; and the second *Replies*, AT VII, pp. 168-169 / AT IX-I, p. 130 / CSM II, p. 119; Cf. Gilson's commentary to the *Discours*, p. 340), does he make use of such terms as *concursus*, *concurrere* etc., whilst on the contrary he does make ample use of the terms *conservatio*, *conservare* etc. Descartes uses *concurrisse* in the third *Meditation* (AT VII, p. 50), but only in a rather non-technical sense. See also note 51, below.

[37] Descartes to Mersenne, 21 April 1641, n° 237, AT III, p. 360 / CSM III, p. 180 and *Replies* to the first objections, AT VII, p. 109 / AT IX-I, pp. 86-87 / CSM II, pp. 78-79.

[38] Descartes to *Hyperaspistes*, August 1641, n° 250, AT III, pp. 429-430 / CSM III, pp. 193-194. For the objections from Descartes' Gassendist adversary, see n° 246, AT III, pp. 405-406.

[39] In spite of the fact that Scholastic philosophers and theologians distinguished between conservation and concurrence, one might, in Descartes' defence, allege that an objective, or, in Scholastic terms, "formal" identification of God's *act* of conservation and His *act* of concurrence is not altogether problematic. The Conimbricenses in fact seem to regard God's conserving and concurring influence as more or less the same thing, where they write: "si effecta causarum secundarum,

9.4.2 Concurrence in the Cartesian World

Descartes' notion of divine concurrence differs radically from the Scholastic idea and not merely because of a careless confusion of Scholastic terminology. The way in which God concurs in a mechanistic Universe makes all the difference. Divine concurrence in its Scholastic sense either consists only in God's co-operation along with the secondary cause, or in both God's co-operation and a previous motion by God, initiating, as it were, the work of the operative faculty of the natural object itself.[40] The Prime Cause either only helps, or He both "starts up" *and* helps the secondary causes. Both of these possibilities are missing in the clockwork Universe described by Descartes. In fact, in the Cartesian world, divine concurrence is bound to be reduced to the continuous creation or conservation of the Universe from one moment to the next, since there is no longer any room for a divine co-operation.

In an attempt to free Descartes of the charge of occasionalism, Daniel Garber has reinterpreted Descartes' idea of divine conservation by distinguishing between God as a "modal cause" of action and God as a "substantial cause" of being.[41] The distinction is a modern way of describing the difference between concurrence and conservation, a distinction moreover, which matches the one Descartes himself draws between a *causa secundum fieri* and a *causa secundum esse*. But although the distinction is illuminating, Descartes himself applies it in order to describe God as a *causa secundum esse*, that is, as a substantial cause of being.[42] A hermeneutical transformation will therefore be required if we are to accept that Descartes in fact saw

---

vt fuit à nobis superius demonstratum, sine influxu Dei ea proximè tuentis nec momento consistere queunt, certè nec produci sine eodem influxu poterunt." *Commentarii in octo libros physicorum* I, p. 276. However, the argument is meant to establish that there is *as much reason* to accept concurrence as there is to accept conservation. Accordingly, *idem influxus* may well be read as "a similar influx". In any case, the question of whether *conservatio* and *concursus* are to be identified is not raised. Contrary to Descartes, God's conservation and concurrence are discussed as two different things, distinguished with respect to their effects.

[40] See hereafter, section 9.5.2.

[41] Daniel Garber, "How God Causes Motion: Descartes, divine Sustenance, and Occasionalism", in *The Journal of Philosophy* 84 (10), January-December 1987, pp. 567-580. The issue is further elaborated in Garber's impressive recent study: *Descartes' Metaphysical Physics*, Chicago (The University of Chicago Press) 1992, chapter 9, pp. 263-305.

[42] AT VII, p. 369 / CSM II, pp. 254-255. Cf. Daniel Garber, *Descartes' Metaphysical Physics*, p. 265.

God as a *causa secundum fieri*. Garber quotes two passages from Descartes' correspondence with Henry More dating from April and August 1649, in order to prove the latter position.[43] The first passage indicates that God may move matter in the same way as an angel or a human being does.[44] This is a convincing argument for regarding God as a "modal cause" in Garber's sense. However, the argument does not make clear that God has to continue this type of movement in order to sustain the more or less independent movement of secondary causes. In other words, although it is clear that God *may* move matter in the same way as human beings and angels do, there is no indication that this is the way in which God concurs. The second passage quoted by Garber is the more convincing one. Descartes there says:

> I consider 'matter left to itself and receiving no impulse from anything else' as plainly being at rest. But it is impelled by God, conserving the same amount of motion or transference in it as He put there from the first.[45]

The passage indicates that if left to itself, matter would be completely inert if God did not conserve the motion which he attributed to it in the beginning. Garber therefore opts for the "divine impulse view" rather than the "cinematic", or "motion-picture view" of Cartesian concurrence.[46] However, as Garber himself interprets this idea, it is in fact by sustaining the original *impulse* of motion, rather than by acting as a source of motion, that God "moves" inert matter. Conserving the

---

[43] Daniel Garber, *idem*, pp. 277-278 and Descartes to Henry More, 15 April 1649, AT V, pp. 340-348 / CSM III, pp. 371-375 and Descartes to Henry More, August 1649, AT V, pp. 401-405 / CSM III, pp. 380-382.

[44] AT V, p. 347 / CSM III, p. 375 and Garber, *Descartes' Metaphysical Physics*, p. 281.

[45] AT V, p. 404 / CSM III, p. 381 and Garber, *Descartes' Metaphysical Physics*, p. 282.

[46] Garber, *op. cit.*, pp. 278 ff. The cinematic view is, basically, the view that God would conserve the world by continually re-creating it in discrete moments of time, or even "atemporal instants". Garber attributes the cinematic view to Martial Gueroult in particular. See Garber, *op. cit.*, p. 267: "Just as when watching a movie, our mind creates continuous action from a sequence of frozen frames, Gueroult claims that the Cartesian continuum of duration and motion is imposed by us onto a world of discrete and timeless instants. For references to Gueroult and others, see *idem*, p. 361, note 10. Garber convincingly argues that there is neither conclusive textual evidence for this view, nor for the opposite view of Jean-Marie Beyssade that the idea of discrete instances is merely an abstract representation of the objective reality of a time-continuum. Cf. Garber, *idem*, pp. 269-273, and p. 361, note 12.

original impulse is far from actively co-operating with worldly objects, or from continually propelling matter in a "modal" way.[47] Although I prefer to interpret Descartes' notion of a *concours divin* in terms of a theory representing God as a *causa secundum esse*—rather, that is, than to leave the matter undecided—I agree with Daniel Garber that Descartes himself at least saw no reason for detailed explanation.[48]

Contemporaries, however, did. As Voetius' ally Jacobus Revius, who was one of the first to write against Descartes' *Principles*, would clamour in 1650:

> I am ashamed, for the love of God, I am ashamed about your ignorance, Descartes. That such a great philosopher as yourself has not learned to distinguish between *conservation* and *concurrence*![49]

Hot-headed debates in Leiden subsequent to those in Utrecht may have fuelled Revius' sarcasm.[50] Nevertheless, the point is clear enough.

---

[47] God may well once, in the beginning, have caused matter to move and since have caused both matter and motion to be "substantially" conserved, "leaving it"—to quote the *Discours*—"to act according to the laws He established." (AT VI, p. 42; translation from CSM I, p. 132.) Although I agree with Garber that there is reason to assume that Descartes accepted a distinct causality of secondary causes, I do not see that he really distinguished between the substantial causation of a conserving God and a modal causation of God pressing secondary causes—or matter—to act. In any case, Descartes ignores the sharp distinction made in Scholastic texts between conservation and concurrence.

[48] "Despite Descartes' clear commitment to the doctrine that God must continually support his creation, it is not at all clear just how he thinks God actually performs this most remarkable of his feats. Nor, I suspect, did Descartes think that there was any particular need for detailed explanation or defense." Cf. Garber, *Descartes' Metaphysical Physics*, p. 265. Garber seems in the end to leave the whole matter undecided: "[the objects of physics] are created and sustained by God, and, *perhaps*, subject to his continual push [...]." (Garber, *idem*, p. 293; my italics.)

[49] Jacobus Revius, *Statera Philosophiæ Cartesianæ qua Principiorum ejus falsitas, & dogmatum impuritas expenditur ac castigatur* [etc.], Lvgdvni Batavorvm, Ex Officina, Petri Leffen, [1650]. Revius (1586-1658) was both a minister of the Curch and a poet, but first and foremost a theologian. In his polemics against Arminians and Cartesians, he was, according to the Arminian leader Johannes Wtenbogaert, "as bitter as soot, as sharp as a lancet, or razor-blade." Revius held the important position of *regent* in Leiden University's *Statencollege* for bursary students of theology. On Revius see *Jacobus Revius: a Dutch Metaphysical Poet*, Trans. with an introd. by Henrietta ten Harmsel, Detroit (Wayne State U. P.) 1968.

[50] The *Statera* were written despite the explicit prohibition by the Leiden University Curators to write anything either for or against Descartes. Revius however, was provoked by Adriaan Heereboord, the professor of philosophy who had "explicitely and excessively praised Descartes" in the *Letter to the Curators*

Descartes does not get away with his all too easy application of Scholastic notions. As Revius explains, God necessarily acts alone when He acts as a creating and conserving cause, whilst He co-operates with secondary causes in concurring with their operations, which is clearly something else. Identifying the two is a downright blasphemy as regards the miracle of Creation.[51]

According to the *Discours*, moreover, the Cartesian Universe is said to be "left to act according to the laws" God once "established".[52] Scholastic authors could agree that Nature had been invested with all sorts of causes that had regular effects. The difference between the Cartesian and the Scholastic views is therefore not that the Cartesian Universe is endowed with laws of Nature. The difference is that, in the Cartesian world, substantial forms are absent. Accordingly, there are no individual sources for action. Neither, then, is there room for a concurrence of God by which He either stimulates or helps the secondary causes along. Revius likewise puts forward the question how the First Cause is supposed to concur with a merely material cause, when concurrence can only take place between causes "of the same order".[53] Really, how could First and secondary causes concur in order to produce a certain action when there is in fact no active force (*vis agendi*) in Descartes' purely material world?[54]

A present-day reader might conclude that Revius is mistaken in his search for motive forces and would be better off if he were to accommodate his theological idea of co-operation to the insights of the New Philosophy. There had, however, been no shortage of materialistic philosophies before Descartes'. Theologians like Revius and Voetius had many reasons for accepting the Aristotelian philosophy and for interpreting theological doctrines along Aristotelian

---

that accompanied the first volume of his *Selected Disputations* (1650). Cf. Theo Verbeek, *Descartes and the Dutch*, pp. 78-79.

[51] Jacobus Revius, *Statera*, pp. 73: "quomodo ulla in creationis miraculum injuria statui possit causa secunda in conservando, atque ita in creando quoque, concurrere cum prima?" Revius finds the same error in the *Principles*: "Eodem errore, Princ. p. 19. §. LI. dicis *omnes substantias* (præter Deum) *non nisi* ope CONCVRSVS *Dei existere posse*. ib. §. LII. substantiam corporem creatam definis *esse rem, quæ solo Dei* CONCVRSV *eget ad existentiam*, quid est à concursu creationem, & ex creatione concursum facere, si hoc non est?" Cf. Descartes, *Principia* I 51, AT VIII-I, p. 24 / CSM I, p. 210 and I 52, AT VIII-I, pp. 24-25 / CSM I, p. 210.

[52] Descartes, *Discours de la Méthode*, AT VI, p. 42 / CSM I, p. 132.

[53] Revius, *Statera*, pp. 72.

[54] Revius, *Statera*, pp. 73.

lines. On the other hand, the theological problems posed by Cartesianism far outweighed its possible, but as yet uncertain, "scientific" merits. Indeed, with respect to *concurrence*, what did Cartesianism have to offer but the return to the fatalistic idea of a God operating in a Universe of objects having no causal efficacy of their own? The Cartesian type of concurrence could not but lead to the kind of "occasionalism" of which, as we saw in chapter 6, Voetius was not particularly fond. The idea of a God acting as the only source of motion was associated with the teachings of the Mutakallimûm. From the theological point of view, instead of new insights, only old errors were to be gained from a New Philosophy such as Descartes'. Since he cannot at the time have known very much about Descartes notion of concurrence, Voetius must have guessed at what was to come of it. But his guesses were right. As a result of the loss of individual causality by which the mechanical Universe is defined, the concurrence of God with His Creation will either be of an occasionalist type, or be reduced to a divine conservation through time, in which case Voetius would expect external spiritual motors to accompany physical objects in motion.

9.5 Voetius on Concurrence

In the second *Thesis* of the essay in defence of the forms, Voetius warns that once the ideas of the New Philosophy are carefully considered, there would be no more room for a divine concurrence.[55] This was a serious accusation, since, to Voetius, the concept of concurrence was an all-important one.

9.5.1 Man's Fixed Hour of Death

In Scholastic philosophy, the term *concursus* may refer to any form of co-operation.[56] Accordingly, in his disputations *De Creatione*, Voetius at some point puts forward the question whether or not, in the first few days of the existence of the Universe, formerly created substances *concurred* with the creation of what came into being at a later moment in time. For instance, did the Earth actively *concur* in the formation of plants on the Third Day of Creation? The question was whether or not

---

[55] Voetius, *Narratio*, p. 40 / *Select. Dispp.* I, p. 873 / *Responsio*, p. 12 / *Querelle*, p. 106.

[56] For examples of this type of ordinary use of the term *concurrere*, see for instance the quotations in chapter 4, notes 21 and 41, above.

God let the Earth co-operate in the formation of things coming to be at a later stage of His Creation.[57]

Co-operation of some sort was generally what a "concurrence" of causes came down to. Mostly, however, the role of the Supreme Cause was involved. An interesting example—and one that has managed to haunt public debate in the Netherlands up to the present day—is the question of medical care in the light of divine providence. In 1634, Voetius was asked to give his opinion on the matter by the famous physician Johan van Beverwijk.[58] Voetius wrote a *Dissertatio Epistolica de Termino Vitæ* for the occasion, which, in 1669, was incorporated in the fifth part of his selected disputations.[59] In it, he defends the view that the end of every human life is

> fixed, estalished, certain and immovable; and this because [it] is immutably determined by God.[60]

However, although God as the "cause of all causes" has pre-established man's hour of death, this does not mean, according to Voetius,

> that [man] must rest, or look after his life in a more careless way, or neglect intermediaries (*media*), but [on the contrary, that he must] do what he is able to according to the prescription of the divine will. [...] For if God should think it fit to provide effects for the preservation of

---

[57] Indeed, is it not written in *Genesis* 1:11 that "God said, Let the Earth bring forth grass"? Nevertheless, Voetius held that such concurrence could only be passive, and not active, as Cajetan had thought. See Voetius, *Select. Dispp.* I, pp. 662-663. Voetius' arguments are, that the "active and plastic forces" of plants are found in the seed and not in the Earth at all, so why should they be found there at the moment of creation? Moreover, the Earth passively supplies matter for the formation of animals: should not the same then be said with regard to the formation of plants?

[58] For further details, see my: "New Philosophy to Old Standards; Voetius' Vindication of Divine Concurrence and Secondary Causality", in *Nederlands Archief voor Kerkgeschiedenis / Dutch Review of Church History* 71 (1991), pp. 61 ff.

[59] For bibliographical details, see "New Philosophy to Old Standards", p. 62, notes 8 and 9. The text of the *Dissertatio Epistolica* in the fifth volume of the *Disputationes Selectæ* has separate pagenumbers. In what follows, I shall therefore refer to it as "*De Termino Vitæ*".

[60] Voetius, *De Termino Vitæ*, p. 11.

> life, He will provide them by means of the intermediaries He Himself prescribed; if not, then will He do what is good in His eyes.[61]

Voetius compares the case to that of a soldier, who, unaware of the outcome of his strivings, is no less bound to fight the enemy according to the orders of the supreme command.

Man's destiny may therefore be regarded as both mutable and immutable, depending on the view one takes. In respect of the "infallible prescience and unchangeable decree" of the Prime Cause, our destiny is "wholly immutable and even necessary".[62] In respect of the ordinary causes, however, that is, in respect of those causes which are the "most proximate", man's fate must be regarded mutable. Accordingly, our time of death may be explained in respect of the efficacy of either of two sets of causes. However, a complete explanation would involve the *concurrence* of both. In medical terms, this means that the production, conservation and cessation of life can and must be attributed to both God and Nature, i.e. to both Prime and secondary causes.[63]

---

[61] Voetius, *De Termino Vitæ*, p. 14. Voetius defends his view by referring to *Deuteronomy* 29:29: "The secret things belong unto the LORD our God: but those things which are revealed belong to us and our children for ever, that we may do all the words of this law."

[62] "By necessity of the effect, not logical necessity", Voetius adds; *De Termino Vitæ*, p. 13.

[63] Voetius, *De Termino Vitæ*, p. 11: "Ut omnis ἐνέργεια, omne apotelesma (i.e., ἀποτέλεσμα; completion, fulfillment), ita & vita hominis, atque adeo vitæ motus & terminus pendet à concursu duplicis causæ, primæ scilicet & secundæ. Illa est *Deus*, ista *Natura* & *Cura* seu *Ars*." *Ars* and *Cura* are named besides Nature as indirect secondary causes of the preservation of health, since, in medical treatment, natural processes depend on deliberate human action. As for the natural causes of the preservation or cessation of human life, we must, according to Voetius, distinguish between *natura insita* and *natura assita*. The *natura insita* is "the initial warmth and humidity (in the equilibrium of which Aristotle saw [the cause of the conservation of] life) and the necessary proportion, power (*vigor*) and good quality (*bonitas*) of both; which kind of nature some call *Balsamum Naturale*." *Natura assita* stands for the "aliments and medicaments with which we are amply provided by the mineral, vegetable and animal worlds. On top of which come those that the physicians call Non-Naturalia." See *De Termino Vitæ*, p. 12. Note that *non-naturalia* stands for the six factors of nourishment, retention and evacuation, air, sleeping and waking, exercise, and passions of the soul. All of these could, according to contemporary medical theory, influence the equilibrium of the bodily humours and hence cause a disease. Voetius may have consulted Johan van Beverwijk, who offered similar ideas in his popular *Schat der Gezondheid*, or *Treasury of Health* of 1636. Cf. Johan van Beverwijk / Lia van Gemert (ed.), *De schat der gezondheid*, Amsterdam (Querido) 1992, pp. 9-16.

On the other hand, the relation between the two types of causes concurring to produce a certain effect is one of subordination. Compared to the First Cause, the organic and moral causes must be regarded as "accessory"[64] and inferior. For as, in the example of the prolongation of life, the outcome of medical treatment is still dependent on God's Will, so in every other action do secondary causes only act in virtue of the First and Prime Cause, as His instruments.

Voetius refers to the authority of the great Aquinas and his followers as a source for his ideas on man's fate.[65] With respect to the question of concurrence, however, the reference to "the followers" of the Angelic doctor conceals a serious theological controversy which, from the end of the sixteenth century onwards, had divided Scholastic philosophers and theologians.

9.5.2 Voetius' Defence of Premotionism

In relation to theological discussions concerning God's grace, providence and foreknowledge, Jesuit philosophers had presented a theory of concurrence in which God, as Prime Cause, only operates and acts *simultaneously* with secondary causes.[66] This theory was opposed to the one generally advocated by the Dominican party. The

---

[64] Voetius uses the Greek term συναίτιος, which occurs in Plato, *Timæus* 46^{c}7 and 76^{d}7, in the sense of "being an auxiliary cause of". Cf. the *Lexique* by Éduard des Places S.J.; Platon, *Œuvres Complètes*, Paris (Soc. d'Éditions 'Les Belles Lettres') 1964, Tôme XIV, p. 478. Aristotle uses the term *inter alia* in the sense of "concomitant cause" / "part-cause" in *De Anima* II 4, 416^{a}14 and *Ethica Nicomachea* III 7, 1114^{b}23; of "extra cause", in *De Generatione Animalium* V 3, 782^{a}26, and "concomitant condition" in *Metaphysica* IV 5, 1015^{a}21. Cf. Bonitz, *Index Aristotelicus*, Berlin 1870 / reprint Graz (Academische Drücke und Verlagsanstalt) 1955, p. 722. Nowhere however, does Aristotle use the concept in order to describe the relation between prime and secondary causes.

[65] Voetius, *De Termino Vitæ*, p. 13: "Quomodo jampridem satis distinctè hoc enunciârunt Scholastici. Thomas I. quæst. 116. art. 3. *Fatum, ut in secundis causis est mobile, ut autem à divinâ præscentia est immobile, non necessitate absolutâ, sed conditionali.*" And, *idem*, p. 12: "ut passim tradit Scholasticorum princeps Thomas, & qui eum sequuntur."

[66] For historical details concerning the debates within the Roman Catholic Church on the question of divine foreknowledge and secondary causality, see William Lane Craig, *The Problem of Divine Foreknowledge and Future Contingents From Aristotle to Suárez*, (Brill's Studies in Intellectual History Volume 7), Leiden (E. J. Brill) 1988, and my own "New Philosophy to Old Standards: Voetius' Vindication of Divine Concurrence and Secondary Causality", in *Nederlands Archief voor Kerkgeschiedenis / Dutch Review of Church History* 71 (1991), pp. 58-91.

Dominicans claimed that God not only operates simultaneously with secondary causes, but that He also produces a "previous motion" or "physical predetermination" in every secondary cause in order for it to act.[67] In the latter view—which has become known as *premotionism*—God also activates the virtues of secondary causes. If, for instance, a fire is to heat a bowl of water, God has first to activate the virtues of the fire in order for it to produce its action. Next, He will also concur along with the action of the fire in order to produce the desired effect.

From a philosophical point of view, it does not matter much which of the two interpretations is presented in a more detailed explication of the notion of concurrence.[68] However, the choice for or against the premotionist standpoint was of the utmost theological importance. According to the Dominicans, if God only co-operated with secondary causes, He would be determined in His action by the secondary cause. This was of particular relevance in the context of divine grace. For if man, as a secondary cause, were not to be predetermined to act in a certain way, he might, as it were, "provoke" divine co-operation.[69]

Such a view was thought to be dangerously Pelagian, not only by Dominicans, but also by Calvinists such as Voetius. According to the latter, to say that divine concurrence does not predetermine the secondary cause, but only works *in effectum* would involve doing

---

[67] The famous Jesuit author Francisco Suárez offers various other names by which God's previous activation of secondary causes became known. See his *Meditationes metaphysicæ*, in *Opera Omnia*, Paris, Apud Lodovicum Vives, 1861, Vol. XI, p. 26: "Hoc autem prævium, quidam [...] applicationem causæ secundæ ad agendum, alii motiones, alii complementum virtutis, quo in actu primo consummatur, alii excitationem virtutis activæ, alii denique determinationem ejusdem causæ ad suam actionem vocant." Suárez himself however rejects the idea of God effecting a previous and necessary motion in the cause itself, "quid nec fides docet, nec ratio suadet, nec mens fere capere potest."

[68] The Jesuit philosophers of Coimbra for instance, in their commentary on Aristotle's *Physics* II, first deal with the question of God's concurrence in general, only to present their co-operationalist interpretation of concurrence in the next *Quæstio*. Conimbricenses, *Commentarii in octo libros physicorum* I, *Quæstiones* XII-XIIII, pp. 273-282.

[69] See iner Specht, "Zur Bezeichnung unzureichender Zweitursachen bei Francisco Suarez", in *Philosophisches Jahrbuch* 68 (1959), p. 393. Ra Specht explains that, according to Suárez, "eine Zweitursache keine echte Kausalität auf die Erstursache ausüben kann". Still, "die menschliche Entscheidung wird zur Occasio der gesetzmäßigen Mitwirkung: sie provoziert die Mitwirkung des an die von ihm selber geschaffene hypothetische Notwendigkeit gebundenen Gottes und affiziert ihn, damit er wirkt."

> God's Greatest Majesty the greatest of injustices, [making] man, in acting, not subordinate and secondary to God, but His equal and companion.

Voetius refuses to compromise in any way concerning the total dependence of all natural and mental operations on God. Where the Jesuits by their co-operationalist view of divine concurrence try to find a *concordia* between divine grace and human free will, Voetius retorts that

> We do not go for such a concord, which subjects God to man, Creator to Creation.[70]

Voetius thus took a firm stand with regard to the question of divine concurrence, as did other Calvinist writers of his day.[71] Their interest in the question of concurrence may be explained by the controversies on Arminianism which divided the Dutch Reformed Churches in the first decades of the seventeenth century. By 1609, the year in which Jacobus Arminius died, the Catholic debates were re-enacted in a Protestant version by the Arminian Petrus Bertius and the Contra-Remonstrant leader Franciscus Gomarus at Leiden University.[72] In fact, not only did the question of concurrence occupy the minds of professors in the years in which Voetius studied, it might even, as the case of Adriaan Heereboord's corrections of his teacher Burgersdijk proves, become a test-case for Calvinistic orthodoxy in later years.[73]

Having become one of the leading Calvinist authorities by the time he faced the New Philosophy of Descartes, Voetius must still have

---

[70] Voetius, *Select. Dispp.* I, p. 306.

[71] The English Puritan William Twisse, the Polish professor of theology at Franeker Johannes Maccovius and Gisbertus Voetius' son Paulus for instance, all defended premotionism. See my article on "Franco Petri Burgersdijk and the Case of Calvinism Within the Neo-Scholastic Tradition", in E. P. Bos / H. A. Krop (edd.), *Franco Burgersdijk (1590-1635), Studies in the History of Ideas in the Low Countries*, Amsterdam / Atlanta, GA (Rodopi) 1993, pp. 37-65. For further details on Voetius' views on divine concurrence, see my "New Philosophy to Old Standards", pp. 67-78 especially.

[72] See my "Franco Petri Burgersdijk and the Case of Calvinism Within the Neo-Scholastic Tradition", pp. 48-50. Being a student in the *Statencollege* at the time, Voetius was even temporarily expelled by Bertius in 1609, presumably because of his intense devotion to the anti-Arminian cause. See e.g. the extremely biased, but no less usefull account by the reverend C. Steenblok, *Gisbertus Voetius: zijn leven en werken*, Gouda (Gereformeerde Pers) 1976, pp. 17-19.

[73] "Franco Petri Burgersdijk and the Case of Calvinism Within the Neo-Scholastic Tradition", *passim*.

been very much on his guard against Jesuit, Arminian, or generally "Pelagian" doctrines of concurrence. The fact of the matter is, however, that Voetius' remarks concerning divine concurrence in the essay "On the Natures and Substantial Forms of Things" does not in any way refer to the controversies which had divided both Catholic and Protestant Churches for half a century. Presumably, the conflict of causality between the Aristotelian and Cartesian world-views overruled the theologically important opposition between co-operationalism and premotionism. As we shall see, even the whole idea of joining theology to a philosophical framework was at stake.

9.6 Causality in the Face of a Materialist Ontology

A striking element in Voetius' essay on substantial forms is that it discusses the subjects of divine concurrence and of "occasionalist" ideas even though Voetius cannot have known much about Descartes' thoughts on divine concurrence, nor of any the occasionalistic ideas that were to result from it.[74] Apparently, the philosophy of the forms and the idea of divine concurrence were intimately related.

9.6.1 Theological Questions with Regard to Aristotle's Physics

In the essay "On the Natures and Substantial Forms of Things", Voetius warns that the mechanistic ideas of the new philosophers might bring people to accept the "absurd opinion"

> 'That there are no secondary causes fitted with a causality of their own, but that only the first and universal cause acts in the presence and [given a certain] disposition of the secondary causes.' For the absurdity of this position, see the Scholastics and the modern metaphysicians and theologians, where they discuss the question of providence and God's concurrence with creation. The consequences with which they may be confronted are the following: 1. That there would be no divine concurrence nor any movement of the prime mover which is accommodated to the natures and properties of secondary causes, whether these are necessary or contingent. 2. That there is no intrinsic motor in created substances, or substantial principle of motion which is internal and proper [to the thing in

[74] For a critical account of present-day interpretations of Descartes' *concours ordinaire de Dieu* and its relation to occasionalism, see Daniel Garber, *Descartes' Metaphysical Physics*, Chicago (The University of Chicago Press) 1992, chapter 9, pp. 263-305. See also notes 46 and 47, above, and chapter 6, note 36.

> question]: for the disposition of the movable to move in virtue of its quantity, shape, position, is neither an activity nor a causality of an efficient cause, but only a necessary condition and a *causa sine quâ non.*[75]

At this very point, Voetius comes up with his list of alternative "external motors" of

> the Platonic-Vergilian world-soul, or intelligences, or God, or atoms, or heavenly globules.[76]

As is the case in the Renaissance commentaries on Aristotle's *Physics*, likewise in Voetius' essay, do we find the idea of composite substances, of the efficacy of secondary causes and of divine concurrence all as part of one and the same body of related topics.

Voetius refers to "the Scholastics and the modern metaphysicians and theologians". I have previously argued that he might here have a specific tradition of premotionist thinkers in mind, since, with regard to the question of concurrence, both Voetius and his son Paul saw themselves a representatives of a specific tradition following Aquinas.[77] The discussions concerning a "physical predetermination"

---

[75] Voetius, *Narratio*, p. 40 / Select. Dispp. I, p. 873 / *Responsio*, p. 12 / *Querelle*, p. 106.

[76] Voetius, *ibidem*. See section 6.2, above.

[77] See my "New Philosophy to Old Standards", p. 65 note 24 and p. 87, note 89, where I refer to Voetius' disputation *De Potentia Dei*. Voetius writes: "[...] by the same sort of trifle one will be able to say that a created thing can operate without divine concurrence and predetermination (against which the Thomists dispute constantly and truthfully, as can be seen in [the works of] the Thomistic Theologians Cumel, Rispolis, Bañez, Alvarez, Sylvius etc., and the Philosophers John of St. Thomas and the Complutenses in their Thomistic Philosophy)." Voetius, *Select. Dispp.* I, pp. 411-412 (second series). Gisbertus' son Paul Voet argued that: "of the Scholastics, one should especially study those who have declared war on the Jesuits, such as the Dominicans, whom, since they choose a safe way in many cases, we often praise and follow." Paulus Voet, *Prima Philosophia Reformata*, Trajecti ad Rhenum 1657, p. 2. Whether or not Aquinas himself takes a premotionist position in his works, has for centuries been a question of great dispute amongst Christian theologians. See my: "New philosophy to Old Standards", pp. 65-67 and notes 22-28 especially. In the *Dictionnaire de Théologie Catholique*, E. Vansteenberghe gives the following, conciliatory and, to my mind, only possible conclusion on the matter: "Malgré les efforts déployés de part et d'autre pour pénétrer la pensée de saint Thomas, il ne semble évidemment démontré, ni qu'il a été adversaire de la prédétermination physique, ni qu'il en a été partisan." Vansteenberghe, *op. cit.*, Tôme X-II, col. 2184. There simply was no such distinction in Aquinas' days.

would without any doubt have come to Voetius' mind whenever he mentioned divine concurrence.[78] The fact is, however, that Voetius' references to the premotionist tradition occur where he deals with the question of "divine concurrence *and predetermination*".[79] No mention of premotion or predetermination is made in the passage of the essay on substantial forms in which divine concurrence is brought up against Descartes. Voetius in this case does not seem to regard the matter of premotionism of much relevance. In the essay, Voetius is not defending premotionism: he is defending the forms against the mechanical philosophy in which, as he sees it, divine concurence will in any case be dropped.

Accordingly, when referring to "the Scholastics and the modern metaphysicians and theologians", Voetius may just as well at this point have Jesuit thinkers such as Ruvius, Suárez or the Conimbricenses in mind, whom he otherwise quotes approvingly in defence of the forms.[80] The commentaries of these authors display a grouping of arguments which exactly parallels Voetius'.[81] But why did

---

[78] Note that Voetius had been involved in explaining the Scholastic views concerning divine concurrence and grace to Calvinists less familiar with them. Voetius for instance defended Johannes Maccovius, whose orthodox, but very Scholastic way of explaining the "physical causes" of divine grace, had led to a reprimand of the Polish theologian at the Dordrecht Synod. See my "Franco Petri Burgersdijk and the Case of Calvinism Within the Neo-Scholastic Tradition", pp. 39-40 and p. 48, note 40 especially.

[79] See footnote 77, above. (My italics.)

[80] Voetius praises Suárez in particular. See *Narratio*, p. 42 / *Select. Dispp.* I, p. 874 / *Responsio*, p. 21 / *Querelle*, p. 108: "Rationes, quæ à formarum assertoribus adferri solent primum pernoscendæ, deinde perspicuè & solidè solvendæ. Hanc in finem accuratè legendus est junioribus aliquis ex recentioribus disputationibus physicis aut metaphysicis; quales Zwarez, Pererius, Ruvius, Toletus, Conimbricenses, Complutenses, Ariaga, Mendoza &c. Ex omnibus plenissimè & subtilissimè hanc causam agit Zwarez." Voetius also regarded Antonio Rubio, or Ruvius (1548-1615) as a "subtle writer on questions of physics." See above, chapter 6, note 46 and Voetius, *Excercitia et Bibliotheca Studiosi Theologiæ*, 1644, p. 382. It is worthy of note moreover, that, for general questions of physics, Voetius in fact places the Conimbricenses before the Complutenses. See *idem*: "II. Controversiarum scriptores; quales I. *ex recentioribus Scholasticis Theologis* imprimis commendandi *Conimbricenses*, *Ruvius*, Toletus, Pererius; quorum duo priores subtilitate posteriores facilitate & perspicuitate præcellunt. Post illos prodiere Masii, Mendoza, Oviedo, Ariaga, Complutenses, Johannes à S. Thoma."

[81] See sections 6.2.1.2 and 9.3, above. Note that some elements of Voetius' argumentation in the passage concerning concurrence and the efficacy of secondary causality are in fact missing from the texts of premotionist thinkers, whilst they may be found in Jesuit commentaries. Thus, the *Disputationes in octo libros*

they, and why did Voetius, connect divine concurrence (1) with the occasionalism of a Prime Mover operating without the help of secondary causes, (2) with the question of accidental properties versus substantial principles, and, finally, (3) with the question of a need for external motors?

The answer is, that all of these questions are linked up with a single concept of causality that Scholastic theologians found in Aristotle. Because of the way in which Aristotle had framed the question of physical change in *Physics* II, physics was regarded to be the discipline in which causal sources of action were to be pointed out. The restricted task of finding causal virtues and powers enabled theologians to introduce various theological notions in the context of the issue of answering to the *why* of natural change. The question, for instance, why water which is placed on a fire becomes hot, may be answered by referring to the calefactory virtues of the fire. However, the same example may be used in a more absolute sense in relation to the maintenance and effectiveness of the virtues themselves. This is exactly what we find in the Scholastic commentaries, where arguments regarding divine conservation and concurrence are formulated in terms of God preserving the natures of, for instance, fire and water and activating the natural virtues of fire, which, without His concurrence, would remain ineffective.

It should be stressed that this alliance of theology and physics was not thought of in terms devaluing the discipline of physics in any way. Although it is admitted that "the Peripatetics" did not themselves see the necessity of a divine concurrence of secondary causes, since they treated secondary causes in an "absolute" way,[82] their philosophical explanation of causality is nonetheless accepted as what we would call the "scientific" one. For Voetius and for other Renaissance

---

Physicorum Aristotelis of the Complutenses, who generally follow Thomas rather than Aristotle, do not mention Aquinas' list of alternative external movers such as the atoms, intelligences and the world-soul, which, as we have seen, does occur in the *Commentarii in octo libros physicorum Aristotelis Stagiritæ* of the Jesuit Conimbricenses. A Jesuit author like Antonio Ruvio for instance also deals with both divine concurrence and the efficacy of secondary causality. See his *Commentarii in octo libros Aristotelis de Physico*, Madrid 1605, pp. 294 ff.

[82] Conimbricenses, *Commentarii in octo libros physicorum* I, p. 278: "At Peripatetici caussas secundas, vt par est [that is to say, comparable with the Platonic error of not accepting secondary causes as real causes at all], absolutè, & propriè caussas vocant: quia licet à caussa prima dependant (quod Platonicis vt eis caussarum nomen non absolutè concederent, persuasit) in suo tamen genere perfectam caussalitatem obtinent."

Aristotelians, the "sane and sober philosophy" of Aristotle not only sustained the Christian dogmas of providence and grace; it also prevented the introduction of such unphilosophical notions as are presented in Platonic and magical philosophies.

9.6.2 A Carnival of Principles

A fine illustration of Voetius' allegiance to the Scholastic version of Aristotelian principles occurs in his dissertation *De Termino Vitæ*, "On the End (or Duration) of Human Life". After having argued that "the infallible and certain" moment of death "follows from the predefinition of divine will", Voetius puts forward the question how it could be otherwise. If man's last hour does not depend on God's will and absolute decree, it either depends only on itself, or on some other principle. Thus, it might only depend on the "goodness, force (*vigor*), or necessity of nature". And, if not on these natural causes, then maybe death would depend on fortune.[83]

Voetius thus first links the question of the cause of death to the issues relating to causal determinants as dicussed by Aristotle in book II of the *Physics*. Natural causes are, according to Voetius, indeed part of the story. As Aristotle had said, natural forces aim at "good" ends—in this case, the preservation of life—by a hypothetical necessity. If the natural causes cease to have the desired effect in the case of upholding the bodily functions, living things will die. Yet as we know, Voetius' own solution to the problem of how man's end of life is fixed ultimately involves the concurrence of these natural causes with the Prime Cause. There are hence two levels of causality concurring to establish the same result.

Whereas there might be something to say for Nature as the cause of death—as long as it is not taken in an absolute way[84]—this does not count for any of the other alternatives Voetius lists as possible substitutes for divine volition. Again like Aristotle in *Physics* II, Voetius mentions the possibility of fortune as a natural principle. It is rejected. But whereas Aristotle discusses *only* fortune and chance, Voetius knows a great many other possible alternatives:

---

[83] Voetius, *De Termino Vitæ*, pp. 46-47.

[84] See the quotation from the Conimbricenses in note 82, above, concerning the mistake that the Peripatetics have made by regarding secondary causes as causes "absolutè & propriè [per se]".

> either Stoic fate, or the two vessels of Homer's Rhapsody,[85] or the Pythagorean συστοιχειώσεις as discussed in Plutarch's book on the opinions [of the philosophers],[86] or the atoms of Epicurus, or the movement of atoms in a slanting direction, when [their motion] swerves over a minimal distance, [a solution] which, according to Cicero's *De Fato*, Epicurus concocted in order to save the freedom of the will as being exempt from the perpendicular weight and impact or motion of atoms.[87]

Thus, along with mythological determinants of fate, and the "opposite pairs" of the Pythagorean philosophy, Voetius adds the principles of the Hellenistic schools to the Aristotelian dump of rejected δόξαι.[88]

---

85 The two vessels, wine-barrels, or urns "of Homer's Rhapsody" which Voetius refers to, are those from which Zeus ladles good or evil. See Homer, *Iliad* XXIV, 527-528. Whom Zeus gives from both urns sometimes meets with evil and sometimes with good. Those who receive an unmixed share of evil however, will lead a miserable life, roaming the world in hunger. Plato alludes to the Homeric idea of Zeus distributing prosperity and misery at *Republic* II, 379 D, but dissaproves of the idea of God being the cause of evil.

86 Voetius refers to Plutarch's "de placit. libr." What must be meant, is the Περὶ τῶν Ἀρεσκόντων τοῖς Φιλοσόφοις, or *Placita Philosophorum* as it was known in its Latin version. This spurious work occurs along with a Latin translation by Xylander in the second volume of Plutarch's *Opera Omnia*-edition of 1599, or Πλουτάρχου Χαιρωνέως τὰ Σωζόμενα Πάντα / *Plvtarchi Chæronensis Qvæ Exstant Omnia, cvm Latina interpretatione Hermanni Crvserij: Giulielmi Xylandri, et Doctorvm Virorum* [etc.], Francofvrti, Apud Andreæ Wecheli heredes, Claudium Marnium, & Ioannem Aubrium, 1599. However, I have not found the term συστοιχειώσεις in this work, although Pythagoras' principles are discussed at p. 876 / Plutarch, *Moralia* Vol. V, fasc. 2, pars 1, ed. Jürgen Man, Leipzig (Teubner) 1971, pp. 55-56. In the Pythagorian school, some ten elements of the συστοιχία, or "series of related pairs" such as odd and even, male and female etc., are held to be the principles of things. Cf. Aristotle, Metaphysics I 5, 986^{a}23. The term συστοιχειώσεις also occurs in Iamblichus, where it is used in the sense of the "orderly arrangement" of things which reflects the nature of God. Iamblichus, *Adhortatio ad Philosophiam*, ed. Kiessling, Leipzig (F. C. G. Vogelius) 1813, p. 62.

87 Voetius, *Select. Dispp.* V, p. 47 and Cicero, *De Fato* X 22: "itaque tertius quidam motus oritur extra pondus et plagam, cum declinat atomus intervallo minimo (id appellat [scil., Epicure] *elachiston*), and *idem*, X 23: "Hanc Epicurus rationem induxit ob eam rem quod veritus est ne, si semper atomus gravitate ferretur naturali ac necessaria, nihil liberum nobis esset, cum ita moveretur animus ut atomorum motu cogeretur."

88 Hellinistic philosophical themes deeply influenced sixteenth and seventeenth-century thought. Through the intermediate of authors such as Sebastien Basso and David Gorlæus, the revival of ancient non-Aristotelian ideas had troubled Scholastic Aristotelians as Voetius long before Descartes' appearance on the

This however is not all. There may be other principles supposedly governing man's fate in his last hour:

> geniuses either good or bad, or minds, such as the מלאך המות of the Jews, that is, the angel of death and the frightful Lilith,[89] also known with the Gentiles as Erynnyes and Stryges,[90] or the Helec and Alcocoden of the astrologers, and similar positions and arrangements of stars (*astrorum* σχέσεις *&* διαθέσεις), *ferdaria*, as they say, of planets, and time-rulers, which useless art (*mataiotechna*) Archangelus of Burgonovo touches on in his Commentary on the doctrines of the Cabbalists and Scaliger in his Commentary on Manilius.[91]

---

philosophical stage. For a recent oversight of the influence of the Stoic and Epicurean Schools in post-classical times, see Margaret J. Osler (ed.), *Atoms,* Pneuma, *and Tranquillity: Epicurean and Stoic Themes in European Thought*, Cambridge (C. U. P.) 1991. From the early stages of the Reformation onward, the Stoic idea of fate and the Epicurean concept of liberty were overshadowed by discussions on predestination and free will. Stoic and Epicurean concepts might still be used to illustrate the theological discussions. However, former identifications of classical and Christian ideas had to be abandoned. See for instance Concetta Bianca's remark concerning the ill-fate of Coluccio Salutati's book on the ideas of fate and fortune: "Lo scoppio della polemica luterana doveva infine segnare un netto e definitivo abbandono: l'interpretazione storiografica nata nel periodo della controriforma non poteva non rivolgere i propi strali verso un'opera che già nel titolo, De fato et fortuna, e ancor più per alcuni capitoli sul libero arbitrio e la predestinazione, richiamava temi teologici difficili e controversi." Coluccio Salutati, *De Fato et Fortuna*, a cura di Concetta Bianca, Firenze (Leo S. Olschki) 1985, p LXXIX.

89 The *horrid Lilith* is a female night demon that seduces men and kills children. See e.g. Mircea Eliade (ed.), *The Encyclopedia of Religion*, Vol. 8, New York (Macmillan) 1987, pp. 554-555 and *Encyclopaedia Judaica*, Jerusalem (Keter) 1971, vol. 11, cols 245-249.

90 *Erynnyes*, or Erînyes: spirits of punishment. See e.g. *The Oxford Classical Dictionary*, edited by N. G. L. Hammond and H. H. Scullard, Oxford (At the Clarendon Press) 1970^2, pp. 406-407. *Strix, strigis* is from the Greek στρίγξ, meaning screech-owl, a bird thought to harass small children and suck their blood. For referendes, see: Charlton T. Lewis and Charles Short, *A Latin Dictionary*, Oxford (At the Clarendon Press) 1879, pp. 1766-1767.

91 The *hyleg*, or place of life, is a celestial place which is determined by an intricate set of astrological rules and, together with the *alcochoden*, enables the astrologer to determine the length of life. The *alcochoden* is not a place, but a planet, to be chosen on the basis of various other considerations. See J. D. North, *Chaucer's Universe*, Oxford (Clarendon Press) 1990^2, pp. 214-217. In his *Dogmata selectiona, obscurioraque Cabalistarum*, the Minorite Archangelus de Burgonovo explains amongst many other things, that mankind was originally

Alternative Platonic, magical and heretical ideas end the list:

> Plato's numbers or ideas or Great Year[92]; or the evil God of the Manicheans[93], or the aeons of Valentine[94]; Helena or Selene of Simon

---

offered a life of 1000 years. God reduced this to 120 years however, probably for the reason that man could not be expected to resist sin for a whole millennium. Therefore people now live short lives, although some may live very long according to his or her "Hilec, & Alcocodem" as, according to Archangelus, the Astronomers put it. See Pavlvs Ricivs (et. al.), *Artis Cabalisticæ: hoc est Reconditæ Theologiæ et Philosophiæ, scriptorvm:* Tomus I, Basileæ, Per Sebastianvm Henricpetri, 1593, p. 855. Josephus Justus Scaliger criticizes the translation of the Arabic *Hilleg* and the Greek ἀφέτης as "prolonger of life". *Hilleg*, according to Scaliger, should rather be translated as ζωοποιός, or life-bringer. See Scaliger's commentary to Marcus Manilius (a Roman astrologer who died c. 22 A.D.): *Astronomicωn Libri Quinque*, Lvtetiæ, apud Mamertum Patissonium Typographium Regium, in officina Roberti Stephani, 1629, p. 78. Ἀφέτης being a name for certain heavenly bodies, the *Hilleg*, *Hilec*, *Helec*, or *Hyleg* was also interpreted astrologically as a life-prolonging sign, though, as we have said, it originally refers to a certain determinable *place* in the heavenly constellation at the time of birth. A χρονοκράτωρ, or "time-ruler", may be any divinity or genius affiliated with a certain position on the Zodiac, or a "mode of Signs". There are, according to Scaliger, various ways of determining the "modes of Signs", for instance according to the place, or according to the conjunction of stars. *Ferderia* (or *frideria*) are planetary chronocratories, of which Scaliger says that they are only accepted in Arabic astrology. See further: A. Bouché-Leclercq, *L'Astrologie Grècque*, Paris (Ernest Leroux) 1899. Bouché-Leclercq quotes the following from a Latin edition of Albohazen Haly: "Est autem Fridaria seu *chronokrateia* certus quidam annorum terminus et notus, in quo planeta gubernans vitam nati dat et infert et bonum vel malum pro sui natura." Cf. A. Bouché-Leclercq, idem, p. 491, note 1, and *Bibliographie*, p. XVII. Finally, Scaliger also mentions *Alcochoden*, the Lord of the year, which the Persians called *Salchodai*. Scaliger, *In Manilium*, p. 200.

92 The view that Plato held his Ideas to be numbers, is Aristotle's. Cf. *Metaphysics* I 6, 987*b*18-25. For a discussion, see W. D. Ross' introduction to *Aristotle's Metaphysics*, Oxford (At the Clarendon Press) 1924, pp. lxvii-lxxi. Plato's "Great Year" occurs in *Timaeus* 39 D as the period after which all planets take up the same relative positions they had before. If this period may be identified as the period of 36,000 years of the created Universe, the period of one human life (100 years) is exactly a day in the Cosmic year. For details on this extremely complex subject-matter in Plato, see James Adam's "Appendices to Book VIII" of *The Republic of Plato*, Cambridge (At the university press) 1902, vol. II, pp. 288-306 and A. E. Taylor, *A Commentary on Plato's Timæus*, Oxford (At the Clarendon Press) 1928 / reprint New York & London (Garland) 1987, pp. 216-220. On both the astronomical conception and religious appreciation of the Great Year, see J. D. North, "Chronology and the Age of the World", in W. Yourgrau / A. D. Beck (edd.), *Cosmology, History, and Theology*, New York (Plenum Press) 1977, pp. 307-333, and: J. D. North, *Chaucer's Universe*, Oxford (O. U. P.) 1988, pp. 35-37.

the Magician[95]; or Hesiod's Pandora[96]; or the great mother Mylitta, or Alytta of the Gentiles[97]; or Nassub of the Turcs, that is, Ctusara, the

---

[93] The dualistic doctrine of Manichaeism parried the problem of theodicy by teaching the doctrine of double roots, or principles, the one a good root of God and Light, the other an evil root of matter and darkness, and the cause of man's pitifull Earthly condition. On Manichaeism, see, e.g., the *Encyclopedia of Religion*, New York (Macmillan) 1987, IX, pp. 161-171.

[94] According to the fourth-century *Epistle to Rheginos* concerning the resurrection, Christ after having died, "transformed [himself] into an imperishable Aeon", that is to say, into a higher heavenly sphere. Αἰών, which in its normal Greek usage may mean life, lifetime, era, or derivatively something like "fate", was in Gnostic writings the name for mediators between God and men, who had emanated from the First Divinity (itself a supreme Αἰών) and could also be represented as heavenly spheres. Valentinus (c. 110 - c. 160) was a prominent preacher of Gnostic ideas in Rome. The *Epistle to Rheginos*, although influenced by Valentinian Gnosticism, is probably not by Valentinus himself. See Malcolm Lee Peel (ed.), *The Epistle to Rheginos, A Valentinian Letter on the Resurrection* (*New Testament Library*), London (SCM Press) 1969.

[95] Simon Magus, "the father of all heresy", is reported in *Acts* 8:9-24 to have offered the apostles Peter and John money in exchange for the powers of the Holy Spirit. Helena is the graceful Earthly manifestation of the divine *Ennoia*, or first act of thought. She plays an important part in the Gnostic mixture of Christian and pagan ideas that Simon later taught in Rome at the time of Claudius. Cf. *Paulys Real-Encyclopädie der Classischen Altertumswissenschaft*, II. Reihe, III. Band, cols. 180-183. The story of Simon in Rome is told by Justin Martyr (c. 100 - c. 165 A. D.), who describes Simon and Helena as a Samaritan couple of whom the man acted as the Supreme God himself and the woman, whom he had found in a brothel, as his first emanation. See St. Justin, *Apologies*, introduction, texte critique, traduction, commentaire et index par André Wartelle, Paris (Études Augustiniennes / Institut Catholique de Paris) 1987, p. 132. Helena was worshiped in Simon's homeland as an omnipotent goddess, goddess of fecundity, tutelary deity and distributor of immortality. Cf. L. H. Vincent, O. P., "Le culte d'Hélène à Samarie", in *Revue Biblique* 45 (1936), pp. 227-232 esp. *Selena*, the Goddess of the Moon, was in Gnostic writings often associated with Helena. The beautiful Gnostic Helena, an incarnation of the divine spirit, made use of her graciousness to incite strife and discord on Earth not unlike the Homeric Helena had brought about in the war between Greeks and Trojans. As Goddess of the Moon, Helena or Selena also influences the sway of misfortune and prosperity on Earth.

[96] In *The Works and the Days* 42-105, Hesiod tells the story of Pandora, the first woman, who was sent by Zeus as punishment for the fact that Prometheus' had stolen fire from the Supreme God and had given it to man. Against the warning of Prometheus, his brother Epimetheus accepted the gift and realised his mistake only too late. As a Greek Eve, Pandora opened the box that held all the sorrows and distress that since have bothered us. A larger version of the story occurs in Hesiod's *Theogony*, in which, however, Pandora is not named. For further references, see Hesiod, *Theogony*, Edited with Prolegomena and Commentary by M. L. West, Oxford (Clarendon Press) 1966, pp. 305-308.

Goddess of Fortune[98]; or the Platonic-Paracelsan world-soul[99], or rather the macroscopic and microscopic harmony of the universe with its signs and images[100]; or that Spirit of the Universe and common motor of the Stoics which Basso has recently transformed into the Platonic world-soul in his *Natural Philosophy against Aristotle* [...][101]; or those lamps of life and death, [that is to say] the magic

---

[97] *Alitta* or *Alilat* is the Arabic and *Mylitta* the Syrian pendant for the Mother-Goddess Aphrodite. Cf. *Paulys Real-Encyclopädie der Classischen Altertumswissenschaft*, Bd. I, Stuttgart (J. B. Metzler) 1894, col. 1483 and Bd. XVI-I, Stuttgart (J. B. Metzler) 1933, col. 1073.

[98] I am uncertain as to what Voetius here refers to. The goddess Nassub may have affiliations with the Persian *Nasu*, a demon visiting corpses and infecting the family of the dead. Cf. Hans Wilhelm Haussig (ed.), *Götter und Mythen der Kaukasischen und Iranischen Völker*, *Wörterbuch der Mythologie* Band IV, [...] (Klett-Cotta) 1986, pp. 411-412. In any case, worship of the Goddess of Fortune is common to many peoples and was especially strong in Roman times. The deity survived in Medieval and Rennaissance Christianity also. Cf. Howard R. Patch, *The Goddess Fortuna in Mediaeval Literature*, London (Frank Cass & Co.) 1967.

[99] Plato's world-soul is discussed at *Timaeus* 34 B ff. See also above, chapter 6, note 45. As Walter Pagel writes: "The search for the hidden invisible spirit which governs and moves visible bodies is the keynote of Paracelsus' natural philosophy. In this it followed one of the main tenets of Platonism as revived by Ficino." Walter Pagel, *Paracelsus: An Introduction to Philosophical Medicine in the Era of the Renaissance*, Basel (Karger) 1982[2], p. 218.

[100] The Paracelsian idea of the harmony existing between macrocosm and microcosm, that is, between the world at large and its microscopic reflection in the human being, leads to the idea that man's higher, active and intellectual functions mirror the activity of the celestial bodies. Astrological signs and images correspond to the microscopic realm. For the Paracelsan idea of the "astral" or sidereal part of human nature, see note 105, below, and, in general: Walter Pagel, *Paracelsus* (see former footnote), pp. 65-82. James Adam, in his commentary on Plato's *Republic*, points to identities in the construction of Soul and of the Universe, adding: "There can be little doubt that if [Plato] speculated on the subject at all, he followed in the path already marked out by the Pythagoreans, framing the Microcosm, as his manner is, on the lines of the Macrocosm." *The Republic of Plato*, edited with critical notes, commentary and appendices by James Adam, M. A., Cambridge (At the University Press) 1902, II, pp. 294-295.

[101] "lib. de natur. pag. 33", Voetius adds. However, I have only found a copy of the 1649 edition of Basso's, or Basson's, work. Sebastianus Basso, *Philosophia Natvralis adversus Aristotelem*, Amstelodami (Apud Ludovicum Elzevirium) 1649. Basso is one of Voetius' main targets in the essay on substantial forms, and, in Voetius' eyes, an author that had put forward ideas very similar to what Descartes and Regius are now presenting as new. See above, section 7.1, esp. note 3. In his *Natural Philosophy against Aristotle*, he attacks the idea of individual forms, favouring medical, anatomical and "mechanical" accounts of bodily operations. He introduces, however, a general animating cause in Nature, interpreted as an

recently invented by the Hermeticist Ernest Burchgrav and deservedly exploded by the great philosopher-physician Daniel Sennert in his book *On the Agreement and Conflict of the Chemists with the Followers of Aristotle and of Galen,*[102] or Socrates' little demon[103]; or that internal man, that is, the *alter ego*, or *Sofia* of Paracelsus, which the master[104] of these mysteries Osvaldo Crollius in his *Basilica Chymica* calls "the sidereal, Olympic, *Gabalis*, the human genius, the Within (*Penatin*), the shadow of the visible body, the private god, the private man (*hominem umbratilem*), the intimate little man of the Wise, the good demon, Paracelsus' internal *Adech*, the spectre, the light of nature, the foretelling and prophetic *evestrum*, the imagination which includes everything that is to become [and] all the stars and [is itself] every star, that is, the Astral Spirit, or virtues of the Angels etc.[105]

---

amalgam of ancient and Christian ideas. Thus, not only is the *spiritus communis* of the Stoics identified as Plato's world-soul; both are in fact Christianized in such a way that Basso allows himself to reformulate Holy Writ, changing the ultimate physical principle that the Christian Fathers took from *Genesis*, viz. "In principio creavit Deus coelem & terram", to: "In the beginning, God created *fire* and *earth.*" Basso, *op. cit.*, p. 307.

[102] Johann Ernst Burggrav's "lamps", or "lights" of life and death are hypothesized flames governing our fate either by burning steadily and thus securing tranquility of mind and body, or by flickering and smoking and thus causing anxiety and stress. At some disastrous point in time, a sanguine vital spirit blows out both flame and man. Daniel Sennert, in his *De Chymicorum cum Aristotelicis & Galenicis consensu ac dissensu*, merely retorts: "Si hæc non est Pyromantia, nescio, quid pyromantia sit. Et an non talis artifex Deo æquiparatur, qui solus corda & animi affectus scrutatur, & cui soli dies vitæ & mortis nostræ noti sunt?" Daniel Sennert, *De Chymicorum*, in *Opera* Tomus I, Parisiis, Apud Societatem, 1641, 1056^{a}E-1056^{b}A.

[103] In the *Syposium*, it is explained that a demon is a spirit that stands between God and man, and is the intermediate both of sorcery from below and of influence from above. Cf. Plato, *Symposium* 202 E - 203 A. For a scholarly and elegant account of the interpretation and *fortuna* of Socrates' demon in the Renaissance, see M. A. Screech, *Montaigne and Melancholy: The Wisdom of the* Essays, London (Duckworth) 1983 / London (Penguin) 1991.

[104] Ἐπόπτης: the highest grade of mystical initiation.

[105] Voetius' quotation is rather misleading. Using far too many commas and changing the order of words, he presents Crollius' list of names for the sidereal side of man as a nonsensical mess. I have rearranged the quotation at certain points in order to make more sense of it. In accordance with the view that every aspect of the macrocosm is reflected in the human microcosm, Oswald Crollius argues that man is made of two parts. From the Earth—and, mediately, from flesh and blood—man has acquired a physical, elementary, visible and tangible body; from the stars of the

Thus ends the long list of principles on which the fatal hour of death might depend if it does not depend on God's absolute decree. Voetius argues that it is not worth going into the matter any further, seeing that these are the alternatives. In fact, all these principles, says Voetius,

> are of no weight at all, so that to reveal today these delirious insanities which have long been disproved, is in fact more than enough to refute them.[106]

Voetius goes on to discuss related topics concerning the fixed end of human life, which need not concern us here.[107] The whole topic of

---

firmament however, an invisible, insensible, sidereal one. The latter is the active part: "instrumentum invisibilis, & incorporeus Astralicus [...] movet, gubernat & efficit artificia." There are many names for it: "Hinc sit, ut Sydereus, Internus, Olympicus, Incorporeus: Homo seu Gabalis idem sit cum Firmamento Astrorum ut jam sæpius dictum [...]. [Sydereum] corpus vocatur Hominis Genius, quia à Firmamento oritur, dicitur etiam Penatis, quod penes nos, seu nobiscum nascatur, umbra visibilis corporis, Lar domesticus, Homo umbratilis, Familiaris Sophorum Homunculus, Dæmon seu Genius bonus, Internus Paracelsi Adech, Spectrum, Naturæ Lumen, Evestrum præsagiens & Propheticum in Homine: Vocatur etiam Imaginatio, quæ in se claudit omnia Astra, estque omnia Astra, & eundem cum Coelo retinet cursum, Naturam atque Potentiam. Iam Astra [...] nihil aliud sunt quam Angelorum virtutes." Osvaldus Crollius, *Basilica Chymica*, Francofurti, impensis Godefridi Tampachij, 1611, pp. 35-36. According to Paracelsus, all bodies have *spirits*, which themselves are "astral bodies". The human intellect acquires real knowledge of the outside world only through a traffic of spirits, in which the *Evestrum* functions as a sort of intermediate spiritual universe. Imagination plays an important role in the Paracelsian philosophy since it is through the imagination that we can communicate with the spirits of material bodies and thereby influence them. Cf. Walter Pagel, *Paracelsus*, pp. 117-125 and 355. Human spirits may also be influenced themselves. See Walter Pagel, *From Paracelsus to Van Helmont*, London (Variorum Reprints) 1986, XII, p. 439 = "Van Helmont's Concept of Disease", in *Bulletin of the History of Medicine* 46 (1972), p. 439: "Every object is to Paracelsus, by virtue of its life, a 'spiritual being [...]' The spirit, the driving force in the body, cannot be apprehended by grasp or vision. It is liable to contract all diseases and to have and to bear them just as the body does." Since man's microcosmic sidereal part is reflected in the macrocosmic heavens of the stars, the latter may be consulted for prophesizing about sickness and death.

[106] Voetius, *De Termino Vitæ*, p. 47.

[107] Voetius also discusses the possibility that the occurence of death might not depend on any principle, but solely on itself, an opinion he attributes to Aphrodisius. Voetius, *De Termino Vitæ*, p. 47: "siquidem κίνησίν τινα ἀναίτιον" Aphrodisæus lib. 2. de Animâ." However, in that case, an existing thing would acquire its existence from itself, the future would itself determine its μελλουσία, i.e., determine what is to be, and potency and act would similarly be their own

death and of the possibilities of influencing the moment of its advent, may strike the modern reader as rather morbid. It should, however, be seen in relation to daily seventeenth-century life, in which high mortality rates and medical helplessness stirred people's interest in the question of influencing one's personal fate. The widely dispersed belief in magical powers is a consequence of these circumstances which has been noted before.[108] With this in mind, it is striking that Voetius does not discuss any ideas of practical magic. Although Voetius himself had had more than his share in mourning the deaths of relatives, he did not care for magical practices and, as we have seen, calls for pious resignation and submission to God's will without however neglecting medical care.[109] Still, in his list of alternative

---

effects. But since there would in fact be neither potency nor actuality, all existence, future and activity, being dependent solely on themselves, would stem from nothing. For Voetius, this means as much as that there would be something without a source, uncaused and by itself (ἄναρχον, ἀναίτιον, αὔταυτον) beside God, challenging, as it were, his independence. Fortunately, the difference between the two principles is that God is *Everything* and this other principle, by definition, *Nothing*. Again there seems nothing more to say. Yet Voetius indicates what consequences this idea might have. He associates uncaused natural events with "the twofold absurdity of the Manichaeans". This indicates to what extent the idea of ascribing natural change to some causal principle pervades Scholastic thought, if not indeed human thought at large. Even if the option is considered that there be *no* cause for a certain event, this is interpreted to mean that the event must then stem from some principle of nothingness, an independent force matching the selfsufficiency of the Prime Being: a shadow of God, a god of nothingness and evil, in other words, the dark root of Manichaeism. Here Voetius draws the line. In view of such consequences, we must stop and assign a genuine cause to all futurition or potential existence. Since the chain of causes cannot be infinitely continued, we ultimately must refer to God. Voetius next discusses and rejects the Jesuit theory of a conditional type of divine foreknowledge; a question intimately related to the discussions on premotion. See my "New Philosophy to Old Standards", pp. 67-78.

[108] In his classic work on *Religion and the Decline of Magic*, Keith Thomas warns against making "the anachronistic mistake of assuming that contemporaries were as daunted by [poverty, sickness, and sudden disaster] as we should be". Nevertheless, his survey of the medical and social uncertainties threatening sixteenth- and seventeenth-century Englishmen easily convinces us that magical beliefs, which "were all concerned to explain misfortune and to mitigate its rigour", were enlivened by the harshness of the times. Keith Thomas, *Religion and the Decline of Magic*, [...] (Weidenfeld & Nicholson) 1971 / London (Penguin Books) 1991. The quotations are from the 1991 edition, pp. 20 and 24.

[109] Even in the first quarter of the seventeenth century, Voetius had had to put up with the deaths of his youngest brother Paul, his elder brother Dirk and Dirk's wife and children, his colleague Johannes Slatius and his two children and of his own youngest child, all of whom died in various epidemics of the plague in the

factors that might influence the span of life, Voetius includes references to, for instance, Simon the Magician and to a wide range of Paracelsan and Hermetic ideas. He rejects these, but it is important to see that his rejection is not so much due to a rational or Reformist antipathy against the occult. What matters most in order to understand Voetius' list, is to see that it does not contain *means* for overcoming fate, but a range of possible *principles* underlying it. For all its fantastical ingredients, Voetius' long list of misconceptions and superstitions is nothing else than an enumeration of philosophical ἀρχαί. According to Voetius, the list of "absurdities" is merely an extension of Aristotle's well-known summaries of the δόξαι of philosophical opponents. Although all affirm that man's hour of death is fixed by some onderlying cause, they disagree "as to the why" (*in* τῷ διότι).[110]

A similar line of argument occurs in the essay on substantial forms. There, Voetius mentions alternative principles that might govern natural change not only in the context of his vindication of the causal efficacy of individual objects against Platonists, occasionalists and atomists, but also in a direct answer to the arguments of two contemporary critics of the forms. Sebastian Basso for instance, had argued that

> the ancients could easily show in what way, by what [means], and from what, forms arise. For they said that the soul and the form [...]

neighbourhood of Voetius' home town of Heusden alone. Voetius was further to outlive his daughter Maria and his sons Daniël and Paul. Personal grief will not have weakened Voetius' orthodox position in matters of faith, given his conviction that only the Calvinist notion of pre-election may offer hope and consolation to the faithful. For an account of Voetius' ideas of providence and pre-election in the light of religious practice, see C. Graafland, *Van Calvijn tot Barth: Oorsprong en ontwikkeling van de leer der verkiezing in het Gereformeerde Protestantisme*, 's-Gravenhage (Uitgeverij Boekencentrum B.V.) 1987, pp. 232-231. However, Voetius' personal sufferings offered his enemies ample evidence for their malignant view that this must be God's way of pronouncing judgement over the Utrecht theologian. Cf. C. Steenblok, *Gisbertus Voetius*, p. 59.

110 Voetius uses the Greek terminology at *De Termino Vitæ*, p. 11 for instance. Posing the question whether the duration of life is "fixed, established, certain and immovable because it is immutably determined by God", he distinguishes *duæ hic sententiæ*: "una negat, altera ait. Posterior in duas dispescitur, quæ quamvis in τῷ διότι h.e. in assignandâ ratione & causâ certitudinis & immobilitatis dissentiant, in thesi tamen principali seu in τῷ ὅτι conveniunt."

> consist of a certain arrangement and proportion of the parts of an object.[111]

David Gorlæus, moreover, argues that entities should

> not be multiplied without necessity, since the effects of natural objects can be sufficiently explained by, and reduced to, other principles.[112]

Yet, Voetius retorts, if we inquire as to what the ancients have had to offer we are at loss what principles to chose. Indeed, as for the effects of natural causation, the ancients

> that Basso praises and Aristotle refutes in Book II of the *Physics*, explain them in one way, modern philosophers in another.

Voetius says that he does not at this point wish to discuss "the loathsome and long discredited [opinions] of both old and modern Paracelsans and Hermeticists", but merely the Cartesian idea that everything derives from motion, rest, quantity, shape, and position or situation, and that all secrets of nature can be explained by them. However, he adds that most of the alternatives to the Aristotelian theory of causation have some common aspect, the rejection of which leads to the "collapse" of all alternative viewpoints.[113] This common element of all non-Aristotelian philosophies is not further explained, but there is no doubt that it should be sought in the absence of individual and internal principles of change. It is this aspect of Aristotelian philosophy which is presented in the first books of the *Physics* and which is defended by Voetius in its Renaissance form. Alternative principles are all alike in as far as they do not accept Aristotle's individual and internal principle of natural change.

In any case, as far as Voetius is concerned, the alternative philosophical principles of Basso and Gorlæus have nothing new to

---

[111] Voetius, *Narratio*, p. 44 / *Select. Dispp.* I, pp. 875-876 / *Responsio*, p. 23 / *Querelle*, p. 109. See above, chapter 7, note 3.

[112] See above, chapter 7, note 24.

[113] Voetius, *Narratio*, p. 44 / *Select. Dispp.* I, p. 876 / *Responsio*, pp. 23-24 / *Querelle*, p. 109: "Aliter enim veteres, quos laudat Bassonis & refutat Aristoteles lib. 2. *Akroas.* aliter recentiores [effectus rerum naturalium explicant]: quamvis pleræque sententiæ in uno aliquo communi conveniant, quo negato aut refutato, & ipsæ concidunt. Nos rancidas & explosas jampridem tum veterum tum recentium Paracelistarum, Hermeticorum [...] nunc non refodiemus: sed solummodo demonstrari petimus eam quæ hodie aut heri emersit statuentem à quantitate, figurâ, situ seu positurâ, motu, quiete, omnia derivari, omniaque arcana [naturæ] optimè per ea explicari ac demonstrari posse: quod nos negamus."

offer. They are merely footnotes to the long history of Platonist, atomist and magical attempts of finding alternatives to the Aristotelian ἀρχαί. However, says Voetius,

> We shall not here unearth the loathsome and long discredited [opinions] of both old and modern Paracelsans and Hermeticists [...], but merely aim to describe the [theory] which has suddenly emerged and in which it is held that everything derives from quantity, shape, position or situation, motion [and] rest, and that all secrets of Nature can be perfectly explained and demonstrated by them—which we deny.[114]

Instead of Basso and Gorlæus, it is in fact Regius and Descartes whom Voetius is interested in. Thus, Voetius implicitly argues that it is the Cartesian principles of mechanicism which may be added to the long list of oddities that the history of philosophy and magic has had to offer.

9.6.3 The Danger of Cartesianism

If we are to evaluate Voetius' primary reaction to Cartesianism, we must keep in mind that to the judgment of Scholastic philosophers, the principles of the mechanical philosophy by definition belong to the list of philosophical misconceptions. Their philosophical merit is, however, not what troubles Voetius most. The Cartesian idea of a world-machine is also blasphemous, since it is based on a conception of causality which it is hard to bring into line with theology. The God of Christianity allows natural causes to operate in submissive concurrence to the action of the Supreme Being. Given the postulate that natural change should be explained by the ascription of action to causal powers responsible for the change, the Aristotelian philosophy was, in Scholastic eyes, seen as the middle course between the occasionalism that introduced a God without an active Nature (a Prime Cause without secondary causes), and the materialist view of an independent Nature, the operations of which would remain inexplicable if it were not for animistic or other types of external motors which alternative philosophies would introduce.

Secondary causality and divine concurrence therefore went hand in hand. In fact, according to Voetius, the idea of divine concurrence either in its co-operative or its premotionist form, could only make

[114] Voetius, *Narratio*, p. 44 / *Select. Dispp.* I, p. 876 / *Responsio*, p. 23 / *Querelle*, p. 109.

sense in the case in which the Prime Agent can exert His influence on an independent causal faculty of the secondary agent. In the medical case for instance,

> The relation between God and man is that God, through His providence, carries out His decree and that in doing so, He uses man as [His] means and instrument.[115]

But the instrument must *act* itself. In fact, so crucial was the idea of the substantial form, that Voetius thought it was the only way to avoid assisting forms and powers besides God. This means that the theory of divine concurrence is dependent on a theory of causality in which there is something for the Prime cause to co-operate with. The idea of concurrence is thus restricted to a philosophy in which individual sources of action are accepted. This is exactly what Aristotelianism had to offer: an ontology of individual natures which are at the same time individual centres of activity.

9.7 Conclusion

In this light, Cartesianism could only be regarded as a step backwards. The clockwork conception of Nature reintroduced a type of materialism which not only denied any activity of secondary causes, but also the idea of a divine Hand helping individual virtues to become effective. The question whether God, in concurring with secondary causes, immediately excited their action by way of a previous movement, or only co-operated with them in order to produce their effects, was of great theological importance. The question of divine concurrence as such was, however, no less related to a religious world-outlook. If the natural virtues of secondary causes were rejected, how could natural motion be understood? A re-introduction of separate intelligences or, even worse, of the God of "occasionalism" might be expected. That would either undermine God's omnipresence, or lead to a crude fatalist philosophy like that of the Islamic Kalâm. And what was the need for such alternatives, when the sane and sober philosophy of the Schools taught a view of Nature which agreed with both common sense and Christian faith? In any case, if David incited the faithful to "sing unto the LORD,"

> Who covereth the Earth with clouds, who prepareth rain for the Earth, who maketh grass to grow upon the mountains,[116]

---

[115] C. Steenblok, *Gisbertus Voetius*, p. 32.

it was not merely for making the Earth or for preparing its effects by way of creating well-directed and well-functioning independent clockwork. Neither is the Lord praised for His continued creation of every one of the Earth's parts from every moment to the next. When we praise Him for preparing the rain and for making grass to grow upon the mountains, we praise Him at least for His *immediate* co-operation with every motive source existing in the Universe. Such help can only be given if there are faculties in need of it. Such faculties were thought to be provided by the "Natures and Substantial Forms of Things" which Voetius set out to vindicate against the New Philosophy. According to Aristotle's Renaissance interpreters, the concurring God of Christianity could not do without the philosophy of internal principles of motion.

---

[116] *Psalms* 17:8 and Conimbricenses, *Commentarii in octo libros physicorum* I, p. 276.

CHAPTER TEN

CONCLUSIONS: THEOLOGY BETWEEN VITALISM AND MECHANISM

In reaction to Voetius' essay on substantial forms, Descartes wrote the following words to his Utrecht advocate Henricus Regius in January 1642:

> I should like to explain in what way automata are also works of Nature and how it is that men, in making them, do nothing other than apply activity to what is passive (*applicare actiua passiuis*), as they also do when they sow wheat or breed mules; which does not bring about any essential difference, but only such as are laid down by Nature.[1]

Reading these lines, one cannot but think that Descartes is fully aware of the theoretical difficulties the example of mules raised for an essentialist interpretation of the Aristotelian theory of forms. Is he making fun of the Scholastic idea of a limited set of natural species ordained in the beginning by God? Even if he is not, the breeding of mules still forms a distinctive example of an occurrence of human craftsmanship in which "no essential difference is brought about." For Descartes, the occurrence of mules offers no problem in itself. The mule is just as much an automat, or a clockwork of divine arrangement as are, for instance, the human body, comets and wheat. No "first root of muleness" need trouble the mechanical philosopher.

The quotation forms an interesting summary of what is at stake in the Utrecht Crisis in yet another way. Descartes maintains that in breeding mules, mankind "applies activity to what is passive". The point is clearly that in fabricating clockwork one makes use of natural properties in no other way than one does in the case of agricultural production. In both instances, natural means are applied in such a way

---

[1] Descartes to Regius, January 1642, AT III, p. 504 / *Rodis-Lewis*, p. 88. In the *Responsio*, Regius transforms the argument slightly, saying that applying activity to what is passive, sowing wheat, breeding mules etc., brings about only "a difference introduced by ***reason***": "*à ratione*", instead of "*à naturâ* inductam". Regius, *Responsio*, p. 18. It may be just a slip of the pen. The argument in any case remains the same: according to Descartes and Regius, there is no essential difference between works of art and works of Nature.

as to produce results that would not follow did man not intervene. The phrase *applicare actiua passiuis* might be reformulated in Aristotelian terms as applying activity to what is conditionally, or hypothetically necessary. In *Physics* II 9, Aristotle explains that there is necessity in Nature in the sense that, in order for certain results to come about, particular—especially material—conditions are *necessary*. Likewise, in sowing wheat there are conditions on which a successful harvest depends. These are primarily material conditions, such as the nature of the soil in which the wheat is supposed to grow. However, whilst according to Aristotle it is Nature which, in teleological causation, makes use of the necessary conditions for natural change, in Descartes' example it is the human agent, in this case the farmer, who applies activity to what is conditional. This makes all the difference. Descartes' formulation is entirely clear as to the point that what is applied by human activity is, in itself, *passive*. This is exactly where the Cartesian and Aristotelian positions diverge.

Descartes illustrates the idea that natural and artificial objects are of one and the same kind by arguing that human practice is identical regarding both: whether we breed mules or build automata, we merely "apply" passive conditions in such a way as to bring them to the desired result.[2] For Aristotle, the possibility of such "active" administration of what is "passive" in human conduct, was a reason to search for a similar kind of active principle in Nature too. Yet according to Descartes, to do so is to confuse mental and physical realms. In the "passive" clockwork of Nature, only mankind is actively involved. Mankind, that is, and of course God, whose splendour is reflected in the incomparable way in which He has applied the "passive" mechanical properties of matter. It is in the same passage that Descartes emphasizes the *gradual* difference between art and Nature:

> since those few wheels in a clock can in no way be compared with the innumerable bones, nerves, veins, arteries etc. of the smallest animal.[3]

As we have noted before, the paradox of mechanicism is that the machine-metaphor was defended on the very grounds on which it was

---

[2] Note that by "applying activity to what is passive", Descartes does not merely mean "applying materials", but, especially, applying the regularities that pertain in Nature. Living creatures as well as automata and clockwork, both when broken and when working properly, all according to the same regularities, or "laws of Nature". See also above, section 8.2.2.2.

[3] Descartes to Regius, January 1642, AT III, p. 504 / *Rodis-Lewis*, p. 88.

rejected by adversaries. Nature's splendour works as well for denigrating man as for praising God as the *artifex mundi*.

What I have tried to make clear in this book is that even if Descartes had convinced Voetius of his sincere intention to praise the glory of the Creator, this would not nearly have sufficed to win over the theologian to the mechanicist's cause. At the heart of Voetius' criticisms of the New Philosophy lies the conviction that the mechanical viewpoint cannot account for the causal relations between God and Creation. Neither Descartes nor Regius seems to have been fully aware of this fact.

10.1 God and Nature

Defending the mechanistic interpretation of Nature against Voetius objections, Descartes, in his letter to Regius, goes on to criticize Voetius' quotations from the Bible.[4] Descartes tells his Utrecht friend that Voetius' way of quoting the Bible is only a malicious trick. The Biblical verses referred to do not prove anything at all. Accordingly, Descartes advises Regius to quote them in full, so that everybody may see for themselves.[5]

Now the verses Voetius refers to may at first sight seem to be rather inappropriate.[6] However, *Psalm* 104:29, *Numbers* 16:22 and 27:16, and *Habakkuk* 2:19 relate of the breath of living man, which

---

[4] Descartes to Regius, January 1642, AT III, p. 504 / *Rodis-Lewis*, p. 88. Cf. Voetius, *Narratio*, p. 41 / *Select. Dispp.* I, p. 874 / *Responsio*, p. 13 / *Querelle*, p. 107. The quotations are from *Psalms*, *Numbers*, *Hebrews* and *Habakkuk*. See note 6, below.

[5] Descartes to Regius, January 1642, AT III, p. 504 / *Rodis-Lewis*, p. 88.

[6] In the King James version, the verses read as follows. *Psalm* 104:29 "Thou hidest thy face, they are troubled: thou takest away their breath, they die, and return to their dust." *Psalm* 104:7 "At thy rebuke [the waters that stood above the mountains] fled; at the voice of thy thunder they hasted away." *Psalm* 104:14-15 "He causes the grass to grow for the cattle, and herb for the service of man: that he may bring forth food out of the Earth; And wine that maketh glad the heart of man and oil to make his face to shine, and bread which strengtheneth man's heart." *Numbers* 16:22 "And they fell upon their faces, and said, O God, the God of the spirits of all flesh, shall one man sin, and wilt thou be wroth with all the congregation." *Numbers* 27:16: "Let the LORD, the God of the spirits of all flesh, set a man over the congregation." *Hebrews* 11:9-10 "By faith [Abraham] sojourned in the land of promise, as in a strange country, dwelling in tabernacles with Isaac and Jacob, the heirs with him of the same promise: For he looked for a city which hath foundations, whose builder and maker is God." *Habakkuk* 2:19: "Woe unto him that saith to the wood, Awake; to the dumb stone, Arise, it shall teach! Behold, it is laid over with silver, and there is no breath at all in the midst of it."

can be taken away by God; of God as the God of spirits; and, finally, of dumb and inanimate materials. Clearly, Voetius' intention is to take these verses as Scriptural proof for the view that *animata* should be sharply distinguished from *automata*, which would already explain Voetius' reference to them in the context of comparing art and Nature. The other verses generally praise God's glory and benevolence, but in what way exactly? *Hebrews* 11:9-10 refers to Abraham's faith in a destiny promised to him by God. What is promised is a "city which hath foundations, whose builder and maker is God." According to the Latin, this is a city *cuius artifex et conditor Deus*—which is an accepted way of referring in philosophical terms to God as designer and creator of the world. Of course, this could be read in Cartesian terms, making God a "mechanical craftsman". As far as Descartes and Regius are concerned, it merely depends on which view one takes. And again, the paradoxical situation is that both proponents and adversaries of mechanicism may point to God's Creation and appreciate it differently.

But now look at the other texts mentioned. In *Psalm* 104, verses 14-15[7] expound that God produces (Latin: *producens* [*te*]) grass and herbs for the sevice of man:

[7] There has been some doubt about which verses Voetius refers to. The 1648 edition of Voetius' essay does not mention *Psalm* 104, verses 7, and 14-15 at all, but mentions *Psalm* 104, 29 and *Psalm* 7:14-15 respectively. Voetius, *Select. Dispp.* I, *Index Sacræ Scripturæ*. The reason for this must be, that the text refers to "Ps. 104 29. & 7. 14. 15.", which would indeed be an akward way of referring to *Psalm* 104:7, 14, 15, 29. *Psalm* 7:14-15 however, although it does exist ("Behold, he travaileth with iniquity, and hath conceived mischief, and brought forth falsehood. He made a pit, and digged it, and is fallen into the ditch which he made"), does not seem to make any sense in this context, since Voetius is not here discussing the attitude of the mechanicists, nor does he use the verses referred to in a rhetorical way, comparing his rivals with the unrighteous of the Biblical verse. The verses are supposed to directly contest the mechanical view of natural objects and artifacts: "quod quomodo cum Psal 104 29. & 7. 14. 15. Numer. 16. 22. & 27. 16. Hebr. 11. 9. 10. Habac. 2. 19. satis conveniat, fateor me nondum videre." Voetius, *Narratio*, p. 41 / *Select. Dispp.* I, p. 874 / *Responsio*, p. 13 / *Querelle*, p. 107 (the *Responsio* does not refer to *Habbakkuk*). *Psalm* 7:14-15 differs from all the other in not mentioning the topics of Nature, art, life, spirit, unliving Nature etc. in any way. Theo Verbeek accordingly discusses only *Psalm* 104:29 in his notes to the text of the *Narratio*, leaving *Psalm* 7:14-15 aside. See *Querelle*, p. 471, note 70. *Psalm* 104:7 and 14-15 on the other hand, *do* take up the subject-matter of Nature against art. Yet the decisive argument in favor of *Psalm* 104:7, 14-15 and 29, is the fact that it is these verses which Regius in his *Responsio* replies to, even though Regius himself repeats the akward notation of "Psal 104 29. & 7. 14. 15." Since the text of Voetius' essay was publicly defended in the tumultuous

> He causes the grass to grow for the cattle, and herb for the service of man: that he may bring forth food out of the earth; And wine that maketh glad the heart of man and oil to make his face to shine, and bread which strengtheneth man's heart.

God causes natural phenomena for the well-being of mankind. The verse endorses exactly the kind of final causality that we examined in chapter 3 and that, as we saw in chapter 5, Descartes was willing to accept as a sign of piety, but refused to admit in physics.[8]

In December 1641, Voetius could surely not have known of Descartes' intentions to completely ban such final considerations. What matters, however, is not so much that we estimate the measure of Voetius' knowledge of Cartesianism as that we explain what Voetius could have meant by the quotations neither Descartes nor Regius took seriously. It is very significant that they illustrate the opposition between artificial and natural objects not merely by pointing to the spiritual properties of living things, but by referring to what Martin Schoock would call "relative finalism", that is, the description of natural phenomena as forming part of a meaningful and purposeful network of interdependent causes working for man's benefit alone. The Scholastic idea of natural causation simply includes this aspect, and Voetius' Biblical quotation neatly represents it as much as it does another aspect, namely that of concurrence. As we saw in chapter 9, according to Scholastic theory, God not only planned the way in which secondary causes work and co-operate with each other; God also works and co-operates along with them Himself. This is reflected in the remaining quotations to which Voetius refers:

---

disputationes of 18, 23 and 24 December 1641 by Lambertus vanden Waterlaet, and since, in his later reaction to Regius, Vanden Waterlaet does not oppose to Regius' choice of verses, but only to his interpretation, we may safely conclude that it is in fact the verses 7 and 14-15 of *Psalm* 104 that Voetius intended. Cf. Regius, *Responsio*, pp. 13 and 18-19. Regius refers to the numbering *secundum Hebrœos*, which is the same as that of King James Version, which we have cited. Regius' quotations however, are from the Septuagint. As was said above, Regius does not refer to the *Habbakkuk*-text, nor does he quote *Psalm* 104:15 in full. See further: Lambertus vanden Waterlaet, *Prodromvs sive Examen Tutelare Orthodoxœ Philosophiœ Principiorum* [etc.], Lvgdvni Batavorvm, Excudebat Wilhelmus Christiani, 1642, pp. 62-64.

[8] Descartes, *Principia* III 3, At VIII-I, p. 81 / CSM I, p. 249. See section 5.2.2, above.

> Thou hidest thy face, they are troubled: thou takest away their breath, they die, and return to their dust.[9]

And:

> At thy rebuke [the waters that stood above the mountains] fled; at the voice of thy thunder they hasted away.[10]

Apart from referring to living Nature, these verses are in fact a literal restatement in Biblical terms of the philosophical idea of a divine sustenance, by which God conserves and co-operates with Nature. If He were to cease to do so, in other words, if God were to cease helping Nature with his existential and productive *influx*, Nature would be lost.

The conclusion must be, that far from being irrelevant, Voetius' quotations summarize every single aspect of the Scholastic position quite sharply. In fact the whole complex of ideas of individuality, finality and concurrence, that is, the complete subject-matter of what Scholastic commentators read into Aristotle's *Physics* I and II, is present in the few Biblical verses with which Voetius confronts the New Philosophers. Although he may not yet have known all the ins and outs of Cartesian philosophy, Voetius clearly saw what would be at stake with regard to the old philosophy if the concept of substantial form were to be rejected.

Theology, according to Voetius, needed the Aristotelian concept of the forms. In his defence of the premotionist view of concurrence and in his attacks on the Jesuit theory of God's conditional foreknowledge, the Utrecht theologian generally reduced rather than affirmed the autonomy of secondary causes.[11] His position may seem inconsistent. The seventeenth century would eventually see examples of thinkers embracing the mechanical view for exactly the reason that it might be employed to lay the physical foundation for the theological idea of divine sovereignty.[12] To Voetius, the loss of individual causation must

---

9 *Psalm* 104:29.

10 *Psalm* 104:7. See also above, note 7.

11 See my: "New Philosophy to Old Standards: Voetius' Vindication of Secondary Causality and Divine Concurrence", *passim*.

12 It was not only the occasionalists who shared this idea. Pierre Bayle for instance, held that action attributed to other than intelligent forces should be rejected precisely because it would make redundant any activity of God. Cf. E. Labrousse, *Pierre Bayle: Hétérodoxie et Rigorisme*, The Hague (Nijhoff) 1964, pp. 219-247. J. E. McGuire, in his article on "Boyle's Conception of Nature", has argued that there is a direct link between voluntarist theories of divine causation on

have seemed more troublesome than the gain of a mechanical sovereignty of Creator over Creation. Only in this way could ethical and doctrinal theses be explained in philosophical terms. In the essay on forms, Voetius asks how, if one rejects qualities, there can be any *habitus*:

> I should like to see how those who deny the existence of natural faculties or potencies could defend [the concept of] *habit*—the necessity of which is urged by Scripture and reason alike—against the Atheists, Sceptics and infidels, granted that they do not also compose

---

the one hand and the acceptance of nominalism and mechanical philosophy on the other. Cf. J. E. McGuire, "Boyle's Conception of Nature", in *Journal of the History of Ideas* 33 (1972), pp. 523-542. An example may be found in Pierre Gassendi, whose support of the Epicurian philosophy cannot be separated from his voluntarist theology. See e.g. Margaret J. Osler, "Fortune, fate, and divination: Gassendi's voluntarist theology and the baptism of Epicureanism", in Margaret J. Osler (ed.), *Atoms, Pneuma, and Tranquillity: Epicurean and Stoic Themes in Euopean Thought*, Cambridge (C. U. P.) 1991, pp. 155-174. Yet the theological position is not always straightforwardly linked to the philosophical. As Martial Gueroult argues, Malebranche for instance, did not share the idea of the Mutakallimûm that all action must be attributed solely to divine Will. In this case, the positions of such divers thinkers as Aquinas and the Mutakallimûm are interpreted along the same line, whilst the occasionalistic position of Malebranche is in fact seen as the one that vindicates the efficacy of Nature! See Gueroult, *Malebranche*, Tôme II, pp. 109-110: "A la question: 'Pourquoi le feu sèche-t-il le ligne' il refuse d'accord avec saint Thomas de répondre: 'Parce que Dieu le veut.' Il s'oppose radicalement à leur affirmation que tout découle de la simple volonté de Dieu 'sans aucune autre raison'." Gary B. Deason, in his "Reformation Theology and the Mechanistic Conception of Nature", universalizes the idea of mechanistic tendencies in Protestant thought and traces the influence of a Reformist theology emphasizing divine sovereignty on what culminated in the eighteenth century as a "natural religion". It is striking that such an interpretation of the mechanization and "Epicurization" of philosophy along Reformist lines is so completely at odds with Voetius' position. Or is it? Gary Deason concludes that the "Universal Ruler" of Boyle and Newton is in fact a long way away from Luther's peronal God: "Why the difference?", Deason asks, and answers that the difference of questions led to a difference of meaning between the Reformers and the mechanists: "While the technical dimensions remained the same, the context changed, and so did the meaning." Gary B. Deason, "Reformation Theology and the Mechanistic Conception of Nature", in David C. Lindberg and Ronald L. Numbers (edd.), *God and Nature: Historical Essays on the Encounter between Christianity and Science*, Berkeley (University of California Press) 1986, pp. 167-191, the quotation is from p. 187. Voetius however, did not wait to see the meaning of divine concurrence change, but acted as soon as he saw the danger.

> it out of that pentagonal prison of motion, rest, quantity, position and shape.[13]

The habitual disposition of rational creatures that have faith, that sin, etc., may only be explained in philosophical (we might say "scientific") terms, when philosophy accepts concepts that may apply to both the realms of inanimate and of rational beings.[14] Voetius saw the philosophy of forms and qualities as the only basis for a theology in which both natural and moral action could be explained in philosophical terms. Such a theology would, moreover, escape the animistic fallacy as long as it was based on Aristotelian foundations. Voetius' theology, which in this aspect is straightforwardly Thomistic, is a theology requiring a philosophical basis that at once vindicates the total sovereignty of God and the individual efficacy of secondary causes. It is not mechanistic, since it accepts individual forms. But neither is it animistic. Indeed, it is rather the mechanical theories that are in danger of having to accept spiritual or mystical external motors of natural change. The co-ordination in causal terms of all secondary causality, both conscious and unconscious, living and dead, and the whole idea of a premotionist divine concurrence, needed a philosophy that accepts individual forms and natures.

10.2 Explaining Physical Change

All this was challenged by Descartes. Yet neither Voetius nor Schoock interpreted Regius' and Descartes' ideas as offering anything radically novel. On the contrary, Cartesianism was presented with arguments that had been raised against other non-Aristotelian philosophies and was seen as merely a new variation on age-old Platonic, mystical and atomist ideas. Voetius' and Schoock's attitude toward Cartesianism may be compared to that of Daniel Sennert with regard to the Hermetic magician Ernest Burchgrav, which we mentioned in the previous chapter. Commenting upon Burchgrav's conception of an inborn "lamp of light and death", Sennert does two things. First, he

---

[13] Voetius, *Narratio*, p. 41 / *Select. Dispp.* I, p. 873 / *Responsio*, pp. 12-13 / *Querelle*, pp. 106-107.

[14] As Theo Verbeek remarks, religion and science were themselves examples of "habitual dispositions": "[d'après] la philosophie scolastique, la science et la foi etaient des 'habitus'ou 'habitudes'." See *Querelle*, p. 471, note 68, where Verbeek also points to Jean-Luc Marion's analyses of the Aristotelian use and of Descartes' rejection of the concept of *habitus*. Jean-Luc Marion, *L'Ontologie grise de Descartes*, Paris (Vrin) 1971, pp. 25-30.

ridicules the philosophical merit of this freely invented hypothesis. Secondly, he criticizes the idea in the light of divine administration.[15] Roughly speaking, when confronted with the rise of Cartesian philosophy, Schoock takes the first strategy, Voetius the second.

As we have noticed, Schoock is often concerned, in the *Admirable Method of the New Philosophy of René Descartes*, with making suspect the Cartesian method of hypothesizing corpuscular models of explanation. Schoock's project in much of the *Admiranda* is in fact to point out the connexion between the invention of mechanical models on the one hand and, on the other, the subjective criterion of evidence that is introduced to make such models acceptable. Schoock thereby exposes Cartesian epistemological, methodological and metaphysical ideas as so many instruments for making his corpuscular theories acceptable. This is not without reason. In fact, Descartes uses the same argument regarding the divine guarantee for the truth of what is clearly and evidently perceived in his metaphysical and in his physical works.[16]

Our analysis of Cartesian thought in the light of Scholastic commentaries has revealed other important aspects of Cartesian philosophy. The distrust of textual authorities that Descartes displays in the *Discours*—and which Schoock interprets as an easy way out of long-standing discussions on the right choice of physical principles—may tell us more about the changing appreciation of textual authority that divides the old from the new, the Scholastic from the Modern. A seventeenth-century "modern" could well, and abundantly, make use of ancient thought, but contrary to his typical Medieval or Renaissance counterpart, a *modernus* of the early modern period would think his philosophical ability excelled over that of former discussions in a way which made the older texts no longer a sufficient background for defining his own position. Schoock's criticisms may, in other words, explain aspects of Cartesian philosophy which—after centuries of having been read as context-free metaphysical meditations—may elude the present-day reader.

By re-evaluating the impact of the Cartesian plea for the reduction of all physical principles to mechanical ones, we have brought together such paradoxical aspects of Cartesian method as, on the one hand, the obvious use of observation and experiment in the works of Regius and Descartes, and, on the other, the defence of *a priori*

[15] See chapter 9, note 102, above.

[16] See above, chapter 7, notes 53 and 59 especially.

deductions in physics and the rejection of commonsense reliance on sense-perception. These aspects are less paradoxical once they are confronted with the traditional philosophy and its methodology. Cartesian *a priori* deductions, far from playing down the importance of experienced phenomena, function as a self-sufficient alternative to rival principles of external causation. The Cartesian *a priori* is a chain of natural causes in the Universal clockwork—not an intuitive deduction of any kind. Also, as a rule of method, the Cartesian rejection of the senses does not involve a rationalistic ban on observation and experience. The *détachement des sens* should rather be interpreted in the light of conflicting ideas concerning the ontological status of sensible qualities and the contrast between Descartes' mechanistic interpretations of natural processes on the one hand and the commonsense explanations of the School philosophy on the other.

So much for understanding Descartes. In order to understand his adversaries, one must first realize that Martin Schoock regarded the five Cartesian principles of motion, rest, quantity, position and shape as so many hypotheses not proved by any fact. This point is already brought forward by Voetius, who, in the essay on forms, confronts the New Philosophy time and again with the limitations of mechanical explanations, and in particular with the fact that there are so many natural phenomena we are unable to explain:

> why are some persons in such a way affected by the presence of cats (which they do not know to be present), that they almost faint; and [why are] others, when they unknowingly eat a piece of cheese, agitated in such a way that they contract a rather serious and dangerous illness from it?[17]

Voetius is pessimistic about being able to explain such phenomena as allergies and sicknesses, and is especially pessimistic about describing them in terms of particle motion. Moreover, after writing a paragraph on the merits of accepting one's learned ignorance, he concludes the essay on the forms with the unassuming recognition that "there is so much that we do not know".[18]

---

[17] Voetius, *Narratio*, p. 46 / *Select. Dispp.* I, p. 878 / *Responsio*, p. 28 / *Querelle*, p. 111.

[18] Voetius, *Narratio*, p. 51 / *Select. Dispp.* I, p. 881 / *Querelle*, p. 115 (not in the *Responsio*). For a discussion of "learned ignorance", see Theo Verbeek, "From 'Learned Ignorance' to Scepticism; Descartes and Calvinist Orthodoxy", in Richard H. Popkin / Arjo Vanderjagt (edd.), *Scepticism and Irreligion*, pp. 31-45.

Yet neither Voetius nor Schoock is concerned with holding back scientific progress. Nor does the idea of explaining unusual events such as earthquakes, plagues and comets as effects of natural causes—and not as portents or signs of divine wrath—in any way play a role in the Utrecht Crisis. On the contrary. Descartes' thought has not without reason been typified as an emancipation from occultism.[19] Yet in his attempt to ban Hermetic, Paracelsan and other magical ideas from philosophy, Voetius was in fact much more fanatical. Of course, the essay *On the Natures and Substantial Forms of Things* is a special Thesis written in defence of "occult qualities". The argument is, however, first and foremost an admonition to accept one's learned ignorance and a caution against intellectual arrogance and *pansofia*.[20] Secondly, Voetius' favours the accepted doctrine of occult qualities, precisely for the reason that preposterous magical hypotheses may thereby be avoided. Had not the very learned Daniel Sennert concluded, in his his book *On the Agreement and Conflict of the Chemists with the Followers of Aristotle and of Galen*, that when philosophers try to explain occult qualities:

> they come up with arguments which are insufficient and ridiculous, or deny what experience in fact confirms[?][21]

Rather than as a support for occultism, Voetius' defence of occult qualities should be read as a defence of the "sane and sober" qualitative philosophies of Aristotle and Galen, and as a warning against alternative principles of natural change. From Voetius' point of view, it is rather the mechanical philosophy that may be compared, not only with the ancient philosophers Aristotle refuted in *Physics* II, but with the modern Paracelsans and Hermeticists as well. In fact, all non-Aristotelian causal explanations amount to more or less the same thing.

---

[19] See e.g. John Cottingham, *Descartes*, Oxford (Basil Blackwell) 1986, pp. 22-23.

[20] Voetius, *Narratio*, p. 46 / *Select. Dispp.* I, p. 877 / *Responsio*, p. 28 / *Querelle*, p. 111: "melius convenit [sc. sententium de qualitatibus occultis] cum doctâ ignorantiâ [...] quam sententia opposita, quæ pansofiam, & quidem absque difficultate, sperare & promittere videtur: Quod non potest non esse suspectum. Videantur interim dicta scripturæ Iobi 38. & 39. & 26. & 42. Eccles. 1. Ps. 92."

[21] Voetius, *Narratio*, p. 46 / *Select. Dispp.* I, p. 877 / *Responsio*, p. 28 / *Querelle*, p. 111 and Daniel Sennert, *De Chymicorum*, in *Opera* Tomus I, Parisiis, Apud Societatem, 1641, cap. 8, "Hypomnemata Physica".

In the last chapter, we quoted Voetius' list of alternative moving principles. Voetius refers to a book by Plutarchus as "de placit. libr." Although the reference is troublesome,[22] the title is interesting in itself. The term *placita* refers to the opinions of the philosophers. As Seneca explains, it is the Latin term for what the Greeks held to be the *dogmata*.[23] The term may mean "principle" in a very broad sense.[24] It may also refer to the philosophical *principles* in the technical sense of causal determinants of natural change. This in fact is what the lists of *placita* of natural philosophy consists of. Not only for Scholastics as Voetius, but for instance for sceptics like Montaigne too, lists of opinions are lists of *principles*: where Aristotle has his matter, form and privation, Plato has his Ideas, his world-soul and his great-and-small, Pythagoras his numbers, Epicurus his atoms, and so on. What divides the philosophers in their opinions concerning natural philosophy, is the motive forces which they introduce. The quest for principles led some, like Montaigne and, to a certain extent, Danæus, to accept the sceptical conclusion that the right principles would never be found. Others, such as Voetius and Gassendi, made a specific choice of principles from the treasury of classical philosophy. Voetius thought Aristotle to be in line with theology as well as physics—Gassendi chose Epicurus. Others still, such as Basso and Zanchius, accepted a conglomerate of principles from the ancient stock, all supposedly matching the ideas of Scripture. It was only Descartes who went his own way.

The introduction of the concept of inertia in the seventeenth-century changed the doxography of natural philosophy for good. No longer could natural philosophy remain the battleground for the philosophic war of the various δόξαι, or *placita*, explaining natural change by referring to either forms, motive forces, geniuses or other principles. The varying opinions of philosophers were exchanged for the mathematical descriptions of dynamics. This development, however, does not play a role in the Utrecht Crisis. As we saw in chapter 5, Descartes did not engage himself in the project of mathematization. His refusal, on the other hand, to discuss substantial forms and real qualities, indicates that Descartes did not want to commit himself to a discussion concerning the right choice of causal principles. His

---

22 See above, chapter 9, note 86, above.

23 Seneca, *Ad Lucilium Epistulæ Morales*, Epistula XCV 10: "nulla ars contemplativa sine decretis suis est, quæ Græci vocant dogmata, nobis vel decreta licet appellare vel scita vel placita, quæ et in geometria et in astronomia invenies."

24 See e.g. Seneca's use of the term at *Epistulæ* LXVI 45 and XCV 37.

physics entirely depends on an image of the material world as a gigantic mechanical device. His causes are external causes. Descartes does show a certain allegiance to the Greek requirement of ἀπόδειξις, or demonstration, by promising to give a full account of material causation—which, in Cartesian physics, would involve the deduction of all natural phenomena from an *a priori* "evolutionary" reconstruction of the world. Descartes' physics is, in other words, still very much dependent on a metaphor visualising causation. His refusal, however, to discuss the ancient principles, and his exclusion of individual motive forces, dramatically changed the concept of causality itself. The "why" of physical change was henceforth answered by referring to the disposition of material parts within the clockwork and to the initial processes to which its functioning could be reduced. Nature, in short, could work independently of individual natures, forms, geniuses or other moving principles.

10.3 Philosophy and Theology

At nearly every stage, Voetius' essay against the New Philosophy takes up the defence of Aristotelian principles in their late-Scholastic interpretation. Moreover, although Voetius' essay does include a thesis on occult qualities; a thesis on corpuscular models, announcing, by the way, the subject-matter of Schoock's *Admiranda*; a thesis elaborating the question of the *ens per accidens* in theological terms; a thesis on Copernicanism; and, finally, a thesis on learned ignorance, all these are like appendices to the theses regarding substantial causation. Having summed up the difficulties with regard to philosophies that do not accept individual essences and individual centres of causation, and having discussed various arguments for and against substantial forms, Voetius says that "as a logical follow-up" (*consectarii loco*), he will add a word on occult qualities. All the other topics follow, but everything points to the fact that Voetius has given his main arguments in Theses II and III, which relate to substantial unity, divine concurrence, habitual dispositions, artifacts and clockwork etc. It was the new theory of causation—rather than, for instance, Copernicanism—that was Voetius' main concern. More important still, it was the theory of causation—again, rather than, Copernicanism—that defined the way in which *Sophia* was subservient to theology.

On 24 November 1641, Henricus Regius publicly defended the propositions of the New Philosophy that provoked the *rector magnificus* Gisbertus Voetius to react.[25] Amongst these, we find, apart from the dangerous thesis concerning the human soul as an *ens per accidens* and some particular theses concerning Copernicanism and mechanical explanations of sensible qualities and occult forces, the general idea that motion, rest, quantity, position and shape form the principles of natural philosophy. Voetius clearly saw the latter question as the most important one. In both versions of his "corollaries" against Regius' disputation, the question of the *ens per accidens* and the Copernican question are mentioned before the question of the forms. The latter however, takes up the better part, defending the "sacred physics of Moses" and its interpretation by Danæus, Zanchius, the commentators on *Genesis* and the commentators on Petrus Lombardus and Aquinas.[26] In the essay on natures and substantial forms, which was presented for discussion on 23 and 24 December 1641, the order was reversed and the defence of the general principles of Scholastic physics became the primary objective.

In this book, I have aimed to point out that Voetius' idea of a physics which—contrary to mechanical philosophy—*is* reconcilable with the *Physica Mosaica*, includes many aspects of Aristotle's conception of Nature and change. Indeed, in comparison with Danæus and Zanchius, Voetius' idea of physics is far more straightforwardly Aristotelian. As shown in chapter 3, Voetius was at once far more confident than Danæus over the possibilities of linking Christian ideas to pagan standpoints and far stricter than Zanchius in his choice of philosophical principles. Zanchius in fact offers the Platonic world-soul as a pagan idea reflecting the Biblical idea of "nature" in much the same way as Sebastian Basso did.[27] And not only are Aristotle's "second nature" and Plato's world-soul put on a par, they are also interpreted by Zanchius as pre-Christian versions of an all-moving *Spiritus Dei*. Although loose ends of Scholastic theory occur on every page of the works of Hieronymus Zanchius and Lambertus Danæus, Protestant thought in fact acquired a much more rigidly Aristotelian character in the hands of Voetius. Voetius excuses himself in the

---

[25] Cf. *Duker* II, pp. 148-153.

[26] *Narratio*, pp. 28-33 / *Querelle*, pp. 97-101. Voetius presented a second version of these "corrolaries", after Regius had complained to Gysbert vander Hoolck, one of the Utrecht Burgomasters, about the biting character of the first. *Ibidem*, and, on Vander Hoolck: *Querelle*, p. 453, note 102.

[27] For Zanchius, see section 3.1.3.2, above, note 34 in particular. For Basson, see above, chapter 9, note 101.

preface to his essay for trespassing across the boundaries of physics. However, although the subject is one which he would otherwise have left to the "Professors of Medicine and Philosophy and other Physicians and Philosophers in this renowned City of Utrecht",

> since I willingly admit that philosophy (and, notably, physics) is your domain and that theologians may be wholly content to use it where it is necessary and prudent,

he nonetheless seems very pleased that the occasion of Regius' disputations offered him

> a most welcome opportunity to elaborate upon the pre-eminent question of substantial forms.[28]

In fact, Voetius must have been very pleased indeed by the opportunity to defend the Aristotelian ideas which may well have been more dear to him and to his fellow-theologians than to the philosophers and men of medicine themselves.

Voetius forms part of the tradition that by sharply distinguishing the *usus* of Scholastic method from its *abusus*, aimed at professionalising the science of theology in the Protestant world.[29] The inheritance of specific and technical notions of Aristotelian philosophy thereby became unavoidable. It may well be regarded as an important external factor for Voetius' attack on Cartesianism, that he had just initiated the project of Aristotelianizing Protestant theology, when Aristotelian philosophy came under attack more and more from within physics itself. The theological disputes in Leiden between Arminius and Gomarus, and the consequent schism within the Dutch churches had, moreover, influenced theological debate up to the point where it was of extreme importance to pronounce upon issues such as God's concurrence with Nature in an extremely precise Scholastic terminology.[30] Finally, the atomism of David Gorlæus, the attack on

---

[28] Voetius, *Narratio*, p. 36 / *Select. Dispp.* I, p. / *Responsio*, p. 5 / *Querelle*, p. 104.

[29] See my "Franco Petri Burgersdijk and the Case of Calvinism Within the Neo-Scholastic Tradition", p. 53 esp. See also Ernst Lewalter's analysis of the German situation, in *Spanisch-Jesuitische und Deutsch-Lutherische Metaphysik des 17. Jahrhunderts*, Hamburg (Ibero-Amerikanisches Institut) 1935 / reprint Darmstadt (Wissenschaftliche Buchgesellschaft) 1967, from which I borrow the idea of using Scholastic methods critically, distinguishing *usus* from *abusus*. *Ibidem*, p. 59.

[30] See above, section 9.5.2, and my "Franco Petri Burgersdijk and the Case of Calvinism Within the Neo-Scholastic Tradition", especially concerning the Maccovius case, pp. 29-30 and p. 48.

Aristotle by Sebastian Basso, and the sixteenth-century revival of Hermeticism were factors that not only troubled any champion of the forms, but had taken up much of Voetius' dialectical energy from his student-years in Leiden onwards.[31] This may explain the fierceness of Voetius' defence of Aristotle at a time when the university was suddenly invaded by "that new Dutch Mechanical Philosophy".[32]

Yet for all the importance of external factors, the internal inconsistency of mechanical and Aristotelian conceptions of philosophy is ultimately what must have troubled Voetius most. Aristotle's natural philosophy had been reshaped along Christian lines. Cartesianism however, had not been inspired by the desire to establish a natural philosophy that was theologically sound. Contrary to someone like Voetius, for whom the subject of physics had no other use than to explain the Mosaic text, Descartes worked outside textual traditions. Moreover, in contrast to both Voetius and other *novateurs* such as Gassendi, he did not develop science along theological lines. In retrospect, Descartes' abandonment of the publication of *Le Monde* might have been worthwhile even had he not followed Copernicus. It was Cartesianism itself that by disengaging philosophical from theological discourse threatened the accepted interdisciplinary links.

Theologians judged philosophical systems primarily in terms of their ability to define the relation between God, the Creator, and Nature, His Creation. When, in 1677, the works of Spinoza came to the fore, in which the God of the Cartesian world-machine was

---

[31] As we have seen in section 9.6.2, above, Voetius compares the views of Basso, Gorlæus, Regius and Descartes to "the loathsome and long discredited [opinions] of both old and modern Paracelsans and Hermeticists". Cf. *Narratio*, p. 44 / *Select. Dispp.* I, p. 876 / *Responsio*, p. 23 / *Querelle*, p. 109. On the revival of Hermeticism in Renaissance thought, see: Frances A. Yates, *Giordano Bruno and the Hermetic Tradition*, London (Routledge and Kegan Paul) 1964 and Brian P. Copenhaver's critical assessment of what has become known as the "Yates thesis": "Natural Magic, hermeticism, and occultism in early modern science", in David C. Lindberg and Robert S. Westman, *Reappraisals of the Scientific Revolution*, Cambridge (C. U. P.) 1990, pp. 261-301.

[32] "Mechanica illa Philosophia nova Hollandica", is how Jean Riolan characterized the new method in physiology of Descartes, Regius and Cornelis van Hogelande. Cf. Étienne Gilson, "Descartes, Harvey et la scolastique", in *Études sur le rôle de la pensée médiévale dans la formation du système cartésien*, Paris (Vrin) 1984^5, p. 96, note 1. Jean Riolan (1577-1657) was professor of medicine in Paris and though an adversary of William Harvey's theory of blood-circulation, respected by the latter as "a very wise man and able anatomist". See Annie Bitbol-Hespériès, *Le principe de vie chez Descartes*, pp. 214-215. On Cornelis van Hogelande, see *Thijssen-Schoute*, pp. 228-230.

internalized into a *Natura naturans* and God identified with Nature itself, the theological reaction was accordingly hostile. For Johannes Bertling, the Groningen philosopher with whom we began our introduction, there were thus, in 1685, even more troubles to be dealt with. They had, however, been foreseen by Gisbertus Voetius. Just before Christmas 1641, the *rector* of Utrecht University tried to save the day in a skilled, but ultimately unsuccessful attempt to restrain the maiden engaged in the household of theology, to prevent her from running away and settling down on her own. According to the Utrecht theologian, the problem with the mechanical philosophy was that it yielded causal metaphors much less appropriate to theological images of divine guardianship. His blackest dreams were nevertheless destined to come true. For all the jubilation with which, in the eighteenth century, Newtonians and physico-theologians might describe God's managing and mending of the clockwork of the Universe, there would never again be that intimate kind of co-operation seen in the past.

BIBLIOGRAPHY

Abbreviations between brackets refer to the libraries in which rare books and prints may be found. The following libraries are included:

AD	Athenaeumbibliotheek, Deventer
BERL	Staatsbibliothek Preußischer Kulturbesitz, Berlin
BL	British Library, London
BN	Bibliothèque Nationale, Paris
BODL	Bodleian Library, Oxford
BTH	Bibliotheca Thysiana, Leiden
CARM	Nederlands Carmelitaans Instituut, Boxmeer
HERB	Bibliothek Evangelisch Theologisches Seminar, Herborn
KB	Koninklijke Bibliotheek (Royal Library), The Hague
KUN	University Library, Nijmegen
MB	Museum Boerhaave, Leiden
PBL	Provincial Library of Frisia, Leeuwarden
THK	Library of the Theological University, Kampen
UBG	University Library, Groningen
UBL	University Library, Leiden
UBU	University Library, Utrecht
UVA	University Library, Amsterdam
VU	Library of the Free University of Amsterdam

Primary Sources

Albert the Great, *De Generatione et Corruptione*, edidit Paulus Hossfeld in *Opera Omnia*, [...] curavit Institutum Alberti Magni Coloniense Wilhelmo Kübel praeside, Münster (Asschendorff) 1980.

Ambrose, *Exameron* in *Sancti Ambrosii Opera* I (*Corpus Scriptorum Ecclesiasticorum Latinorum* Vol. XXXII-I), ed. Carolus Schenkl, Prague / Leipzig / Vienna (F. Tempsky / G. Freytag) 1897.

Aquinas, *In Octo Libros Physicorum Aristotelis Expositio*, cura et studio P. M. Maggiòlo O. P., Turin / Rome (Marietti) 1965.

— *S. Thomæ Aquinatis Opera Omnia* [etc.], curante Roberto Busa, Stuttgart-Bad Cannstatt (Fromann-Holzboog) 1980 in 7 vols.

— *S. Thomæ Aquinatis Summa Theologiæ*, Cura Fratrum eisdem Ordinis, Madrid (Bibliotheca de Autores Cristianos), 1961[3] in 5 vols.

Aristotle, *Aristotelis Opera*, ed. Immanuel Bekker, Berlin (Georg Reimer) 1831.

— *The Physics*, ed. P. H. Wicksteed and F. M. Cornford, Vol. I, Cambridge, Mass. / London (Harvard University Press / Heinemann) 1980[4] [first edition 1929].

— *Physics Book I and II*, translated with introduction and notes by William

Charlton, *Clarendon Aristotle Series*, Oxford (Clarendon Press) 1992.

— *Aristotle's Metaphysics*, a revised text with introduction and commentary by W. D. Ross, Oxford (At the Clarendon Press) 1924.

— *The Metaphysics*, with an English translation by Hugh Tredennick, Cambridge Ms and London (Harvard / Heinemann) 1987-1980 [first edition 1930] in 2 vols.

— *Aristotle's Metaphysics Books* Γ, Δ, *and* E, translated with notes by Christopher Kirwan, Oxford (Clarendon Press) 1971.

Augustine, *De Genesi ad Literam*, in *Corpvs Scriptorvm Ecclesiasticorvm Latinorvm* XXVIII-I, Prague / Vienna / Leipzig (F. Tempsky / G. Freytag) 1894.

Avicenna, *Liber de Philosophia Prima sive Scienctia Divina*, Édition critique de la traduction latine médiévale par S. van Riet, Louvain / Leiden (Peeters / Brill) 1980.

Baillet, Adrien, *La Vie de Monsieur Des-Cartes*, A Paris (Chez Daniel Horthemels) 1691.

Basil, *Homélies sur l'Hexaéméron*, texte grec, introduction et tradition de Stanislas Giet (*Sources Chrétiennes* N° 26 *bis*), Paris (Les Éditions du CERF) 1968.

Basso, Sebastianus, *Philosophia Natvralis adversus Aristotelem libri XII. In quibus abstrusa Veterum Physiologia restauratur, & Aristotelis errores solidis rationibus refelluntur*, Amsterodami (Apud Ludovicum Elzevirium) 1649. [BL, BN, KB, UBL; 1624 edition in BN]

Bertling, Johannes, *Seminarium totius transnaturalis sapientiæ duobus tomis Philosophiam Primam et integram Pneumaticam sive Theologiam Naturalem Nervosa & succincta Methodo complectens*, Groningæ, Typis Catharinæ Zandt, *Viduæ* Huysman, Provinciæ & Academiæ Typographæ Ordinariæ, Anno 1685. [HERB, UBG]

Beverwijk, Johan van / Lia van Gemert (ed.), *De schat der gezondheid* (1636), Amsterdam (Querido) 1992. [first edition Dordrecht 1636 in UVA; 1637 edition in MB, UBL and UBU]

Biel, Gabriel, *Collectorium circa quattuor libros Sententiarum*, ed. Wilfried Werbeck / Udo Hofmann, Tübingen (J. C. B. Mohr / Paul Siebeck) 1975.

Bois, J. du, *Naecktheyt van de Cartesiaensche Philosophie, Ontbloot in een Antwoort Op een Cartesiaensch Libel, Genaemt Bewys* [etc.], T'Utrecht bij Johannes van Waeberge, 1655. [KB]

Bootius, Gerardus and Arnoldus, *Philosophia Naturalis Reformata. Id est, philosophiæ Aristotelicæ accurata examinatio ac solida refutatio. et novæ ac verioris introductio.* Dublinij in Hibernia, Ex Officinâ Typographicâ Societatis Bibliopolarum, 1641. [BL, BN, UBU]

Burgonovo, Archangelus de, *Cabalistarum selectiona, obscurioraque Dogmata, à Ioanne Pico ex eorvm commentationibus pridem excerpta* [etc.], in Pavlvs Ricivs (et. al.), *Artis Cabalisticæ: hoc est Reconditæ Theologiæ et Philosophiæ, scriptorvm:* Tomus I, Basileæ, Per Sebastianvm Henricpetri, [1593]. [UBL; 1569 edition in BL, BN]

— *Apologia pro defensione doctrinæ Cabalæ contra Petrum Garzam.* [...] et Conclusiones cabalisticæ numero LXXI, Basileæ, Per S. Henricpetri, [1600]. [BL, BN, UBL]

Casmann, Otto, *Somatologia, Physica Generalis, seu Commentationvm Disceptationumque Physicarum Syndromvs Problemat. I. De Natvralivm Corporum in Genere Essentia, & Qvalitatibvs Physicis, tum manifestis, tum occultis*, Francofvrti, Apud D. Zachariam Palthenium, 1607. [BN, KB, UBG; 1598 edition in BL]

Cicero, *De Fato*, in Cicero, *De Oratore III, De Fato, Paradoxa Stoicorum, De Partitione Oratoria*, with an English translation by H. Rackham M.A., London / Cambridge, Mass. (William Heinemann / Harvard U. P.) 1942.

Complutenses, *Dispvtationes in octo libros Physicorvm Aristotelis*, Francofvrti ad Mœnum, Sumptibus Casparis ab Rath / Typis Erasmi Kempfferi, 1629. [CARM; Lyon edition 1637 in BN]

Conimbricenses, *Commentarium Colegii Conimbricensis S. J., In Octo Libros Physicorum Aristotelis Stagiritæ*, Lvgdvni, Svmptibvs Ioannis Baptistæ Bvysson, 1594 / reprint Hildesheim (Georg Olms Verlag) 1984.

— *Commentarii Collegii Conimbricensis S. J., In Tres Libros de Anima Aristotelis*, Coloniæ, Sumptibus Hærendum Lazari Zetneri, 1629.

— *Commentarium Collegii Conimbricensis S. J., In Libros Meteororum Aristotelis Stagiritæ*, Coloniæ, Impensis Hærendum Lazari Zetneri, 1631.

Crollius, Osvaldus, *Basilica Chymica continens. Philosophicam propriâ laborum experientiâ confirmatam descriptionem et usum Remediorum Chymicorum Selectissimorum é Lumine Gratiæ et Naturæ desumptorum*, Francofurti, impensis Godefridi Tampachij, 1611. [UBL; 1608 edition in BTH and UBU; various editions in BL and BN]

Cyril of Alexandria, *Contre Julien* I, Livres I et II. Introduction, Texte Critique, Traduction et Notes par Paul Burgière et Pierre Évieux, Paris (Les Éditions du CERF) 1985.

Danæus, Lambertus, *Physica Christiana, sive, de rervm creaturarvm cognitione & vsu, disputatio e Sacræ Scripturæ fontibus hausta, & decerpta*, Genevæ, Apud Petrum Santandreum, 1576. [BN, UBU; 1580 edition in BL, UBL]

— *Physica Christiana Pars Altera, sive de rervm creaturarvm natvra. Qvæ in sex tractatvs pro diervm, quibvs Deus ipse operatus est, numero diuiditur*, in *Opvscvla Omnia Theologica*, Genevæ, apvd Evsthativm Vignon, 1583. [BN, UBG, UBU; 1598 edition in BL]

Descartes, René, *Descartes' Conversation with Burman*, trans. with introduction and commentary J. Cottingham, Oxford (Clarendon) 1976.

— *Discours de la Méthode*, texte et commentaire par Étienne Gilson de l'Académie Française, Paris (Librairie philosopique J. Vrin) 1976[5] [first edition 1925].

— *Entretien avec Burman*, Texte présenté, traduit et annoté par Ch. Adam, Paris (Librairie philosopique J. Vrin) 1975[2] [first edition 1937].

— *Lettres à Regius et Remarques sur l'explication de l'esprit humain*, Texte latin, traduction, introduction et notes par Geneviève Rodis-Lewis, Paris (Vrin) 1959.

— / Schoock, Martin, *La querelle d'Utrecht*, Textes établis, traduits et annotés par Theo Verbeek, Paris (Les impressions nouvelles) 1988.

— *Œuvres de Descartes*, publiées par Charles Adam & Paul Tannery, Paris (Léopold Cerf) 1897-1913 in 12 vols. / Nouvelle présentation, en co-édition avec le Centre National de la Recherche Scientifique, Paris (Librairie Philosophique J. Vrin) 1964-1971 in 12 vols.

— *Œuvres philosophiques*, Textes établis, pésentés et annotés par Ferdinand Alquié, Paris (Garnier) 1963.

— *The Philosophical Writings of Descartes*, Translated by John Cottingham, Robert Stoothoff and Dugald Murdoch, Cambridge (C. U. P.) 1985-1991.

— *Meditations on First Philosophy. With selections from the Objections and Replies*, Translated by John Cottingham with an introduction by Bernard Williams, Cambridge (C. U. P.) 1986.

Empedocles, *The Extant Fragments*, edited, with an introduction, commentary and concordance, by M. R. Wright, New Haven and London (Yale University Press) 1981.

Eustace of St. Paul, *Summa Philosophica quadripartita, de rebus Dialecticis, Moralibus, Physicis et Metaphysicis*, Parisiis, Carolus Chastellain, 1609.

Franzius, Wolfgang, *Historia sacra animalium*, VVittenbergæ, Sumtibus Zachariæ Schureri & Iohannis Gormanni, 1616. [UBU, 1612 edition in BL, 1613 edition in BN, various later editions in BL, BN, KB, UBU and UVA]

Galileo, *Le Opere di Galileo Galilei*, Edizione nazionale sotto gli auspicii di sua Maestà il Re d'Italia, Firenze (G. Barbèra) 1890-1909 in 22 vols. / *Le Opere di Galileo Galilei*, sotto l'alto patronato del presidente della Republica Italiana G. Saragat, Firenze (Barbera) 1968 in 20 vols.

— *Discoveries and Opinions of Galileo*, Translated with an Introduction and Notes by Stillman Drake, Garden City, New York (Doubleday Anchor Books) 1957.

— *Systema Cosmicum*, Lugduni, sumptibus J. A. Huguetan, 1641. [BL, BN,UBL; 1635 and other editions in BL]

Gassendi, Pierre, *Abregé de la philosophie de M. Gassendi, en VIII tomes*, par François Bernier, Lyon, Anisson & Posnel, 1678. [BL, BN, KB; 1684 edition BL, BN, UBL]

Gorlæus, David, *Excercitationes Philosophicæ quibus Vniversa fere discutitur Philosophia Theoretica. Et Plurima ac præcipua Peripateticorum dogmata evertuntur.* In Bibliopolio Commeliniano [Leiden] (Sumptibus viduæ Ioannis

Comelini) 1620. [BL, BN, UBL]

— *Idea Physicæ. Cui adjuncta est Epistola, Cujusdam Anonymi de Terræ Motu*, Ultrajecti, Ex officina Johannis a Waesberge, 1651 [BL] / Tilburg (F. A. H. Peeters) 1986.

Harvey, William, *Excercitatio de Motu Cordis et Sanguinis in Animalibus*, Francofvrti, Svmptibus Gvilielmi Fitzori, 1628.

Hesiod, *Theogony*, edited with Prolegomena and Commentary by M. L. West, Oxford (Clarendon Press) 1966.

— *Works and Days*, edited with Prolegomena and Commentary by M. L. West, Oxford (Clarendon Press) 1978.

Hillers, Henricus, *Bovennatuurkundig Vertoog waar in dat geen, 't welk twyffelagtig is in de vraag aangaande de geschapene Beste Werelt, opgeheldert wort*, onder voorzitting van Nicolaus Engelhard, Te Groningen, By Hajo Spandaw, Boekverkoper in de Gulden straat, in het Oude Zwarte Kruys, 1755. [UBU]

Homer, *The Iliad*, with an English translation by A. T. Murray, in 2 vols, London / New York (William Heinemann / G. P. Putnam's Sons) 1924-1925.

Iamblichus, *Adhortatio ad Philosophiam*, Textum [...] recensuit interpretatione Latina [...] et animadversionibus instruxit M. Th. Kiessling, Leipzig (F. C. G. Vogelius) 1813.

Iunius, Franciscus, *Opera Theologica*, Editio postrema, Genevæ, Apud Samuelem Crispinum, 1613 in 2 vols. [PBL, UBG, UBU; various other editions elsewhere]

Justin Martyr, *Apologies*, introduction, texte critique, traduction, commentaire et index par André Wartelle, Paris (Études Augustiniennes / Institut Catholique de Paris) 1987.

Kant, Immanuel, *Kritik der reinen Vernunft*, Zweite Auflage 1787, in *Kant's Gesammelte Schriften, Werke* Band III, Berlin (Georg Reimer) 1904.

— *Antropologie in pragmatischer Hinsicht*, in *Kant's Gesammelte Schriften, Werke* Band VII, Berlin (Georg Reimer) 1917.

Leibniz, Gottfried Wilhelm, *De rerum originatione radicali*, in *Die philosophische Schriften von G. W. Leibniz*, herausgegeben von C. I. Gerhardt, Berlin (Weidmann) 1875-1890 / reprint Hildesheim (Olms) 1965, Bd. VII, Berlin (Weidmann) 1890.

Luther, Martin, *D. Martin Luthers Werke, Kritische Gesamtausgabe* Weimar (Hermann Böhlaus Nachfolger) 1883-1985.

Maimonides, *Guide des Égarés* (ed. S. Munk), Paris 1856-1866 / reprint Paris (Maisonneuve et Larose) 1960.

Manilius, M., *Astronomicωn Libri Quinque* (ed. Josephus Justus Scaliger), Lvtetiæ, apud Mamertum Patissonium Typographium Regium, in officina Roberti Stephani, 1579. [BL, BN, KB; various later editions in BL, BN, BTH, PBL, UBL, UBG, UVA]

Meineke, August (ed.), *Fragmenta Comicorum Græcorum*, Berlin (Georg Reimer) 1857 / reprint Berlin (Walter de Gruyter & Co.) 1970.

Melanchthon, Philippe, *Liber de Anima / Initia Doctrinæ Physicæ*, in *Scripta Phil. Melanthonis ad Historiam Profanam et Philosophiam spectantia, Corpus Reformatorum* (ed. Carolus Gottlieb Bretschneider) Vol. XIII, Halle (C. A. Schwetschke und Sohn) 1846.

Mersenne, Marin, *Correspondance du P. Marin Mersenne*, commencée par Mme. Paul Tannery; publiée et annotée par C. de Waard et R. Pintard, Paris (Éditions du C.N.R.F.) 1933-1988.

Montaigne, Michel de, *Les Essais*, Édition conforme au texte de l'examplaire de Bordeaux [...] par Pierre Villey, sous la direction et avec une préface de V.-L. Saulnier, Paris (P. U. F.) 1924, Nouvelle édition 1965 / 1992² in 3 vols.

Pererius Ventinus, Benedictus, *Commentaria et Disputationes in Genesim* I-IV, Colonia Agrippinæ, Ex officina Antonij Hierati, 1606. [UBL; 1601 edition in BL, BN; various later editions in BN, 1612 edition in KUN]

Plato, *Œuvres Complètes*, Paris (Soc. d'Éditions 'Les Belles Lettres') 1964.

— *Timæus and Critias*, Translated with an introduction and an appendix on *Atlantis* by Desmond Lee, London (Penguin Books) 1983.

— *The Republic of Plato*, edited with critical notes, commentary and appendices by James Adam, M. A., in 2 vols., Cambridge (At the University Press) 1902.

Plutarch, Τὰ Σωζόμενα Πάντα [*sive, Opera*] *Qvæ Exstant Omnia, cvm Latina interpretatione Hermanni Crvserij: Giulielmi Xylandri, et Doctorvm Virorum* [etc.], Francofvrti, Apud Andreæ Wecheli heredes, Claudium Marnium, & Ioannem Aubrium, 1599. [AD, BL, BN, PBL, UBG, UBL; later editions in BL, BN, KB]

— *Moralia* Vol. V, fasc. 2, pars 1, ed. Jürgen Man, Leipzig (Teubner) 1971.

Regius, Henricus, *Disputatio medica Ia[-IIIa] de Illustribus aliquot Quæstionibus Physiologicis*, Vltraiecti, Ex Officinâ Ægidii Roman, Academiæ Typographi, 1641. [BERL]

— *Physiologia, sive Cognitio Sanitatis*, Vltraiecti, Ex Officinâ Ægidii Roman, Academiæ Typographi, 1641. [HERB, BL]

— *Responsio, sive Notæ in Appendicem ad Colloraria Theologica-Philosophica* [etc.], Ultraiecti, Apud Ioannem à Doorn, 1642. [BN, UBU]

— *Fundamenta Physices*, Amstelodami, Apud Ludovicum Elzevirium, 1646. [BL, BN, UBL, UBU, UVA]

Revius, Jacobus, *Statera Philosophiæ Cartesianæ qua Principiorum ejus falsitas, & dogmatum impuritas expenditur ac castigatur, Et Ad virulentam Epistulam, præfixam selectarum Disputationum Adriani Heereboortii volumini primo, respondetur*, Lvgdvni Batavorvm, Ex Officina, Petri Leffen, [1650]. [BERL, BN, HERB, UVA]

— *Jacobus Revius: a Dutch Metaphysical Poet*, Trans. with an introd. by Henrietta ten Harmsel, Detroit (Wayne State U. P.) 1968.

Rupert of Deutz, *Libri XLII de operibus sanctæ Trinitatis*, Colonia, Frans Birckman, 1528. [UBG, UBL, UBU]

Ruvius, Antonio, *Commentarii in octo libros Aristotelis de Physico auditu, vna cum dubijs, & quæstionibus hac tempestate agitari solitis*, Madriti, Apud Ludouicum Sanchez, 1605. [PBL; various later editions in BL, BN; 1620 edition in KUN]

Salutati, Coluccio, *De Fato et Fortuna*, a cura di Concetta Bianca, Firenze (Leo S. Olschki) 1985.

Schoockius, Martinus, *De Natura Soni, & Echus*, in *Lusus Imaginis Iocosæ sive Echus a Variis Poetis, variis Linguis & numeris exculti*, Ultrajecti, Ex Officina Ægidii Roman, Acad. Typog., 1638. [BL, BN, KB, PBL, UBL, UBU, UVA]

— *Admiranda Methodvs Novæ Philosophiæ Renati des Cartes*, Vltraiecti, Ex Officina Joannis *van* VVaesberge, 1643. [BL, BN, UBG, UBL, UBU, UVA]

— / Wittichius, Christophorus [*respondens*], *Disputatio Physica de Aqua*, Groningæ, Typis Augustini Eissens, Typographi, 1646. [UBG]

— / Vriesen, Bernhardus [*respondens*], *Disputatio Physica de Glacie*, Groningæ, Typis Augustini Eissens, Typographi, 1646. [UBG]

— *Physica Generalis*, Groningæ, typis J. Cöllenii, 1660. [BN, UBU]

— *Physica Cælestis*, Amstelodami, apud P. Vanden Berge, 1663. [BL, BN, BODL]

Schoockius, Martinus, see also Descartes, René.

Seneca, *Ad Lucilium Epistulæ Morales*, with an English translation by Richard M. Gummere, in 3 vols, London / New York (William Heinemann / G. P. Putnam's Sons) 1917-1925.

Senguerdius, Arnoldus, *Collegium Metaphysicum in qvo Metaphysica generalis breviter proponuntur XVI. disputationibus publice ventilatis in inclyta Academia Ultrajectina*, Ultraiecti, Ex Officinâ Ægidii Roman, Acad. Typograph., 1640. [UBU]

Sennert, Daniel, *De Chymicorum cum Aristotelicis & Galenicis consensu ac dissensu*, in *Opera* Tomus I, Parisiis, Apud Societatem, 1641. [PBL, UBL, UBU; original edition as *De Chymicorum cum Aristotelicis & Galenicis concensu ac dissensu*, Wittebergæ, hæred. Z. Schürer, 1619 in BL, BN, UVA, various later editions in BN]

Spinoza, Baruch, *Opera*, im Auftrag der Heidelberger Akademie der Wissenschaft herausgegeben von Carl Gebhardt, Heidelberg (Winter) 1925-1928 / reprint heidelberg (Winter) 1972-1987 in 5 vols.

Suárez, Francisco, *R. P. Francisci Suarez e Societate Jesu Opera Omnia*, editio nova a M. Andre [et a Carolo Berton] juxta editionem Venetianum [...] accurate recognita, Parisiis, Apud Lodovicum Vives, 1856-1866 in 28 vols.

Testimonium Academiæ Vltrajectinæ, et Narratio Historica quà defensæ, quà exterminatæ novæ Philosophiæ, Rheno-Trajecti, Ex Typographiâ Wilhelmi Strickii, 1643. [BL, BN, BTH]

Timpler, Clemens, *Physica sev Philosophia Natvralis Systema Methodicvm*, Hanoviæ Apud Hæredes Guilielmi Antonii, 1613. [BN, UBG; 1605 edition in BL, BN]

[Valentinus:] Malcolm Lee Peel (ed.), *The Epistle to Rheginos, A Valentinian Letter on the Resurrection, New Testament Library*, London (SCM Press) 1969.

Vergil, *P. Vergili Maronis Æneidos Liber Sextus*, with a commentary by R. G. Austin, Oxford (Clarendon Press) 1977.

— *P. Vergilius Maro Æneis Buch VI*, erklärt von Eduard Norden, Stuttgart (B.G. Teubner) 1957[4] [first edition Leipzig (Teubner) 1903].

Voet, Paulus, *Prima Philosophia Reformata*, Trajecti ad Rhenum, ex officinâ Johanni à Waesberge, 1657. [BL, BN, UBL, UBU]

Voetius, Gisbertus, *Dissertatio Epistolica de Termino Vitæ, ad Ampliss. Clarißimumque Johannem Beverovicium*, in *Select. Dispp.* V, 1669, seperate paging.

— *Excercitiæ et Bibliotheca, Studiosi Theologiæ*, Rheno-Trajecti, Apud Wilhelm Strick, Anno 1644. [KB, THK, UBG, UBU; second edition Ultrajecti, Ex officina Johannis à Waesberge, 1651 in BN, BN, UBL, VU]

— *Selectarum Disputationum Theologicarum Partes V*, Utrecht and Amsterdam 1648-1669; I-III, Ultraiecti, Joh. à Waesberge 1648-1659; IV, Amstelodami, Jo Janss. à Waesberge 1667; V, Ultrajecti, Ant. Smijtegelt, 1669. [I-IV: THK, UBG, UBL, UBU; I-V: BL, BN, THK, UBG, UBU]

— *Sermoen van de Nuttigheydt der Academien ende Scholen mitsgaders der Wetenschappen ende Consten die in de selve gheleert werden: Gedaen inde Dom-kercke tot VTRECHT den 13. Martij Ouden stijls, des Sondaechs voor de Inleidinghe der nieuwe Academie aldaer* [etc.], Tot Utrecht, Bij Ægidius ende Petrus Roman, Druckers van de Universiteyt, woonende inde korte Nieu-straat, 1636. [BL, KB, PBL, UBU]

— *Tersites Heautoutimeros Hoc est, Remonstrantium Hyperaspystes, catechesi, et Liturgiæ Germanicæ, Gallicæ, & Belgicæ Denuo insultans, Retusus* [etc.], Ultrajecti, Ex Officinâ Abrahami ab Herwiick & Hermanni Ribii, 1635. [BL, BN, UBL, UBU, UVA]

Voltaire, *Lettres philosophiques sur les Anglais*, ed. Gustave Lanson, Paris (Didier) 1964.

Waterlaet, Lambertus van den, *Prodromus, sive Examen Tutelare Orthodoxæ Philosophiæ Principiorum, Contra Fictitium quoddam hujus temporis eorum pentagonum Tum vel Præcipue Adversum eum modum, quo illud, non ita pridem prætendit Clariss. D. Henricus Regius, Med. Doct. & Profess. in Responsione ad Corrollaria Theologico-Philosophica sub præsidio Clariss. Eximii, ac Celeberrimi D. Gisberti Voetii triduanâ disputatione ventilata*, Lvgdvni Batavorvm, Excudebat Wilhelmus Christiani, 1642. [BN, BODL]

Zanchius, Hieronymus, *Opera Theologica* III, *De Operibus Dei intra Spacium Sex*

Dierum Creatis, Genevæ, Excudebat Stephanus Gamonetus, 1613. [AD, BN, UBG, UBL; original edition as *De Operibus Dei intra Spacium Sex Dierum Creatis*, Neostadii Palatinorum, typis M. Harnisii in BN; 1597 edition in BL; 1605 *Opera*-edition in AD, UBU]

Reference Works

Astius, Fredericus, *Lexicum Platonicum, sive Vocum Platonicarum Index*, Berlin (Hermann Barsdorf Verlag) 1908².

Ausführliches Lexikon der Griechischen und Römischen Mythologie [...] herausgegeben von W. H. Roscher, Leipzig (Teubner) 1884-1921 in 12 vols.

Bonitz, Hermann, *Index Aristotelicus*, Berlin 1870 / reprint Graz (Academische Drücke und Verlagsanstalt) 1955.

Dictionnaire de Théologie Catholique, contenant l'exposé des doctrines de la théologie catholique: leurs preuves et leur histoire, commencé sous la direction de A. Vacant et E. Mangenot; continué sous celle de E. Amman, Paris (Letouzey et Ané) 1899-1972.

Encyclopaedia Judaica, editor in chief: Cecil Roth, Jerusalem (Keter) 1971.

The Encyclopedia of Religion, editor in chief: Mircea Eliade, New York (Macmillan) 1987.

Gilson, Étienne, *Index Scolastico-Cartésien*, Paris (Vrin) 1979 [first edition as Paris Dissertation, 1912].

Haussig, Wilhelm (ed.), *Götter und Mythen der Kaukasischen und Iranischen Völker, Wörterbuch der Mythologie* Band IV, Stuttgart (Klett-Cotta) 1986.

Jöcher, Christian Gottlieb, *Allgemeines Gelehrten-Lexicon* [etc.], Leipzig (Johann Friedrich Gleditschens Buchhandlung) 1750-1751 / Reprint Hildesheim (Olms) 1960-1961.

Lalande, André, *Vocabulaire technique et critique de la philosophie*, Paris (P. U. F.) 1991¹⁷ [first edition 1926].

Lewis, Charlton T., and Short, Charles, *A Latin Dictionary*, Oxford (At the Clarendon Press) 1879.

Lexicon für Theologie und Kirche, begründet von Michael Buchberger [...], völlig neu herarbeitete Auflage [...] von Josef Höfer und Karl Rahner, Freiburg (Herder Verlag) 1957-1968.

Molhuysen, P. C., Blok, P. J. and Fr. K. H. Kossmann (eds), *Nieuw Nederlands Biografisch Woordenboek*, Leiden (A. Sijthoff's Uitgevers-maatschappij) Leiden 1911-1937 / reprint Amsterdam (Israel) 1974 in 10 vols..

The Oxford Classical Dictionary, edited by N. G. L. Hammond and H. H. Scullard, Oxford (At the Clarendon Press) 1970².

Paulys Real-Encyclopädie der Classischen Altertumswissenschaft, Neue Bearbeitung herausgegeben von Georg Wissowa u.a., Stuttgart (J. B. Metzler) 1894-1974 in 44 vols.

Thesavrvs Lingvæ Latinæ, Leipzig (B. G. Teubner) 1900-1981.

Tucker, T. G., *A Concise Etymological Dictionary of Latin*, Halle (Max Niemeyer Verlag), 1931.

Critical Studies

Ahsmann, Margreet, *Collegia en Colleges. Juridisch onderwijs aan de Leidse universiteit 1575-1630 in het bijzonder het disputeren*, Dissertation Leiden 1990, Groningen (Wolters-Noordhoff / Egbert Forsten) 1990.

Algra, Keimpe, *Concepts of Space in Greek Thought* (*Philosophia Antiqua* 65), Leiden (E. J. Brill) 1995.

Balme, David M., "Aristotle's Biology was not Essentialist", in *Archiv für Geschichte der Philosophie* 62 (1980), pp. 1-12.

— "The Place of Biology in Aristotle's Philosophy", in Allan Gotthelf / James. G. Lennox (edd.), *Philosophical Issues in Aristotle's Biology*, Cambridge (C. U. P.) 1987, pp. 9-20.

Berkel, K. van, "Descartes in debat met Voetius. De mislukte introductie van het Cartesianisme aan de Utrechtse Universiteit (1639-1645)", in *Tijdschrift voor de Geschiedenis der Geneeskunde, Natuurwetenschappen, Wiskunde en Techniek* 7-I (1984), pp. 4-18.

Bernard Cohen, I., *The Birth of a New Physics*, London (Heinemann) 1960 / London (W. W. Norton & Co.) 1985 / London (Penguin) 1992.

Bitbol-Hespériès, Annie, *Le principe de vie chez Descartes*, Paris (Vrin) 1990.

Boas, Marie, "The establishment of the Mechanical Philosophy", in *Osiris* 10 (1952), p. 412-541.

Bornkamm, Heinrich, "Kopernikus im Urteil der Reformatoren", *Archiv für Reformationsgeschichte* 40 (1943), pp. 171-183.

Boswell, John, *Christianity , Social Tolerance and Homosexuality*, Chicago and London (University of Chicago Press) 1980.

Bouché-Leclercq, A., *L'Astrologie Grècque*, Paris (Ernest Leroux) 1899.

Bréhier, Émile, *La philosophie et son passé*, Paris (Alcan / P.U.F.) 1940.

Brunschvicg, Léon, *Descartes et Pascal: Lecteurs de Montaigne*, New York (Brentano's) 1944.

Buisman, J. W., *Tussen Vroomheid en Verlichting*, Zwolle (Waanders) 1992.

Burtt, E.A., *The Metaphysical Foundations of Modern Physical Science*, London (Routledge and Kegan Paul) 1932^2.

Cassirer, Ernst, *Das Erkenntnisproblem der Philosophie und Wissenschaft der neueren Zeit*, Berlin (Cassirer) 1911.

Charles, David, "Theological Causation in the *Physics*", in Lindsay Judson, ed., *Aristotle's* Physics*: A Collection of Essays*, Oxford (Clarendon Press) 1991, pp. 101-128.

Clarke, Desmond, *Descartes' Philosophy of Science*, Manchester (M. U. P.) 1982.

Copenhaver, Brian P., "Natural Magic, hermeticism, and occultism in early modern science", in David C. Lindberg and Robert S. Westman (edd.), *Reappraisals of the Scientific Revolution*, Cambridge (C. U. P.) 1990, pp. 261-301.

— and Schmitt, Charles B., *Renaissance Philosophy, A History of Western Philosophy 3*, Oxford (O. U. P.) 1992.

Cottingham, John, *Descartes*, Oxford (Basil Blackwell) 1986.

— "Cartesian dualism: theology, metaphysics, and science", in John Cottingham (ed.), *The Cambridge Companion to Descartes*, Cambridge (C. U. P.) 1992, pp. 236-257.

— *The Rationalists, A History of Western Philosophy 4*, Oxford (O. U. P.) 1988.

Craig, William Lane, *The Kalâm Cosmological Argument*, London (The Macmillan Press) 1979.

— *The Problem of Divine Foreknowledge and Future Contingents From Aristotle to Suárez*, (*Brill's Studies in Intellectual History* 7), Leiden (E. J. Brill) 1988.

Crapulli, Giovanni, *Mathesis universalis; Genesi di una idea nel XVI secolo*, Roma (Ateneo) 1969.

Davain, Alfred, *François du Jon (Junius), Pasteur et Professeur en Théologie 1545-1602*, Genève (Slatkine) 1970 [= Reprint Paris 1882].

Deason, Gary B., "Reformation Theology and the Mechanistic Conception of Nature", in David C. Lindberg and Ronald L. Numbers (eds.), *God and Nature: Historical Essays on the Encounter between Christianity and Science*, Berkeley (University of California Press) 1986, pp. 167-191.

Dibon, Paul, *L'Enseignement philosophique dans les universités néerlandaises à l'époque pré-cartésienne (1575-1650)*, Dissertation Leiden 1954, Amsterdam (Institut Français d'Amsterdam) 1954.

Dijksterhuis, E. J., *De Mechanisering van het wereldbeeld; De Geschiedenis van het natuurwetenschappelijk denken*, Amsterdam (Meulenhoff) 1989[6] [first edition 1950].

Du Plessis, P. J., *Teleios: The Idea of Perfection in the New Testament*, Dissertation Kampen (NL) 1959, Kampen (Kok) 1959.

Duker, A. C., *Gisbertus Voetius*, Leiden (E. J. Brill) 1893-1915 / reprint: Leiden (Groen) 1989 in 4 vols.

Fatio, Olivier, *Méthode et théologie: Lambert Daneau et les débuts de la scolastique réformée*, Thèse Genève 1974 / Genève (Éditions Droz) 1976.

Félice, Paul de, *Lambert Daneau (de Baugency-sur-Loire) Pasteur et professeur en théologie 1530-1595*, Paris (Librairie G. Fischbacher) 1882.

Freeland, Cynthia A., "Accidental Causes and Real Explanations", in Judson, Lindsay (ed.) *Aristotle's* Physics*: A Collection of Essays*, Oxford (Clarendon Press) 1991, pp. 49-72.

Gadoffre, Gilbert, "La chronologie des six parties", in N. Grimaldi and J.-L. Marion (eds.), *Le Discours et sa Méthode*, Paris (P. U. F.) 1987, pp. 19-40.

Gapaillard, Jacques, *Et pourtant elle tourne, Le mouvement de la Terre*, Paris (Éditions du Seuil) 1993.

Garber, Daniel, "How God Causes Motion: Descartes, Divine Sustenance, and Occasionalism", in *The Journal of Philosophy* 84 (10), January-December 1987, pp. 567-580.

— *Descartes' Metaphysical Physics*, Chicago (The University of Chicago Press) 1992.

Gueroult, Martial, *Descartes selon l'ordre des raisons*, Paris (Aubier) 1953.

— *Malebranche*, Paris (Aubier, Éditions Montaigne) 1959 in 2 vols.

Gewirtz, Alan, "Experience and the Non-Mathematical in the Cartesian Method", in *Journal of the History of Ideas*, 2^2 (1941), pp. 183-210.

Gilson, Étienne, *Études sur le rôle de la pensée médiévale dans la formation du système cartésien*, Paris (Vrin) 1984^5 [first edition in *Études de philosophie médiévale* 13, Paris (Vrin) 1930].

— "Pourquoi saint Thomas a critiqué saint Augustin", in *Archives d'Histoire doctrinale et littéraire du Moyen Age* 1 (1926), pp. 6-127.

Gotthelf, Alan (ed.), *Aristotle on Nature and Living Things: Philosophical and Historical Studies*, Pittsburgh, Penn. (Mathesis Pulications) / Bristol (Bristol Classical Press) 1985.

Graafland, C., *Van Calvijn tot Barth: Oorsprong en ontwikkeling van de leer der verkiezing in het Gereformeerde Protestantisme*, 's-Gravenhage (Uitgeverij Boekencentrum B.V.) 1987.

Grafton, Anthony, "Joseph Scaliger and Historical Chronology: The Rise and Fall of a Discipline", in *History and Theory* 14 (1975), pp. 156-185.

Gregory, Tullio, "David van Goirle e Daniel Sennert", in *Giornale Critico della Filosofia Italiana* 20 (1966), pp. 44-63.

Grimaldi, Nicolas, *L'expérience de la pensée dans la philosophie de Descartes*, Paris (Vrin) 1978.

Hatfield, Gary, "Metaphysics and the new science", in David C. Lindberg and Robert S. Westman (edd.), *Reappraisals of the Scientific Revolution*, Cambridge (C. U. P.) 1990, pp. 93-166.

— "Descartes' physiology and its relation to his psychology", in John Cottingham (ed.), *The Cambridge Companion to Descartes*, Cambridge (C. U. P.) 1992, pp. 335-370.

Hoeven, P. van den, *Metafysica en Fysica bij Descartes*, Gorinchem (J. Noorduyn en Zoon N.V.) 1961.

Hooykaas, R., "The Reception of Copernicanism in England and the Netherlands", in *The Anglo-Dutch Contributiuon to the Civilization of Modern Society; An Anglo-Dutch Symposium*, London (Oxford U. P.), pp. 33-44.

— *G. J. Rheticus' Treatise on Holy Scripture and the Motion of the Earth*, Amsterdam / Oxford / New York (North-Holland Publishing Company) 1984.

Judson, Lindsay (ed.), *Aristotle's* Physics*: A Collection of Essays*, Oxford (Clarendon Press) 1991.

Kahn, Charles H., "The Prime Mover and Teleology", in Allen Gotthelf (ed.), *Aristotle on Nature and Living Things*, pp. 183-205.

Kalmar, Martin, *Some Collision Theories of the Seventeenth Century: Mathematicism vs. Mathematical Physics*, PhD. dissertation John Hopkins University, Ann Arbor (University Microfilms International) 1981.

Katz, David S., "Vossius and the English Biblical Critics", in Popkin / Vanderjagt, *Scepticism and Irreligion*, pp. 142-184.

Koyré, Alexandre, *From the Closed World to the Infinite Universe*, Baltimore and London (John Hopkins U. P.) 1957.

— *Newtonian Studies*, London (Chapman & Hall) 1965.

Kretzmann, Norman, Anthony Kenny and Jan Pinborg (edd.), *The Cambridge History of Later Medieval Philosophy*, Cambridge (C. U. P.) 1982.

Labrousse, E., *Pierre Bayle; Hétérodoxie et Rigorisme*, The Hague (Nijhoff) 1964.

Lang, Helen S., *Aristotle's* Physics *and Its Medieval Varieties*, New York (State University of New York Press) 1992.

Jerome J. Langford, *Galileo, Science and the Church*, Ann Arbor 1966 / Ann Arbor (The University of Michigan Press, Ann Arbor Paperbacks) 1992^4.

Laporte, Jean, *Le Rationalisme de Descartes*, Paris (P. U. F.) 1945.

Lasswitz, Kurd, *Geschichte der Atomistik vom Mittelalter bis Newton*, Hamburg und Leipzig (Leopold Voss) 1890 / reprint Hildesheim (Georg Olms) 1963 and 1984.

Lennox, James G., "Are Aristotelian Species Eternal?", in Alan Gotthelf (ed.) *Aristotle on Nature and Living Things*, pp. 67-94.

Lewalter, Ernst, *Spanisch-Jesuitische und Deutsch-Lutherische Metaphysik des 17. Jahrhunderts*, Hamburg (Ibero-Amerikanisches Institut) 1935 / reprint Darmstadt (Wissenschaftliche Buchgesellschaft) 1967.

Liebert, Arthur, "Johannes Müller, der Physiologie in seinem Verhältnis zur Philosophie und seiner Bedeutung für dieselbe", in *Kantstudien* 10 (1915), pp. 357-375.

Lindberg, David C. and Numbers, Ronald L. (edd.), *God and Nature: Historical Essays on the Encounter between Christianity and Science*, Berkeley (University of California Press) 1986.

— and Westman, Robert S. (edd.), *Reappraisals of the Scientific Revolution*, Cambridge (C. U. P.) 1990.

Linde, S. van der, *Opgang en Voortgang der Reformatie*, Amsterdam (Ton Bolland) 1976.

Lohr, Charles H., "Metaphysics", in *The Cambridge History of Renaissance Philosophy*, Cambridge (C. U. P.) 1988 / 1990^2, pp. 535-638.

Marion, Jean-Luc, *Sur l'ontologie grise de Descartes*, Paris (Vrin) 1981.

— *Sur la théologie blanche de Descartes*, Paris (P. U. F.) 1981.

McGahagan, Thomas Arthur, *Cartesianism in the Netherlands, 1639-1667; The New Science and the Calvinist Counter-Reformation* (Ph. D. diss. Univ. of Pennsylvania), Ann Arbor and London (University Microfilms International) 1976.

McGuire, J. E., "Boyle's Conception of Nature", in *Journal of the History of Ideas* 33 (1972), pp. 523-542.

McMullin, Ernan, Concepts of Science in the Scientific Revolution", in David C. Lindberg, Robert S. Westman (edd.), *Reappraisals of the Scientific Revolution*, Cambridge (C. U. P.) 1990, pp. 27-92.

Meyer, Susan Sauvé, "Aristotle, Teleology, and Reduction", in *The Philosophical Review* 101 (1992), pp. 791-825.

Miles, Murray, "Descartes' Mechanicism and the Medieval Doctrine of Causes, Qualities and Forms", in *The Modern Schoolman* LXV, January 1988, pp. 97-117.

Mittelstraß, J., "Die Idee einer Mathesis Universalis bei Descartes", in *Perspektiven der Philosophie* 4 (1978), pp. 177-192.

Moravcsik, Julius M., "What Makes Reality Intelligible? Reflections on Aristotle's Theory of *Aitia*", in Judson, Lindsay (ed.) *Aristotle's* Physics*: A Collection of Essays*, Oxford (Clarendon Press) 1991, pp. 31-47.

North, John David, "Chronology and the Age of the World", in W. Yourgrau / A. D. Beck (edd.), *Cosmology, History, and Theology*, New York (Plenum Press) 1977, pp. 307-333.

— *Chaucer's Universe*, Oxford (Clarendon Press) 1988.

Osler, Margaret (ed.), *Atoms,* Pneuma*, and Tranquility. Epicurean and Stoic Themes in European Thought*, Cambridge (C. U. P.), 1991.

— "Fortune, fate, and divination: Gassendi's voluntarist theology and the baptism of Epicureanism", in Margaret J. Osler (ed.), *Atoms,* Pneuma*, and Tranquillity: Epicurean and Stoic Themes in Euopean Thought*, Cambridge (C. U. P.) 1991, pp. 155-174.

Pagel, Walter, *Paracelsus: An Introduction to Philosophical Medecine in the Era of the Renaissance*, Basel (Karger) 1982².

— *From Paracelsus to Van Helmont*, London (Variorum Reprints) 1986.

Parkinson,G. H. R., *Logic and Reality in Leibniz's Metaphysics*, Oxford (Clarendon Press) 1965.

Patch, Howard R., *The Goddess Fortuna in Mediaeval Literature*, London (Frank Cass & Co.) 1967.

Patrides, C. A., *Premises and Motifs in Renaissance Thought and Literature*, Princeton, New Jersey (P. U. P.) 1982.

Pellegrin, Pierre, "Aristotle: A Zoology without Species", in Alan Gotthelf (ed.), *Aristotle on Nature and Living Things: Philosophical and Historical Studies*,

Pittsburgh, Penn. (Mathesis Pulications) / Bristol (Bristol Classical Press) 1985.

Platt, John E., *Reformed Thought and Scholasticism: the Arguments for the Existence of God in Dutch Theology 1575-1650*, Leiden (E. J. Brill) 1982.

Popkin, Richard H., *The History of Scepticism from Erasmus to Descartes*, Assen (Van Gorcum) 1960.

— *The History of Scepticism from Erasmus to Spinoza*, Berkeley (University of California Press) 1979.

— / Vanderjagt, Arjo (edd.), *Scepticism and Irreligion in the Seventeenth and Eighteenth Centuries* (*Brill's Studies in Intellectual History* 37), Leiden (E. J. Brill) 1993.

Reed, E. S., "Descartes' Corporeal Ideas Hypothesis", in *Review of Metaphysics* 35 (1981/'82), pp. 731-752.

[Rodis-]Lewis, Geneviève, *L'Individualité selon Descartes*, Paris (Librairie philosophique J. Vrin) 1950.

— *Descartes et le Rationalisme*, Paris (P. U. F.) 1970.

Röd, Wolfgang, [review of Jean-Luc Marion, *Sur l'ontologie grise de Descartes*,] *Zeitschrift für Philosophische Forschung* 32-1 (1978), pp. 146-149.

Ross, W. D. (ed.), *Aristotle's Metaphysics*, Oxford (Clarendon) 1924 (=1966^4).

— (ed.), *Aristotle's Physics*, Oxford (Clarendon) 1936.

Rothschuh, K., *Physiologie: der Wandel ihrer Konzepte, Probleme und Methoden vom 16. bis 19. Jahrhundert*, Freiburg (Alber) 1968.

Ruestow, Edward G., *Physics at 17th and 18th Century Leiden*, The Hague (Matinus Nijhoff) 1973.

Ruler, A. A. van, *Theologisch Werk*, Nijkerk (G. F. Callenbach) 1973.

Ruler, J. A. van, "New Philosophy to Old Standards: Voetius' Vindication of Divine Concurrence and Secondary Causality", in *Nederlands Archief voor Kerkgeschiedenis / Dutch Review of Church History* 71 (1991), pp. 58-91.

— "Franco Petri Burgersdijk and the Case of Calvinism Within the Neo-Scholastic Tradition", in E. P. Bos / H. A. Krop (eds), *Franco Burgersdijk (1590-1635), Studies in the History of Ideas in the Low Countries*, Amsterdam / Atlanta, GA (Rodopi) 1993, pp. 37-65.

— [review of Sorell, Tom (ed.), *The Rise of Modern Philosophy. The Tension between the New and Traditional Philosophies from Machiavelli to Leibniz*, Oxford (Clarendon Press) 1993] in *Archiv für Geschichte der Philosophie* 77-2 (1995), forthcoming.

— "Tradition et Émancipation; la question du 'matérialisme' chez Henricus Regius", paper presented at the Colloque *Tradition et Émancipation*, June, 30-July, 2 1991, Vianden, Luxembourg, publication forthcoming.

Russell, Bertrand, "On the Notion of Cause" (1917), in *Mysticism and Logic, and Other Essays*, London (Allen and Unwin) 1963.

Sakellariadis, Spyros, "Descartes' Use of Empirical Data to Test Hypotheses", in

Isis 73 (1982), pp. 68-76.

Sassen, Ferdinand, "350 jaren wijsgerig onderwijs te Groningen", in *Groninger Universiteitsblad* 15-3 (1965), pp. 68-81.

Schmitt, Charles B., *Aristotle and the Renaissance*, Cambridge, Mass. and London (Harvard U. P.) 1983.

Schouls, P. A., *The Imposition of Method*, Oxford (Clarendon Press) 1980.

Screech, M. A., *Montaigne and Melancholy: The Wisdom of the* Essays, London (Duckworth) 1983 / London (Penguin) 1991.

Sorabji, Richard, *Necessity, Cause, and Blame: Perspectives on Aristotle's Theory*, Ithaca (Cornell University Press) 1980.

Sorell, Tom (ed.), *The Rise of Modern Philosophy. The Tension between the New and Traditional Philosophies from Machiavelli to Leibniz*, Oxford (Clarendon Press) 1993.

Southgate, Beverley, "'Torn between Two Obligations': The Compromise of Thomas White", in Tom Sorell (ed.), *The Rise of Modern Philosophy: The tension between the New and traditional Philosophies from Machiavelli to Leibniz*, Oxford (Clarendon Press) 1993, pp. 107-127.

Specht, Rainer, Specht, "Zur Bezeichnung unzureichender Zweitursachen bei Francisco Suarez", in *Philosophisches Jahrbuch* 68 (1959).

— *Commercium Mentis et Corporis: über Causalvorstellungen im Cartesianismus*, Stuttgart-Bad Canstatt (Fromann) 1966.

— *Rationalismus, Geschichte der Philosophie in Text und Darstellung*, Band 5, Stuttgart (Philipp Reclam jun.) 1986[2] [first edition 1979].

Steenblok, C., *Gisbertus Voetius: zijn leven en werken*, Gouda (Gereformeerde Pers) 1976.

Strong, E. W., *Procedures and Metaphysics*, Hildesheim (Georg Olms Verlag) 1966.

Taylor, A. E., *A Commentary on Plato's Timæus*, Oxford (At the Clarendon Press) 1928 / reprint New York & London (Garland) 1987.

Thijssen-Schoute, C. Louise, *Nederlands Cartesianisme*, avec sommaire et table des matières en français, Amsterdam (Noord-Hollandsche Uitgevers Mij.) 1954 / reprint, bezorgd en van aanvullende bibliografie voorzien door Th. Verbeek, Utrecht (HES) 1989.

Thomas, Keith, *Religion and the Decline of Magic*, London (Weidenfeld & Nicholson) 1971 / London (Penguin Books) 1991.

Van de Pitte, F. P., "Descartes' Mathesis Universalis", in *Archiv für Geschichte der Philosophie* 61 (1979), pp. 154-174.

Vanderjagt, A. J., "Filosofie tussen humanisme en eclecticisme. Van Ubbo Emmius (1547-1625) tot Martinus Schoock (1614-1669)", in G. A. van Hemert / J. Schuller tot Peursum-Meijer / A. J. Vanderjagt (edd.), '*Om niet aan onwetendheid en barbarij te bezwijken': Groningse geleerden 1614-1989*, Hilversum (Verloren) 1989, pp. 31-49.

Vartanian, A., *Diderot and Descartes; A Study of Scientific Naturalism in the Enlightenment*, Princeton (P. U. P.) 1953.

Verbeek, Theo, "From 'Learned Ignorance' to Scepticism; Descartes and Calvinist Orthodoxy", in Richard H. Popkin / Arjo Vanderjagt (edd.), *Scepticism and Irreligion in the Seventeenth and Eighteenth Centuries, Brill's Studies in Intellectual History* 37, Leiden (E. J. Brill) 1993, pp. 31-45.

— "Henricus Reneri (1593-1639)", in H. W. Blom, H. A. Krop and M. R. Wielema (edd.), *Deventer denkers. De geschiedenis van het wijsgerig onderwijs te Deventer*, Hilversum (Verloren) 1993, pp. 123-134.

— "Voetius en Descartes", in J. van Oort, C. Graafland, A. de Groot, O.J. de Jong (eds), *De onbekende Voetius: Voordrachten wetenschappelijk symposium Utrecht 3 maart 1989*, Kampen (J. H. Kok) 1989.

— *Descartes and the Dutch*, Carbondale, Ill. (Southern Illinois University Press) 1992.

— "'Ens per Accidens'": Le origini della *querelle* di Utrecht", in *Giornale Critico della Filosofia Italiana* VI-12 (1992), pp. 276-288.

— (ed.), *Descartes et Regius: Autours de l'Explication de l'esprit humain*, Amsterdam / Atlanta, GA (Rodopi) 1993.

— "Regius's *Fundamenta Physices*", in *Journal of the History of Ideas* 55 (1994), pp. 533-551.

Vermij, Rienk H., *Secularisering en Natuurwetenschap in de zeventiende en achttiende eeuw: Bernard Nieuwentijt*, Amsterdam (Rodopi) 1991.

Vincent, L. H., "Le culte d'Hélène à Samarie", in *Revue Biblique* 45 (1936), pp. 221-232.

Vrijer, M. J. A. de, *Henricus Regius: Een "Cartesiaansch" hoogleraar aan de Utrechtsche Hoogeschool*, 's-Gravenhage (Martinus Nijhoff) 1917.

Weisheipl, James A., O.P., *Nature and Motion in the Middle Ages* (*Studies in Philosophy and the History of Philosophy* 11), Washington D. C. (The Catholic University of America Press) 1985.

Westman, Robert S., "The Melanchthon Circle, Rheticus and the Wittenberg Interpretation of the Copernican Theory", in *Isis* 66 (1975), pp. 165-193.

Wieland, W., "The Problem of Teleology", in Jonathan Barnes, Malcolm Schofield and Richard Sorabji (eds.), *Articles on Aristotle*, Vol. 1, "Science", London (Duckworth) 1975, pp. 141-160.

Woodfield, A., *Teleology*, Cambridge (C. U. P.) 1976.

Wundt, Max, *Die deutsche Schulmetaphysik des 17. Jahrhunderts* (*Heidelberger Abhandlungen zur Philosophie und ihrer Geschichte* 29), Tübingen (Mohr) 1939.

Yates, Frances A., *Giordano Bruno and the Hermetic Tradition*, London (Routledge and Kegan Paul) 1964.

INDEX

www.ingramcontent.com/pod-product-compliance
Ingram Content Group UK Ltd.
Pitfield, Milton Keynes, MK11 3LW, UK
UKHW020822290926
15470UKWH00003B/1316